II Columbian Congress of Electrochemistry (CCEQ 2016) and 2nd Symposium on Nanoscience and Nanotechnology (SNN 2016)

Journal of Physics: Conference Series Volume 786

Bucaramanga, Columbia
4 – 7 October 2016

Editor:

Ely Dannier V. Nino

ISBN: 978-1-5108-3604-4
ISSN: 1742-6588

Printed from e-media with permission by:

Curran Associates, Inc.
57 Morehouse Lane
Red Hook, NY 12571

Some format issues inherent in the e-media version may also appear in this print version.

Printed by Curran Associates, Inc. (2017)

For permission requests, please contact the Institute of Physics
at the address below.

Institute of Physics
Dirac House, Temple Back
Bristol BS1 6BE UK

Phone: 44 1 17 929 7481
Fax: 44 1 17 920 0979

techtracking@iop.org

Additional copies of this publication are available from:

Curran Associates, Inc.
57 Morehouse Lane
Red Hook, NY 12571 USA
Phone: 845-758-0400
Fax: 845-758-2633
Email: curran@proceedings.com
Web: www.proceedings.com

Table of contents

Volume 786

II Colombian Congress of Electrochemistry (CCEQ) and 2nd Symposium on Nanoscience and Nanotechnology (SNN)

4–7 October 2016, Bucaramanga, Colombia

Accepted papers received: 7 December 2016
Published online: 25 January 2017

Preface

Papers

Oral Presentations

Preface

In the present volume of Journal of Physics: Conference Series we publish the proceedings of the "II Colombian Congress of Electrochemistry (CCEQ) and 2nd Symposium on Nanoscience and Nanotechnology (SNN)", that was held from, October 4-7, 2016, at the Bucarica headquarters of the Universidad Industrial de Santander (UIS), Bucaramanga, Colombia. The proceedings consist of 45 contributions that were presented as plenary talks at the event. The abstracts of all participants' contributions were published in the Abstract Book with ISBN 978-958-8819-39-6. The website of the symposium is available at http://cceq.uis.edu.co/.

The scientific program of the II CCEQ and 2nd SNN consisted of 5 Plenary Lecture, 3 Magisterial Conferences, 2 Keynote, 54 Oral and 78 Poster Presentations and 3 Courses with the participation of undergraduate and graduate students, professors, researchers and entrepreneurs from Colombia, Spain, Mexico, Brazil and Venezuela. Moreover, the II CCEQ and 2nd SNN provided a forum of exchange in the research and innovation that enrich the area of electrochemistry, Nanoscience and Nanotechnology of the materials and the industrial applications.

All papers in these Proceedings refer to one from the following topics: New Materials, Thin Film, Surface Physics, Simulation and Diagnosis, Laser and Hybrid Processes, Biomedical Coatings, Preparation/Characterization/Application Nanomaterials, Surface Modification (Ionic Implantation, Ion Nitriding, PVD, CVD), Electrochemistry of Materials (Electrodeposits, Electropolymerization, Nanoelectrochemistry, Semiconductors), Corrosion, Analytical Electrochemistry, Electrochemistry in Mineral Processing and Metals (Extractive Metallurgy), Storage and Conversion Electrochemical Energy and Environmental Electrochemistry and Water Treatment involving Electrochemical Nature Phenomena. The editor hopes that those interested in the area of the science of materials can to enjoy this reading that reflects a wide variety of current issues.

On behalf of the organizing committee of the II CCEQ and 2nd SNN, we are extremely thankful to all authors for providing their valuable contributions for these Proceedings as well as the reviewers for their constructive recommendations and criticism aiding to improve the presented articles. Besides, especially we appreciate the great support provided by the Sponsors and Partners.

The Editor

Ely Dannier V. Niño

doi:10.1088/1742-6596/786/1/011001

Organizing Committee
Ángel Manuel Meléndez Reyes (Chairman)
David Alejandro Miranda Mercado
Sergio Ismael Blanco Vázquez

National Scientific Committee
Enrique Vera López
Henry Reyes Pineda
Walter Pardave Livia
Mauricio Rincón Ortiz
Julio Andrés Pedraza Avella
Darío Yesid Peña Ballesteros
Pedro Luis Delvasto Angarita
Cristian Camilo Viafara Arango
Jorge Andrés Calderón Gutiérrez
Afranio Antonio Cardona Granados
Andrés Giovanni González Hernández
Elcy María Córdoba Tuta
Ana Emilse Coy Echeverría
Martha Eugenia Niño Gómez
Sandra Judith García Vergara

International Scientific Committee
Ely Dannier V. Niño, Spain

Sponsors
- Universidad Industrial de Santander

Partners
- ✓ Foundation of Researchers in Science and Technology of Materials (FORISTOM)
- ✓ Sistemas Automatizados e Industriales División Electro (SAIDE)
- ✓ Centro Colombiano de Tecnología (CECOLTEC)
- ✓ Centro de Materiales y Nanociencias (CMN)
- ✓ Universidad del Quindio
- ✓ Hi-Tech Automatización
- ✓ S&S Ingeniería S.A.S.
- ✓ Vortexcompany
- ✓ DropSens

IOP Conf. Series: Journal of Physics: Conf. Series **786** (2017) 011001 doi:10.1088/1742-6596/786/1/011001

Peer review statement

All papers published in this volume of *Journal of Physics: Conference Series* have been peer reviewed through processes administered by the proceedings Editors. Reviews were conducted by expert referees to the professional and scientific standards expected of a proceedings journal published by IOP Publishing.

Fabrication of pH sensitive nanovalves using smart surface coated nanopores

A M Nieto-Soto[1], D K Diaz-Maldonado[1] and F A Rios Angarita[1]
[1] Universidad Industrial de Santander, Bucaramanga, Colombia

E-mail: fabianra@uis.edu.co

Abstract. A pH sensitive nanovalve was fabricated using different smart surfaces covalently attached to an anodized aluminium oxide membrane (AAO). The smart surfaces were synthesized using a mixture of aliphatic and aminated silanes. Effect on the contact angle of the aliphatic silane chain length was evaluated. The smart surface, in conjunction with a nanoporous membrane, allowed the formation of a hydrophobic plug which controlled the transport of the molecule safranine depending on the pH of the solution. It was demonstrated that mixtures of butyl and methyl-trimethoxysilane with aminopropyl-trimethoxysilane were able to perform as effective nanovalves creating a plug that remained closed at pH>7 and opened up at pH<5.

1. Introduction

Fabrication of smart nanomaterials is an area of active research due to its broad field of applications [1–4]. Especially in medicine, the search of a valve that can be opened and closed upon application of external stimuli to release a molecule of interest constitutes a fundamental need to develop an effective drug delivery system. A number of nanovalves have been proposed in the literature with different degrees of complexity [5,6].

Rios et al. demonstrated the use of smart surfaces as a simple and effective way to control the mass transport through a nanopore [7]. The system was based on the use of an alumina membrane coated with a smart hydrophobic surface. Such a hydrophobic surface was basically a mixture of two silanes, one aminated and one aliphatic, and had the property of being hydrophobic at pH>7 but hydrophilic at pH<5. If the system is modulated properly it could, in principle, be used as an efficient drug release mechanism. Although pH smart surfaces have been known for a while [8,9], the approach proposed here possesses more degrees of freedom allowing tuning of the contact angle response based through the different properties of the species used in the modification.

Once an adequate formulation is obtained, the smart surface can be covalently attached to a nanopore, forming an effective plug to the entrance of an aqueous media when in its hydrophobic state [10]. However, it is difficult to predict the right conditions to reach the "on" or "off" states which makes the experimental evaluation of different formulations necessary. In this work, we report the influence of the aliphatic silane chain length on the response of the smart surface, for a flat substrate, as well as a nanovalve in an alumina membrane. Was verified that the length of the aliphatic chain effectively varies the switching contact angle at different values of pH.

2. Experimental

Anodized aluminium oxide membranes (AAO) were fabricated using soft anodization as described elsewhere [11]. Briefly, high purity aluminium plates (99%, Goodfellow) were used as substrates. Pretreatment of aluminium included sonication in ethanol for 5 minutes, immersion in a solution of sodium hydroxide for 5 minutes and electropolishing in a solution of perchloric acid and ethanol (1:4V/V) at 10°C and 20V for 1min to obtain a mirror finished surface.

Anodization was carried out using an aqueous solution of oxalic acid (0.3M) as electrolyte in a two step method. During the first step, a potential difference of 60V was applied between the working and counter electrode for 2 hours. The first layer of oxide was dissolved with a solution of chromic acid (2%) and phosphoric acid (6%). A second step, at the same voltage, allowed the formation of an ordered alumina membrane which was separated from the aluminium using a solution of CuSO4 and HCl.

AAO and flat glass slides were used as substrates for surface modification as described previously [12,13]. Glass slides were cleaned using piranha solution (35% H2O2-65% H2SO4) for 20min at 70°C, washed with deionized water and dried in an oven for 30min at 115°C. AAO were cleaned in boiling water for 20min and dried in the oven 30min at 115°C. Silanization of the substrates was performed overnight in 2% v/v toluene solution of silane mixtures. The mixture was 2% of APTS and 98% of aliphatic silane: octadecyl-trimethoxysilane (OTS), Butyl-trimethoxysilane (BTS) and methyl-trimethoxysilane (MTS). After modification, membranes were cured at 130°C for 1h.

Membranes were characterized using SEM to determine morphology, pore size and thickness. Contact angle analysis were performed in a Dataphysics SCA20 instrument using the sessile drop method. Droplets were buffered with a 0.1M phosphate buffer for pH 7 and an acetate buffer for pH 5. All measurements were performed at 25°C.

Diffusion experiments were carried out at 25°C in a diffusion cell made of two plastic cuvettes (1cm path) which held the different membranes in the middle. Effective area of transport through the membrane was 12mm2. Solution on both sides had the same pH and ionic strength, but only one reservoir contained safranine dye.

3. Results and analysis

Figure 1 shows SEM images of a typical AAO membrane. Measured pore diameter is ~100nm and the thickness of the membranes is ~50μm. The anodization process produced well ordered pores and, after dissolution of the protecting oxide layer and the remaining aluminium metal, the membranes were modified and used for transport experiments.

Flat glass substrates were used as a mean to evaluate the hydrophobic switching of the smart surface. Figure 2 shows contact angle behaviour of the different mixtures with different pH solutions. At a high pH, both APTS:BTS and APTS:OTS surfaces are hydrophobic, which is opposite of the mixture APTS:MTS. If the formation of a monolayer is assumed, the previous result is expected since the mixture APTS:MTS will leave more of the amino groups exposed on the surface favouring its hydrophilic character. In the case of the other two mixtures, the aliphatic molecules will take a more dominant role giving the surface more of a hydrophobic character.

At pH 5, the surface APTS:OTS remains hydrophobic while the mixtures APTS:BTS and APTS:MTS produce hydrophilic surfaces. It is important to notice however, that all surfaces demonstrated a reduction in their contact angle, which suggests the presence of amino groups. In the case of mixtures made with OTS, the long aliphatic methylene chains produce a more hydrophobic environment, which forbids a hydrophilic contact angle. From a practical standpoint in its use as a pH controlled valve, the mixture APT:BTS shows a better response, since it is hydrophilic at low pH and hydrophobic at a higher pH.

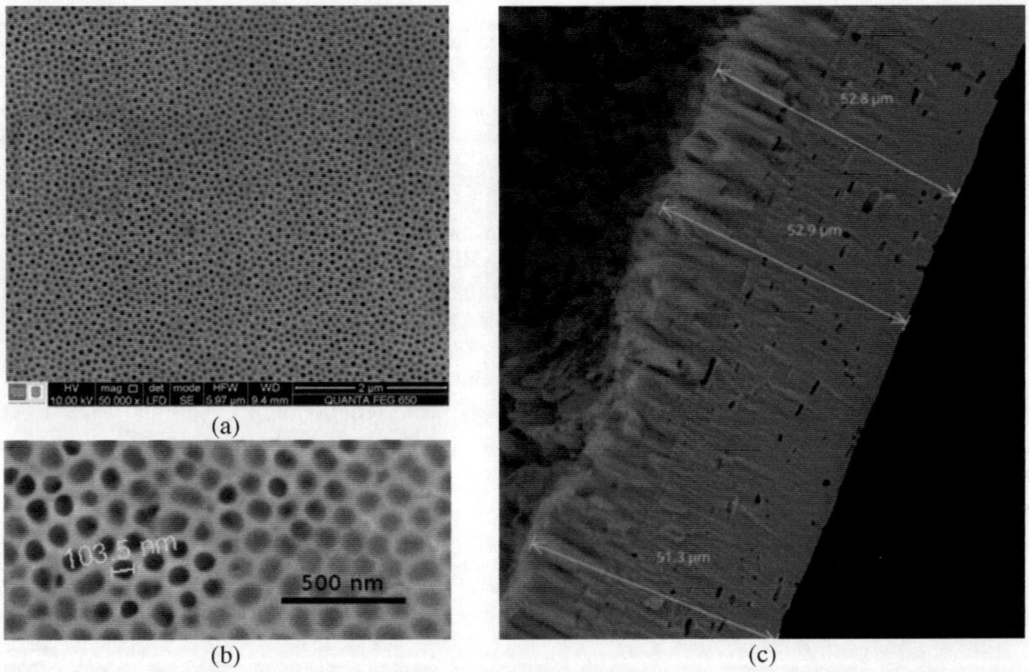

Figure 1. SEM analysis of alumina membranes fabricated using two step soft anodization. (a) Top view of the membrane showing the homogeneity of the membrane (b) Top view zoom. Pore size is ~100nm. (c) Cross-section view. Thickness of the membranes was ~50 μm.

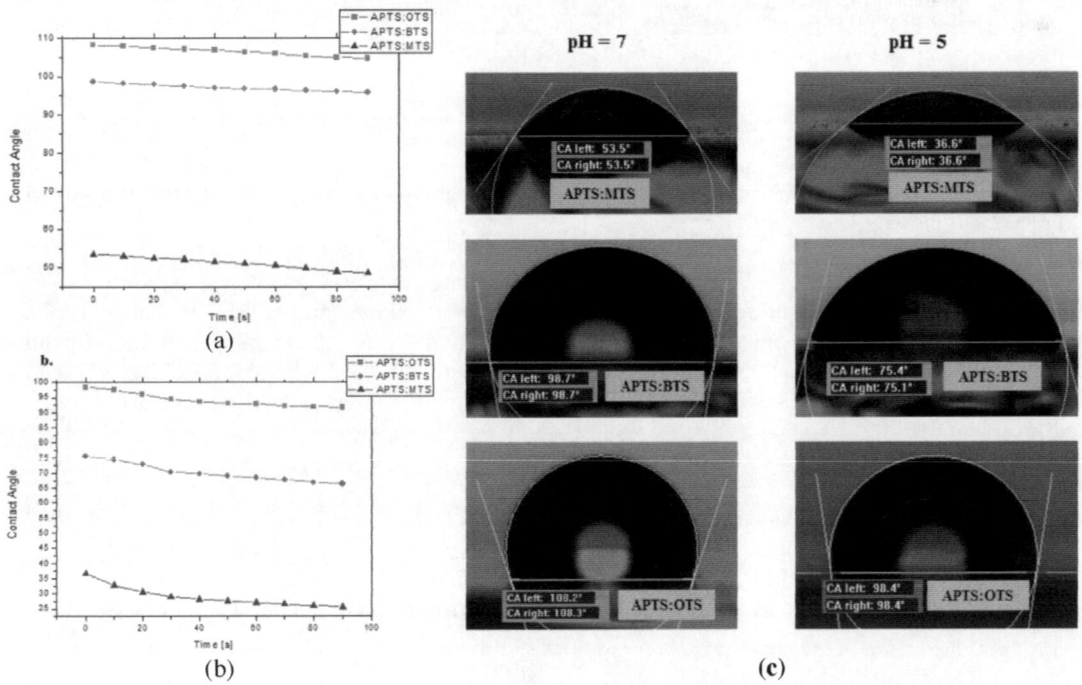

Figure 2. Contact angle measurements. (a) Change of contact angle at pH=7 (b) Change of contact angle at pH =5. (c) Contact angle profiles on different surfaces with solutions of pH 5 and 7.

All three modification were carried out on homemade AAO membranes. Figure 3 shows results for the different transport experiments. At pH 7, the unmodified membrane shows passage of safranine while modifications with APTS:OTS and APTS:BTS block the pores in accordance to the presence of a hydrophobic plug as expected. On the other hand, it was also observed that the mixture APTS:MTS shows a blockage of the pore. There is a possibility that an increase in rugosity on the alumina surface can also produce an increase in the contact angle, thereby producing the hydrophobic plug, although further verification of this fact is necessary.

At pH 5, all membranes allowed the transport of safranine, with the exception of the one modified with APTS:OTS. It can be observed that the surfaces APTS:BTS and APTS:MTS, although they allow safranine passage, show a higher mass transport rate than an unmodified membrane during the first minutes, however this rate decreases afterwards. This behaviour cannot be explained yet since the presence of salt in the solution would eliminate most of the charge effects on the walls. On the other hand, the hydrophobic plug would hinder transfer at the beginning, during the transition to an open state, and hence a smaller rate of mass transport would be expected, not higher. This and other aspects are currently under research in our lab.

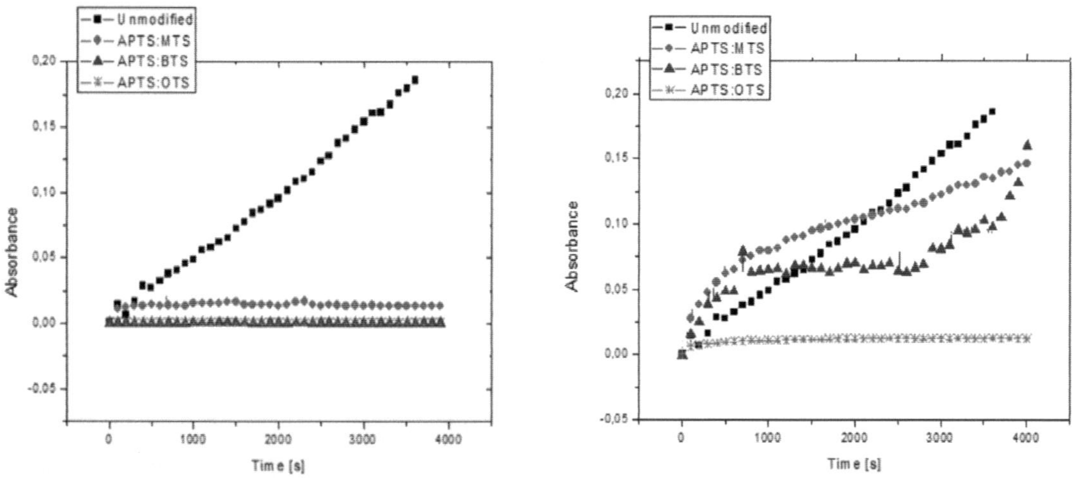

Figure 3. Transport experiment through a nanopore membrane modified with a pH sensitive surface. (a) pH=7. (b) pH=5.

4. Conclusions

We have demonstrated the influence of the length of aliphatic silanes on pH sensitive smart surface modifications. Although all mixtures show pH sensitivity, it is clear that the longer the aliphatic change, the higher the hydrophobic character of the membrane will be. Two mixtures, APTS:BTS and APTS:MTS, performed as an effective pH sensitive valve maintaining an "on" state at pH<5 and an "off " state at pH>7.

Acknowledgments

Authors aknlowledge the financial support from the Office of the Vice President for Research (VIE-UIS) through the grant 1731 as well as the use of the Research Facilities at Parque Tecnologico de Guatiguara. A. M. Nieto and D. K. Diaz acknowledge Colciencias and UIS for a Young Researcher Fellowship. SEM measurements at Tecnoparque SENA Bucaramanga are gratefully acknowledged.

IOP Conf. Series: Journal of Physics: Conf. Series **786** (2017) 012001 doi:10.1088/1742-6596/786/1/012001

References

[1] Colilla M and Gonzalez B 2013 *Biomater. Sci.* **1** 114–134
[2] Wei H, Wang Y, Guo J, Shen N Z, Jiang D, Zhang X, Yan X, Zhu J, Wang Q, Shao L, Lin H and Wei S 2015 *J. Mater. Chem. A* **3** 469–480
[3] Trewyn B G, Giri S, Slowing I I and Lin V S Y 2007 *Chem. Comm.* **43** 3236–3245
[4] Kim K T, Meeuwissen S A, Nolte R J M and Hest J C M 2010 *Nanoscale* **2** 844–858
[5] Giret S and Wong Chi Man M 2015 *Chem. Eur. J.* **21** 13850–13865
[6] Alberti S and Soler-Illia G J A A 2015 *Chem. Comm.* **51** 6050–6075
[7] Rios F and Smirnov S N 2011 *Chem. Mater.* **23** 3601–3605
[8] Holmes-Farley S R, Reamey R H, MCarthy T J, Deutch J and Whitesides G M 1985 *Langmuir* **1** 725–740
[9] Bain C D and Whitesides G M 1989 *Langmuir* **5** 1370–1378
[10] Smirnov S, Vlassiouk I, Takmakov P and Rios F 2010 *ACSnano* **4** 5069–507
[11] Thompson G 1997 *Thin Salid Films* **297** 192-201
[12] Vlassiouk I, Rios F, Vail S A, Gust D and Smirnov S 2007 *Langmuir* **23** 7784-7792
[13] Rios F and Smirnov S 2009 *ACS Appl. Mater. Interfaces* **1** 768-744

Solid-state fabrication of Al/CNTs nanocomposites

O J Toro Cadavid[1], M F Valencia García[2], H V Martínez Tejada[1] and G Suárez Guerrero[1]
[1] Universidad Pontificia Bolivariana, Medellín, Colombia
[2] Universidad EIA, Medellín, Colombia

E-mail: oscar.toro@upb.edu.co

Abstract. In the present work, synthesis of Al-based composites reinforced with carbon nanotubes has been carried out using solid–state manufacturing routes exclusively to consolidate Al–1100/CNTs laminates. Preliminary results regarding consolidation of Al/CNTs nanocomposites in which pressure and temperature are combined in a single step are presented. Preliminary Scanning Electron Microscopy (SEM) and Transmission Electron Microscopy (TEM) characterization allow us to conclude that CNTs were effectively set on the Al matrix.

1. Introduction

In recent years, there has been an increase in the development of composite materials that surpass the properties of traditional metallic, polymeric and ceramic composites, using carbon nanotubes (CNTs) as reinforcement due to their excellent properties [1], such as an average elastic modulus in the range of 1 to 2TPa and a tensile strength from 11 to 63GPa [2].

Currently, incorporation of CNTs on aluminium-based metal matrix nanocomposites (Al-MMCs) has become one of the most active fields of research in order to obtain thermal stability, high strength and stiffness or higher electrical conductivity. Challenges in this area still include poor dispersion of reinforcements in matrices due to agglomeration [3], Van der Waals forces between nanotubes and poor wettability between Al surface and CNTs [4].

Due to chemical incompatibility between CNTs and aluminium matrices, surface modification strategies and complementary routes must be applied in order to guarantee homogenous dispersion and densification of the composite material. In this study, forging of Al/CNTs laminated nanocomposites is proposed. Surface-modified multi-walled carbon nanotubes (m-CNTs) and three different strategies for dispersion and mixing are implemented, including: (I) surfactants, (II) Electro-hydro-dynamic-atomization (EHDA) and (III) dispersion through gas currents.

2. Experimental

Multi-walled carbon nanotubes (MWCNTs) were synthetized via carbon vapour deposition (CVD) following a process previously reported by Hoyos Palacio et al. [5]. With the aim of improving wettability of CNTs on aluminium, a surface modification step was carried out during the growth of CNTs in the CVD process. An ultrasonic nebulizer coupled to a quartz tube was implemented, and liquid aluminium nitrate [($Al(NO_3)_3$] was introduced into the nebulizer as aluminium precursor, being posteriorly supplied through pulses to the CVD tube. The nebulized precursor was dragged to the reaction zone using nitrogen (80cc/min).

2.1. Dispersion of m-MWCNTs
Dispersion of the nanometric reinforcement plays an important role in the fabrication of Al-MMCs. Three kinds of dispersions were carried out as explained below.

2.1.1. Surfactants. For this project purpose, Sodium Dodecyl Sulphate (SDS) was selected, an anionic analytical grade surfactant. According to different authors [6,7], sodium dodecylbenzene sulfonate (NaDDBS) yields the best results for dispersion of MWCNTs. NaDDBS is useful to produce stable dispersions, without any need for chemical modification of CNTs or any sort of polymeric coating application [6]. Nevertheless, among all surfactants, anionic are the most common in terms of industrial volumes. Besides, due to their effectiveness/price ratio, they are also the best qualified [8].

1g of modified MWCNTs (m-MWCNTs) was dispersed in 100ml of deionized water, with a ratio of 1:1 (H_2O: SDS). The solution was sonicated at 32Hz during 30min at room temperature and completely dried in a convection oven at 100°C (Figure 1(a)). Finally, the container was cooled down at room temperature and the nanotubes were extracted from the bottom of the beaker (Figure 1(b)).

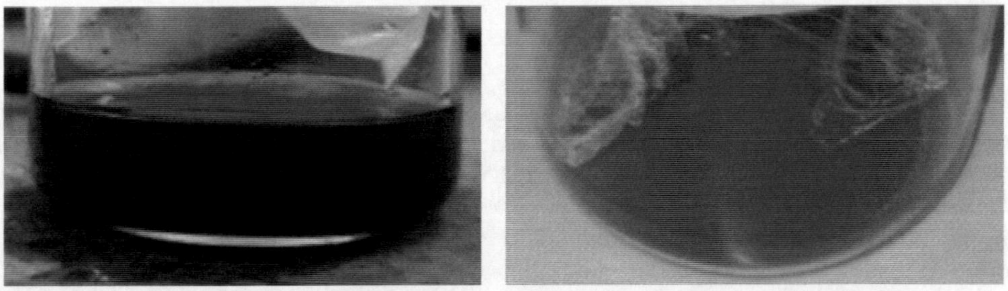

Figure 1. (a) m-MWCNTs dispersed in deionized water and SDS, (b) dry CNTs.

2.1.2. EHDA. In this case, a dispersion of 0.01g of m-MWCNTs in 1ml of an aqueous solution with 0.01g of SDS was prepared (Figure 2(a)). Ideal conditions were obtained for a cone-jet mode (7.5kV, 5cm of distance and a flux of 0.6ml/h). However, after several tests, the nozzle started to show signs of clogging, indicating that the process wouldn't be easily scalable despite reaching minimum dispersions in a matter of hours (Figure 2(b)).

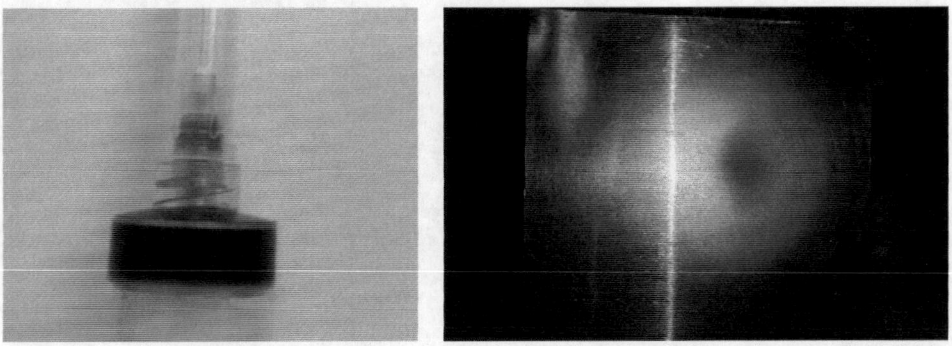

Figure 2. (a) m-MWCNTs dispersed in SDS, (b) detail of aluminium substrate after 50min of EHDA.

2.1.3. Compressed air dispersion. This process was designed in order to disperse great quantities of m-MWCNTs over the Al sheets. This method uses a "cloud" of m-MWCNTs (0.15g) previously dispersed in SDS and completely dried at 100°C inside a tight-lid container with the aid of compressed air at a pressure of 0.5Psi (Figure 3(a)). The appearance of the aluminium sheets after dispersion is presented in Figure 3(b).

Figure 3. (a) Al sheet in the bottom with cloud of m-MWCNTs on top, (b) Al sheet after dispersion.

2.2. Fabrication of Al/MWCNTs laminated nanocomposite

Aluminium 1100 sheets were used for the consolidation of nanocomposites. At first, the sheets were immersed in a solution of nitric acid (5%vol.), washed and cleaned with ethanol and subsequently dried. Then, an aqueous solution of polyvinylpyrrolidone (PVP) was sprayed to mitigate slippage between m-MWCNTs and the supporting matrix. Finally, m-MWCNTs were dispersed over the sheets using compressed air dispersion as explained in section 2.1.3.

The fabrication of the laminated nanocomposite was carried out in a press under 38Ton-force at 500°C during 600s (Figure 4(a)) and then thermally treated at 560°C during 3h. In total, 10 Aluminium 1100 sheets containing 0.25%wt of m-MWCNTs were stacked and then consolidated to obtain the MMC showed in Figure 4(b).

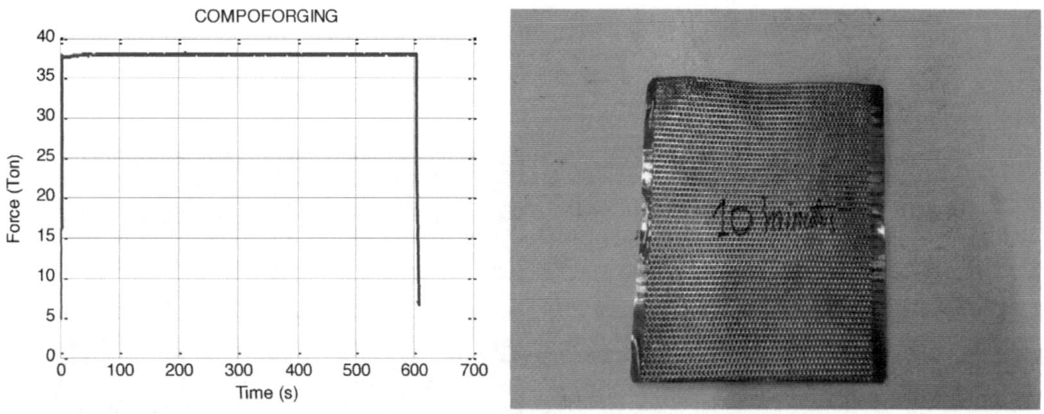

Figure 4. (a) Process of consolidation (time vs. force at 500°C) and (b) Laminated nanocomposite of 10 Al sheets (23x23cm).

3. Materials characterization

Characterization of MWCNTs was carried out using Raman spectroscopy and TEM. Comparison of m-MWCNTs with pristine MWCNTs (p-MWCNTs) from the same CVD furnace was performed in order to identify morphological differences. The consolidated nanocomposite was characterized using SEM and TEM.

IOP Conf. Series: Journal of Physics: Conf. Series **786** (2017) 012002 doi:10.1088/1742-6596/786/1/012002

3.1. Raman spectroscopy

Raman spectroscopy was carried out in a Horiba Jobin Yvon®, Labram HR Raman spectrometer. Figure 5 shows the spectra correspondent to the m-MWCNTs and the p-MWCNTs. In the spectra, D, G and G' bands appear at 1328, 1580 and 2650cm^{-1} respectively, corresponding with the typical theoretical values, which are 1350cm^{-1} for the D band and 1582cm^{-1} for the G band [9,10].

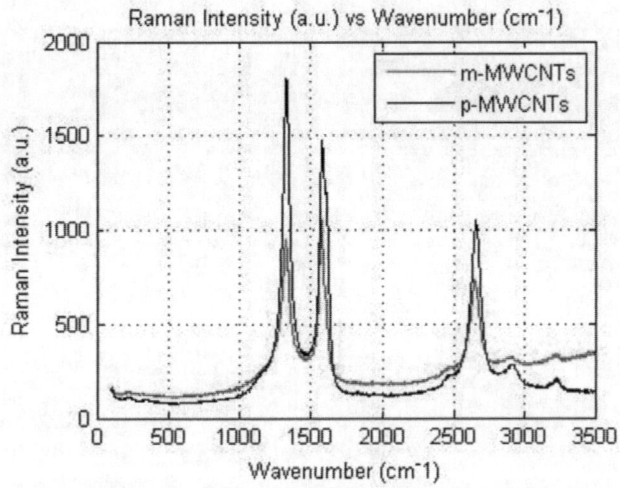

Figure 5. Raman spectra of p-MWCNTs and m-MWCNTs.

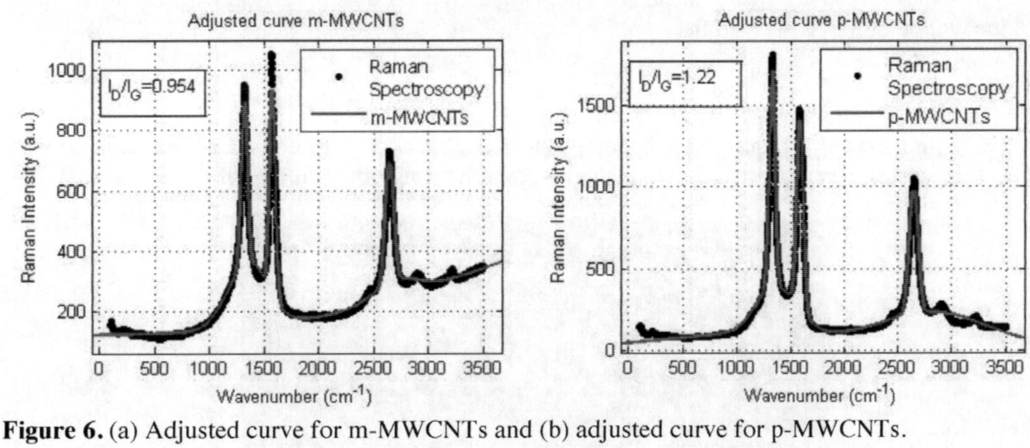

Figure 6. (a) Adjusted curve for m-MWCNTs and (b) adjusted curve for p-MWCNTs.

The I_D/I_G was obtained by adjusting the acquired data to a Gaussian approximation of eight-order through Matlab®'s cftool toolbox (Figure 6). According to S. Liang et al. [11], a higher I_D/I_G indicates the presence of defects or functional groups on the surface of MWCNTs. In this regard, for m-MWCNTs the I_D/I_G ratio is 0.954, which is relatively high. In addition, when compared to the spectra obtained for p-MWCNTs, it can be noted that the potential modification of MWCNTs contributed to the increment in the material structural order, to the extent that the I_D/I_G ratio acquired a value of 1.22 for p-MWCNTs.

3.2. TEM of MWCNTs

A TEM microscope (FEI TF20) operated at 200Kv was used to obtain TEM images. The samples of p-MWCNTs and m-MWCNTs were dispersed in ethanol and sonicated during 30min. For each case, a drop was extracted and dried over a cooper grid before being analysed with the microscope. Figures 7(a) and 7(b) correspond to p-MWCNTs and m-MWCNTs respectively. The average inner diameter for both p-MWCNTs and m-MWCNTs was found to be 9.12nm. Meanwhile, the external diameter was within a range from 30.8 to 46nm. These values confirm that MWCNTs have large diameters, while explaining why the Radial Breathing Mode (RBM) wasn't observed in the samples Raman spectra.

Theoretically, interplanar distance between CNTs averages an estimate of 0.34nm [12], thus the approximate number of concentric nanotubes used varies from 17 to 55.

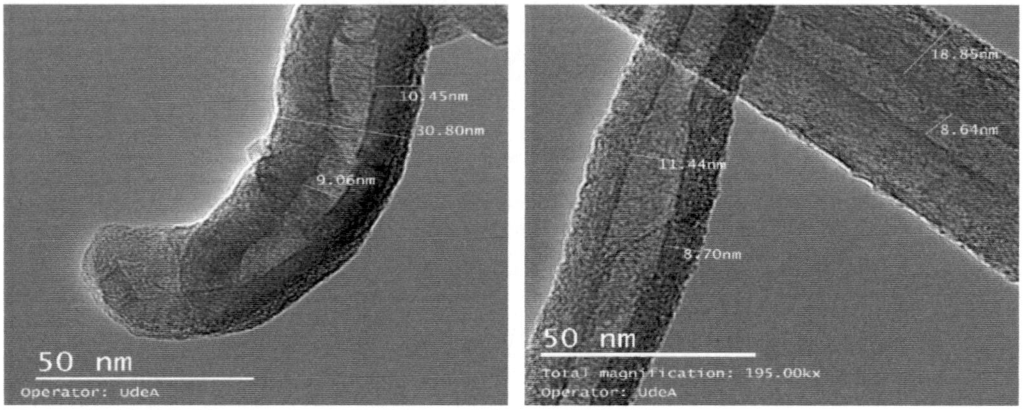

Figure 7. (a) TEM image of p-MWCNTs and (b) m-MWCNTs.

In Figure 7(b), correspondent to the m-MWCNTs TEM, there are no noticeable modifications, probably because the differences in the Raman spectra showed above correspond only to Al atoms embedded in the m-MWCNTs' walls.

3.3. SEM of Al/m-MWCNTs nanocomposite

A SEM microscope (JEOL JSM 5910 LV) with a 3nm resolution was operated at an acceleration voltage of 30kV. Figure 8 shows a laminated nanocomposite consolidated at 500°C with a compressing load of 38Ton-force during 10min, with a previous thermal treatment at 560°C during 3h.

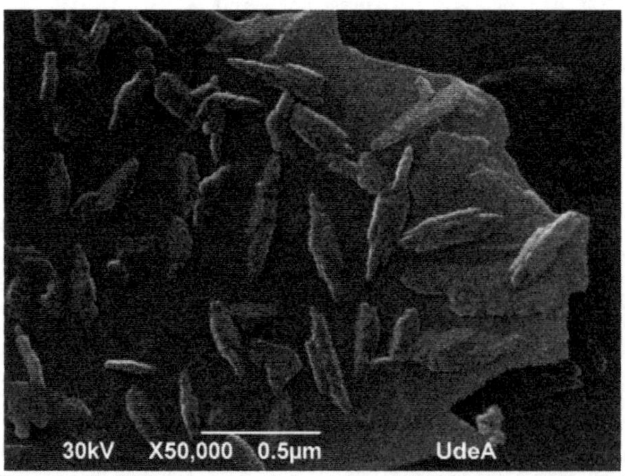

Figure 8. SEM image of the Al/m-MWCNTs laminated nanocomposite.

Homogeneously dispersed MWCNTs can be observed in the matrix, and results are comparable to previous information found for Al/MWCNTs [13] and CNTs/Al-Cu [14] nanocomposites.

3.4. TEM of Al/m-MWCNTs nanocomposite

TEM observation was carried out using a TECNAI F20 SUPER TWIN TMP microscope with a 0.1nm resolution and an acceleration voltage of 200kV. The laminated nanocomposite sample was prepared according to the procedure developed by Yan Liu et al. [15] for transversally observing films deposited on metallic substrates.

Xudong et al. [16] characterized an Al/CNTs composite manufactured by powder metallurgy. In their results, the interface between Al and CNTs had no interfacial sub products, like the Al_4C_3 observed by Z. Y. Liu et al. [17] at higher processing temperatures. Figure 9 shows the interface between the m-MWCNTs and the Al matrix obtained in this study. As can be seen on the image the interface does not exhibit any signs of interfacial sub products and a reinforcement layer of approximately 33.4nm can be noticed.

Figure 9. Interface between the m-MWCNTs and the Al matrix.

4. Conclusions

The CNTs used in this study have typical characteristics of CNTs fabricated via CVD. The differences between pristine and modified CNTs may likely be due to weak bonding related to physical adsorption of aluminium on the surface of the nanotubes.

Infiltration of CNTs into the matrix was lower than expected, mainly because of the temperature used during the consolidation process, which was selected within a range suitable to avoid formation of Al_4C_3 intermetallic compounds at higher temperatures.

In contrast to EHDA, mechanical dispersion through compressed air proved to be a technique useful for the deposition of greater quantities of MWCTNs on Al sheet surfaces.

Acknowledgments

The authors would like to acknowledge PhD. J. F. Pérez's (CINVESTAV) assistance in obtaining samples of m-CNTs.

References

[1] A Esawi and M Farag 2007 Carbon nanotube reinforced composites: Potential and current challenges *Mater. Des.* **28(9)** 2394–2401

[2] E Neubauer, M Kitzmantel, M Hulman and P Angerer 2010 Potential and challenges of metal-matrix-composites reinforced with carbon nanofibers and carbon nanotubes *Compos. Sci. Technol.* **70(16)** 2228–2236

[3] S C Tjong 2013 Recent progress in the development and properties of novel metal matrix nanocomposites reinforced with carbon nanotubes and graphene nanosheets *Mater. Sci. Eng. R Reports* **74(10)** 281–350

[4] Z Y Liu, B L Xiao and Z Y Ma 2011 Fabrication of CNTs/Al composite with enhanced dispersion pre-treatment *18th ICCM International Conference on Composite Materials* (Korea: ICCM)
[5] L M Hoyos-Palacio, A G García, J F Pérez-Robles, J González and H V Martínez-Tejada 2014 Catalytic effect of Fe, Ni, Co and Mo on the CNTs production *IOP Conf. Ser. Mater. Sci. Eng.* **59(1)** 12005
[6] O Matarredona, H Rhoads, Z Li, J H Harwell, L Balzano and D E Resasco 2003 Dispersion of Single-Walled Carbon Nanotubes in Aqueous Solutions of the Anionic Surfactant NaDDBS *J. Phys. Chem. B* **107(48)** 13357–13367
[7] Y-F Hsieh, B-Y Huang, C-P Liu and C-Y Liu 2010 NaDDBS as a dispersion agent for multiwalled carbon nanotubes in capillary EKC separation of nucleotides *Electrophoresis* **31(19)** 3288–3295
[8] S Che, A E Garcia-bennett, T Yokoi, K Sakamoto, H Kunieda, O Terasaki and T Tatsumi 2003 A novel anionic surfactant templating route for synthesizing mesoporous silica with unique structure *Nat. Mater.* **2(18)** 801–805
[9] S K Singhal, R Pasricha, S Teotia, G Kumar and R B Mathur 2011 Fabrication and characterization of Al-matrix composites reinforced with amino-functionalized carbon nanotubes *Compos. Sci. Technol.* **72(1)** 103–111
[10] S K Singhal, R Pasricha, M Jangra, R Chahal, S Teotia and R BMathur 2012 Carbon nanotubes: Amino functionalization and its application in the fabrication of Al-matrix composites *Powder Technol.* **215–216** 254–263
[11] S Liang, G Li and R Tian 2016 Multi-walled carbon nanotubes functionalized with a ultrahigh fraction of carboxyl and hydroxyl groups by ultrasound-assisted oxidation *J. Mater. Sci.* **51(7)** 3513–3524
[12] P Somani and M Umeno 2007 Importance of transmission electron microscopy for carbon nanomaterials research *Mod. Res. Educ. Top. Microsc.* **3** 634–642
[13] H J Choi, S M Lee and D H Bae 2010 Wear characteristic of aluminum-based composites containing multi-walled carbon nanotubes *Wear* **270(1–2)** 12–18
[14] D H Nam, S I Cha, B K Lim, H M Park, D S Han and S H Hong 2012 Synergistic strengthening by load transfer mechanism and grain refinement of CNT/Al-Cu composites *Carbon N. Y.* **50(7)** 2417–2423
[15] Y Liu, R Wang, X Guo and J Dai 2007 A cross-sectional TEM sample preparation method for films deposited on metallic substrates *Mater. Charact.* **58(7)** 666–669
[16] X Yang, E Liu, C Shi, C He, J Li, N Zhao and K Kondoh 2013 Fabrication of carbon nanotube reinforced Al composites with well-balanced strength and ductility *J. Alloys Compd.* **563** 216–220
[17] Z Y Liu, S J Xu, B L Xiao, P Xue, W G Wang and Z Y Ma 2012 Effect of ball-milling time on mechanical properties of carbon nanotubes reinforced aluminum matrix composites *Compos. Part A Appl. Sci. Manuf.* **43(12)** 2161–2168

Nanocomposites fabrication by self-assembly method to modify macroscopic properties

N Lopez-Barbosa[1] and J F Osma[1]
[1] Universidad de los Andes, Bogotá D.C., Colombia

E-mail: jf.osma43@uniandes.edu.co

Abstract. Polymeric nanocomposites have been in the scope of scientists for the last decade due to their multiple applications and simple synthesis. Self-assembly fabrication can be performed through different methods such as layer-by-layer or the controlled growth of nanostructures on a surface. These methods allow fast elaboration of nanocomposites that can be readily integrated in sensors or films. The current work exposes the self-assembly of nanocomposites for the modification of material's macroscopic properties such as hydrophobicity and temperature's resistance on textiles. Hydrophobicity properties of cotton textiles were modified by the application of functionalized silica nanoparticles on their surfaces. Thermal resistance of cotton textiles was achieved by incorporating TiO_2 nanoparticles into the matrix, increasing the scope of their applications. Functionalization was attained by chloro-trimethyl-silane (CTS) and γ-amino(propyl) triethoxy silane (APTES) in organic and inorganic solvents. Wetting phenomena characteristics appeared to highly depend on the synthesis parameters.

1. Introduction

Nanocomposites are an emerging technology used to modify macroscopic properties of different materials. Among others, textiles have been in the scope of researchers to provide them with multifunctional properties such as UV [1], thermal and flame resistance [2], and antibacterial [2–5], antimicrobial and conductive [4,6] properties. Similarly, the need of fabricating hydrophobic surfaces have increased from self-cleaning applications to engineered surfaces for anti-sticking, antifouling and controlled wettability, among others [7]. Superficial hydrophobicity principally depends on the differential interfacial energy between the surface and the water droplet [8], which can be modified by lowering the surface free energy or changing the surface roughens. Several attempts [9–13] have been done to modified the wetting phenomena of cotton textiles by incorporating different nanocomposites onto their surface in the attempt to either decrease the surface energy or simulate structures found on the lotus leaf surface.

Hydrophobic and superhydrophobic surfaces have been used in a great variety of applications such as spray cooling [14], spray painting [15], self-cleaning coatings [16], ink-jet printing and turbines [17], among others. When resting on a solid surface, the shape of a droplet is only determined by the surface tensions acting on the liquid if the effect of inertial forces is negligible. This is the case when the free fall velocity of the droplet is considerably smaller than the surface tension [18]. In such case, the resting droplet can be assumed to have a spherical crown shape [19] and the resting contact angle can be interpreted as the interactions of the solid-liquid, liquid-gas and gas-solid interphases.

In the current study, the wetting phenomena of cotton textiles was modified by SiO_2 nanoparticles functionalization through the attachment of chloro-trimethyl-silane (CTS) and γ-amino (propyl) triethoxy silane (APTES). In addition, the interaction between functionalized SiO_2 and TiO_2 nanoparticles was also studied. Wetting phenomena was studied by means of the sessile drop static contact angle method.

2. Materials and methods

2.1. Materials

SiO_2 and TiO_2 industrial grade nanoparticles were used. Acetone, chloro-trimethyl-silane (CTS) and γ-amino (propyl) triethoxy silane (APTES) were obtained from Sigma Aldrich. Textiles used were 100% cotton without dyes.

2.2. SiO_2 nanoparticles silanization

A SiO_2 nanoparticles solution in milli Q water of 1.66mg/mL was used. 1mL of CTS 2% (v/v) solution was added and left still overnight. The same procedure was performed with APTES in both milli Q water (APTES-W) and acetone (APTES-A) solvents. Functionalized nanoparticles were dried by solvent evaporation method.

The effective bonding between CTS or APTES with SiO_2 nanoparticles was confirmed by FT-IR measurements in an A250/D FT-IR (Bruker, United States).

2.3. Nanoparticles immobilization on textiles

Seven different solutions were attained for immobilization over cotton textiles. All solutions were prepared in milli Q water with a concentration of 1mg/mL. Solution TiO_2 was prepared by dissolving TiO_2 nanoparticles. Solution CTS-W was prepared by dissolving CTS-silica complexes. Solution APTES-W was prepared by dissolving APTES-silica complexes prepared in water. APTES-A was prepared by dissolving APTES-silica complexes prepared in acetone. Solution CTS-TiO_2 was prepared by dissolving both TiO_2 nanoparticles and CTS-silica complexes. Solution APTES-W-TiO_2 was prepared by dissolving both TiO_2 nanoparticles and APTES-silica complexes prepared in water. Finally, solution APTES-A-TiO_2 was prepared by dissolving both TiO_2 nanoparticles and APTES-silica complexes prepared in acetone.

Five 100% cotton textiles were placed in each solution and left still for 10 minutes. Textiles were removed from the solution and dried at room temperature (25°C) overnight. Each of the textiles had a superficial area of 50×50mm.

2.4. Wetting phenomena measurements

Wetting phenomena was analysed through sessile drop static contact angle measurement. 2.5μL of milli Q water were dropped at a height of 3cm on the surface of each textile. Tangential and perpendicular video of the drops was registered for later image processing. Static contact angle measurements were taken 5, 10 and 15 seconds after the droplet touched the textile surface through video analysis by means of the software Tracker. When the three static angle measurements differed considerably, measurements of the contact angle were performed every second for 15 seconds. The experimental setup is shown in Figure 1.

Textiles were placed on a hotplate and heated at 80°C for 1h to study the effect of the presence of TiO_2 nanoparticles. Wetting phenomena measurements were repeated afterwards.

All contact angle measurements were repeated three times in each textile, giving a total of thirty measurements per solution, fifteen prior to heating and fifteen afterwards.

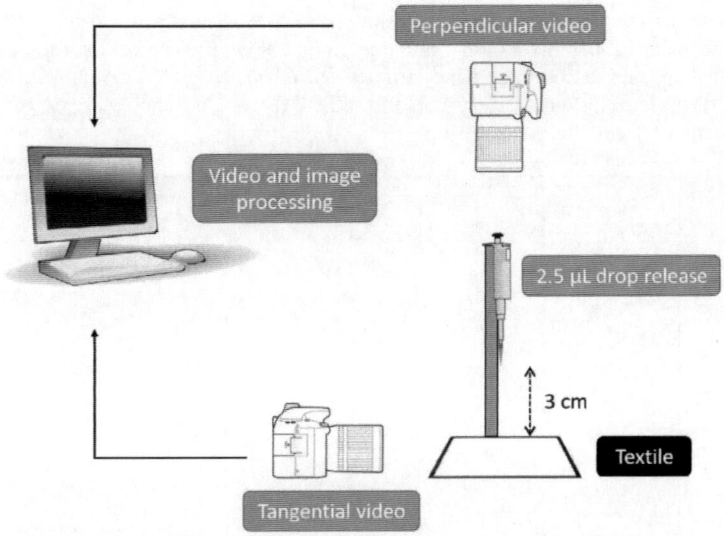

Figure 1. Experimental setup for static contact angle measurements.

3. Results and discussion

3.1. SiO_2 silanization

SiO_2 effective silanization was confirmed through FT-IT measurements of the resulting powder. Results for solutions CTS is shown in Figure 2. Among the peaks, there is a dominant peak at 1059cm^{-1}. This peak is associated to the O-Si-O bond that appears between the SiO_2 nanoparticle and the CTS. The presence of the peak confirms that the silanization was effective.

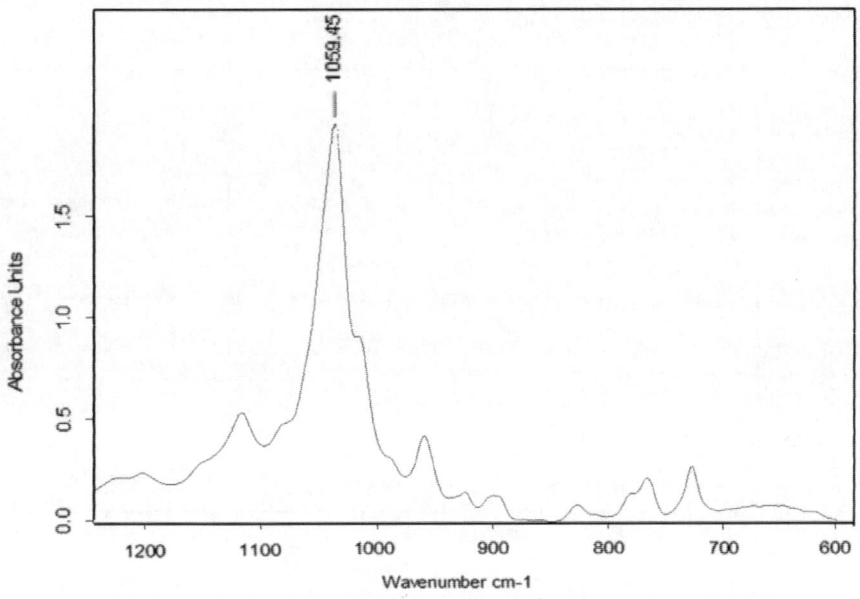

Figure 2. FT-IR spectre for solution CTS.

IOP Conf. Series: Journal of Physics: Conf. Series **786** (2017) 012003 doi:10.1088/1742-6596/786/1/012003

Figure 3 shows the FT-IR spectre of solution APTES-W. The spectre contains two main groups of peaks; one between 1100 and 1000cm^{-1}, and one around 800cm^{-1}. The first group is attributed to the O-Si-O bond that appears between the nanoparticles and APTES, which proves the correct silanization of the sample. The second group corresponds to the Si-CH_3 rocking characteristic of silane. This means that there is an excess of silane in the sample.

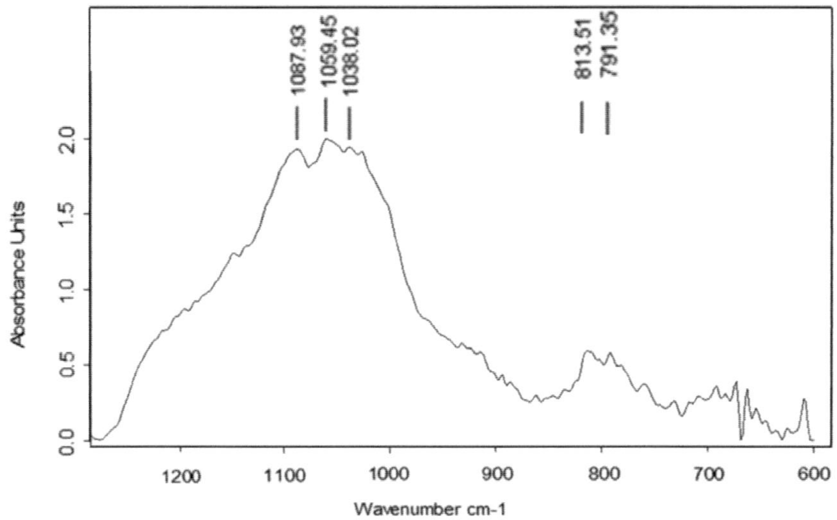

Figure 3. FT-IR spectre for solution APTES-W.

Figure 4 shows two peaks that correspond to the O-Si-O bonding (1087.30cm^{-1}) and the Si-CH_3 rocking of silane (805.29cm^{-1}) in the FT-IR spectre of solution APTES-A. In contrast with the APTES-W spectre, peaks are easier to identify in this sample.

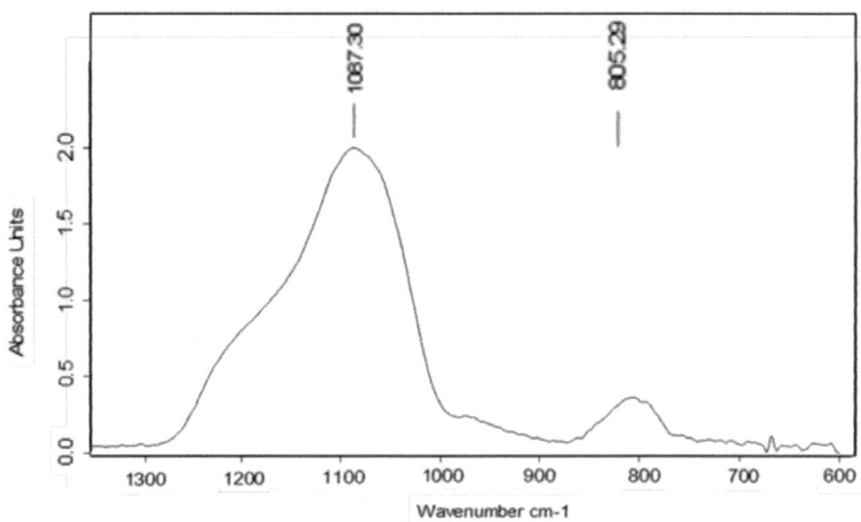

Figure 4. FT-IR for solution APTES-A.

3.2. Wetting phenomena measurements

Contact angle measurements were performed by video analysis in the software Tracker, from three five different textiles from each solution. Tangential videos were used to determine the contact angle between the textile surface and the droplet of water. Perpendicular videos were used to monitor the contact area between the droplet and the textile. The mean contact angle of three measurements per textile prior to heating are shown in Figure 5 and after heating in Figure 6.

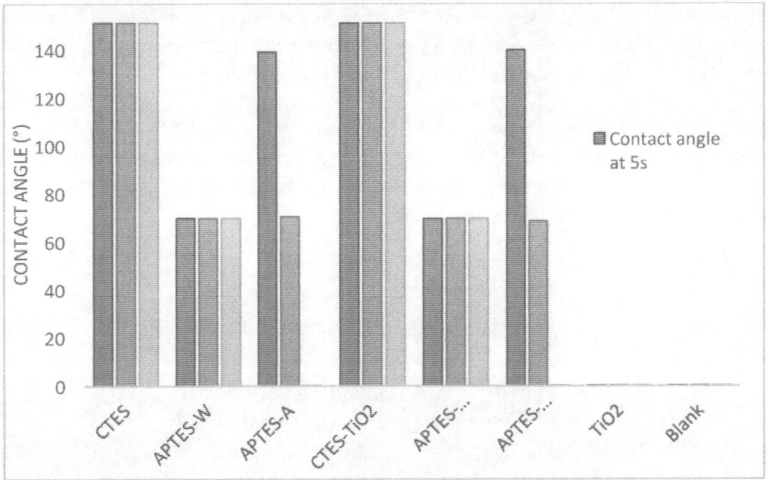

Figure 5. Mean contact angle measurements before heating.

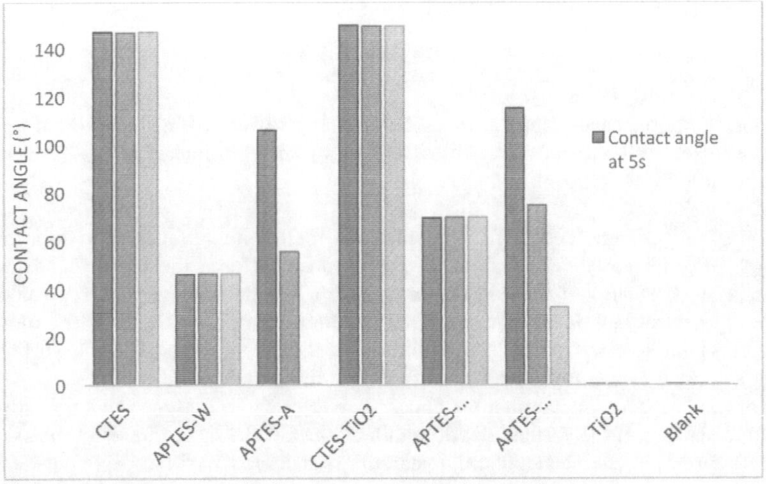

Figure 6. Mean contact angle measurements after heating.

Regardless of the presence of TiO_2 nanoparticles in the treatment, textiles functionalized with solutions CTES and CTES-TiO_2 exhibited a superhydrophobic wetting phenomena with no significant variation after the heating process, as stated in [15]. In contrast, textiles from solutions APTES-W and APTES-W-TiO_2 exhibited a hydrophilic behaviour before the heating process. This is attributed to the hydrogen bonds formed between the water droplet molecules and the APTES superficial amino groups. However, after heating, there is a greater decrease in the contact angle between the droplet and the surface in textiles treated with solution APTS-W than those treated with solution APTS-W-TiO_2. This is attributed to two different phenomena. For instance, it is known that TiO_2 nanoparticles absorb part of the heat before it is transferred to the textile. This allows a greater number of bonds between the silane and the SiO_2 nanoparticles to remain unchanged regardless of the heating process, which is translated in a smaller diminishment upon the contact angle between the water droplet and the textile surface. On the other hand, silanization processes were performed at room temperature. Aissaoui et al. [20] showed that room temperature silanization of SiO_2 surfaces results in a more or less homogeneous layer covered by silane aggregates. Nevertheless, aggregation is sustained by only one chemical anchoring point, which is easily detached during the heating treatment, contributing to the decrease of the contact angle between the droplet and the functionalized surface.

During the video processing, it was observed that the contact angle between the textile surface and the water droplet of textiles from solutions APTES-A and APTES-A-TiO_2 was neither constant before or after the heating process. Because of this, contact angle measurements were performed for fifteen seconds every second after the first five seconds of being the droplet in contact with the textile. The obtained measurements are shown in Figure 7.

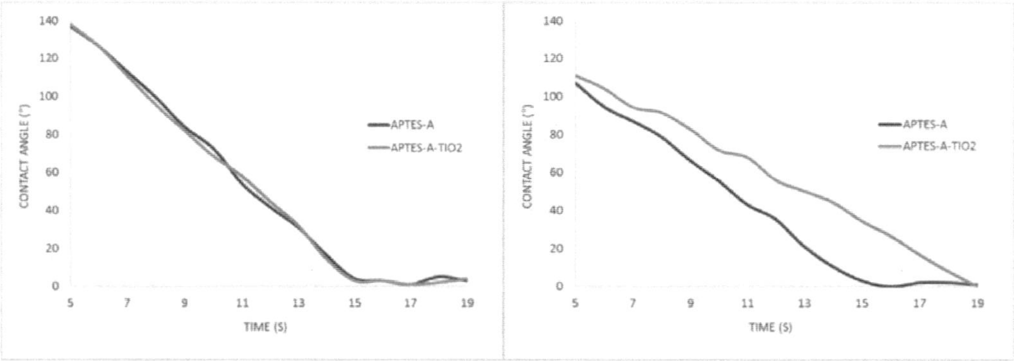

(a) (b)

Figure 7. Contact angle measurements every second after five seconds of being the droplet in contact with the textile surface in textiles from solution APTES-A and solution APTES-A-TiO_2 (a) before and (b) after heating at 80°C for 1 hour.

Although the contact angle between the droplet and the textile surfaces from solutions APTES-A and APTES-A-TiO_2 decreased with time, the contact area between the droplet and the textile was always constant at $10.18mm^2$. If the contact angle decreases, but the contact area remains constant, there must be a loss in the volume of the droplet. A diminishment of the droplet volume is usually attributed to a loss due to evaporation [19]. However, since the textile is a porous surface, a decrease of the droplet volume may be attributed to water going through the textile.

In an ideal system, static contact angles appear because of the presence of three surface tensions between the three phases of the system; solid, liquid and gas. If the three phases in the system exhibit a force balance, there is a chemical potential matching attributed to all components, and there is a temperature matching such that the gas phase represents the saturated vapour of the liquid phase, the system is said to be in thermodynamic equilibrium [18]. Under this circumstance, a decrease of the volume of the droplet would break down the thermodynamic equilibrium, and a resultant force towards the centre of the droplet would arise [19]. As a consequence, the solid-liquid interphase would shrink until the three surface tensions reach a new equilibrium, i.e. a decrease of the volume of the droplet translates in a reduction of the contact area under a steady contact angle. In contrast, a reduction of the droplet volume was translated in a decrease of the contact angle, rather than the contact area.

There are at least three factors that might have been involved in this phenomena. For instance, APTES-A and APTES-A-TiO_2 solutions were prepared with acetone as a solvent. Pan et al. [21] showed through NMR measurements of silica nanoparticles in acetone solution, that additional hydroxyl groups are given to the SiO_2 nanoparticles by the presence of acetone molecules in form of hydrogen bonds. This implies that there is a greater number of bonding sites on SiO_2 nanoparticles immerged in acetone than there is when immerged in water. A higher number of APTES bonds with the nanoparticles could be the reason why, at time 0, APTES-A and APTES-A-TiO_2 appear to be hydrophobic rather than hydrophilic. However, since the hydroxyl groups that result from the presence of acetone are not covalently bonded, they are easily detached from the surface of the nanoparticle when the droplet enters in contact with the textile surface, which results in a decrease of the contact angle. However, if this was the only difference between APTES-W and APTES-A solutions, the contact angle of the last one should stabilized to the contact angle of the first one with time, which did not happen. This can be attributed to the other two factors present in the system. Firstly, the interactions between the water molecules of the droplet and the hydrogen atoms of the amino group are translated in a more hydrophilic behaviour. Finally, there is a possibility that due to the higher density of hydroxyl groups in SiO_2 nanoparticles in acetone solution than those in water, a steric impediment of the APTS groups is created, resulting in a different organization than the one encountered when silanized in water. The change in organization can be translated in non-homogenous distributions that led to non-compact structures, which produces additional surface forces [19]. Also, Choi et al. [22] showed that an incorrect distribution, especially when hydrophobic chains are short, can allow water penetration, which may induce a deformation of the organic layer. This phenomenon is coherent with the fact that the volume reduction is not due to evaporation but because of water going through the textile.

4. Conclusions

Different combinations of functionalized SiO_2 and TiO_2 nanoparticles were tested as mechanisms to modify the wetting phenomena of 100% cotton textiles. It was found that while CTS produced constant hydrophobic surfaces, APTES produced a final hydrophilic behaviour. However, there was a direct relation between the behaviour of the textiles and the solvent in which silica nanoparticles were functionalized when using APTES. This allowed us to create surfaces where the contact angle between the droplet and the textile was constantly steady and hydrophilic, or presented a short-term hydrophobic property that changed to a hydrophilic behaviour with time, depending on the solvent used for the functionalization of the silica nanoparticles. Variable contact angles were attributed to a loss in volume due to the fact that the contact area remained constant. This loss in volume was a result of the droplet going through the textile.

The fact that the contact area of the droplet remained constant with a decreasing contact angle translated in a greater diffusion time of the droplet through the textile when compared with an untreated one. This could be exploited in applications in which interactions between the textile surface and the passing liquid are more efficient if contact time is increased.

In the combined presence of TiO_2 and functionalized silica nanoparticles, the thermal properties of TiO_2 protected the silane bonds of the functionalized silica nanoparticles, therefore preserving the hydrophobic properties of the heated textiles. This showed that it was possible to integrate multiple types of nanoparticles in the same textile to obtain multipurpose textiles.

References

[1] G Broasca, G Borcia, N Dumitrascu and N Vrinceanu 2013 Characterization of ZnO coated polyester fabrics for UV protection *Appl. Surf. Sci.* **279** 272–278

[2] A El-Shafei, M ElShemy and A Abou-Okeil 2015 Eco-friendly finishing agent for cotton fabrics to improve flame retardant and antibacterial properties *Carbohydr. Polym.* **118** 83–90

[3] A Hebeish, M K El-Bisi and A El-Shafei 2015 Green synthesis of silver nanoparticles and their application to cotton fabrics *Int. J. Biol. Macromol.* **72** 1384–1390

[4] W-G Kwak, M H Oh and M-S Gong 2015 Preparation of silver-coated cotton fabrics using silver

carbamate via thermal reduction and their properties *Carbohydr. Polym.* **115** 317–324

[5] Z A Raza, A Rehman, M Mohsin, S Z Bajwa, F Anwar, A Naeem and N Ahmad 2015 Development of antibacterial cellulosic fabric via clean impregnation of silver nanoparticles *J. Clean. Prod.* **101** 377–386

[6] K Kardarian, T Busani, I Osório, H Domingos, R Igreja, R Franco and J Cortez 2014 Sintering of nanoscale silver coated textiles, a new approach to attain conductive fabrics for electromagnetic shielding *Mater. Chem. Phys.* **147(3)** 815–822

[7] C Neinhuis and W Barthlott 1997 Characterization and Distribution of Water-repellent, Self-cleaning Plant Surfaces *Ann. Bot.* **79(6)** 667–677

[8] N D Tissera, R N Wijesena, J R Perera, K M N de Silva and G A J Amaratunge 2015 Hydrophobic cotton textile surfaces using an amphiphilic graphene oxide (GO) coating *Appl. Surf. Sci.* **324** 455–463

[9] A Berendjchi, R Khajavi and M E Yazdanshenas 2011 Fabrication of superhydrophobic and antibacterial surface on cotton fabric by doped silica- based sols with nanoparticles of copper *Nanoscale Res. Lett.* **6(1)** 1–8

[10] N Gogoi, D Rastogi, M Jassal and A K Agrawal 2014 Low-surface-energy materials based on polybenzoxazines for surface modification of textiles *J. Text. Inst.* **105(11)** 1212–1220

[11] M Shateri-Khalilabad and M E Yazdanshenas 2013 Preparation of superhydrophobic electroconductive graphene-coated cotton cellulose *Cellulose* **20(2)** 963–972

[12] L F Hao, Q F An, W Xu and Q J Wang 2010 Synthesis of fluoro-containing superhydrophobic cotton fabric with washing resistant property using nano-SiO2 sol-gel method *Advanced Materials Research* **121** 23–26

[13] C-H Xue, S-T Jia, H-Z Chen and M Wang 2008 Superhydrophobic cotton fabrics prepared by sol–gel coating of TiO2 and surface hydrophobization *Sci. Technol. Adv. Mater* 9(3) 035008

[14] W M Grissom and F A Wierum 1981 Liquid spray cooling of a heated surface *Int J. Heat Mass Transf.* **24(2)** 261–271

[15] A Nakajima, T Miyamoto, M Sakai, T Isobe and S Matsushita 2014 Comparative study of the impact and sliding behavior of water droplets on two different hydrophobic silane coatings *Appl. Surf. Sci.* **292** 990–996

[16] E Gutierrez-Miravete, E J Lavernia, G M Trapaga, J Szekely and N J Grant 1989 A mathematical model of the spray deposition process *Metall. Trans. A* **20(1)** 71–85

[17] C Mundo, M Sommerfeld and C Tropea 1995 Droplet-wall collisions: Experimental studies of the deformation and breakup process *Int. J. Multiph. Flow* **21(2)** 151–173

[18] N Lopez-Barbosa, N Zuluaga, S Jumah and J F Osma 2015 Fabrication of folding polymeric thin films with hydrophilic properties *Novos Saberes* **2(2)** 71–83

[19] X Zeng, G Xu, Y Gao and Y An 2011 Surface Wettability of (3-Aminopropyl)triethoxysilane Self-Assembled Monolayers *J. Phys. Chem. B* **115(3)** 450–454

[20] N Aissaoui, L Bergaoui, J Landoulsi and P Vi 2012 Silane Layers on Silicon Surfaces: Mechanism of Interaction , Stability and Influence on Protein Adsorption *Langmuir* **28(1)** 656–665

[21] V H Pan, T Tao, J-W Zhou and G E Maciel 1999 Hydrogen Bonding between Acetone and Silica Gel, as Studied by NMR *J. Phys. Chem. B* **103(33)** 6930–6943

[22] H Choi and H Liang 2016 Wettability and spontaneous penetration of a water drop into hydrophobic pores *J. Colloid Interface Sci.* **477** 176–180

IOP Conf. Series: Journal of Physics: Conf. Series **786** (2017) 012004 doi:10.1088/1742-6596/786/1/012004

Characterization of the optical non-linear response of the *(E)*-4-(4-dimetylaminophenyl) but-3-en-2-one and *(E)*-4-(4-nitrophenyl) but-3-en-2-one by Z-Scan

K Rodriguez[1]**, A Pérez**[1] **and F Racedo**[1]
[1] Universidad del Atlántico, Barranquilla, Colombia

E-mail: fran@mail.uniatlantico.edu.co

Abstract. We presents the study carried out by the technique Z-Scan, to analyse the nonlinear optical properties of (E)-4-(4-dimetylaminophenyl) but-3-en-2-one and (E)-4-(4-nitrophenyl) but-3-en-2-one, diluted in Ethyl Acetate with concentration levels of [0.02M, 0.08M, 0.23M] and [0.0047M, 0.013M, 0.041M] respectively. The measurements were_performed using a Nd:YAG laser emitting at 532nm, for samples with L_{eff} =1mm thickness, and a automated scanning of 10cm symmetric to the lens focus, the iris diameter was 1mm, the samples were also characterized by an UV-Vis Spectroscopy. We calculated the nonlinear refractive index (η2), the nonlinear absorption coefficient (β) and the the third-order nonlinear optical susceptibility (χ3) of the two researched compounds. The results show a combination of thermal response and nonlinear self-defocusing and self-focusing, which make their application interesting as new optoelectronic materials.

1. Introduction

The aromatic ketones α,β-unsaturated are known because they have in their structure a benzene ring, separated by three carbon atoms, two are connected by a double bond and the third is part of a carbonyl group. The presence of the carbonyl group and the olefin unit together confer the butenone high reactivity, placing them as an interesting class of molecules to study both in projects in basic science, applications in areas such as medicine [1-3,] agriculture and industry._Likewise, some small aromatic ketones have been used in the development of frequency amplification devices due to its SHG high activity (Second Harmonic Generation) [4].

The Z-Scan technique [5-6] is used to monitor the transmittance of the sample according to its Z position, in relationship with the position of the focal plane of the main lens [7-8]. The scan range depends on the parameters of the beam and the thickness of the sample. A critical parameter is the length of the diffraction of the focused beam, which is defined as πω02/λ for a Gaussian beam where the waist of the beam ω0 is the waist of the beam in the focal plane (1/e2). The nonlinear optical effects arise due to the nonlinear properties of the sample such as changes in the index of refraction and the absorption coefficient that takes place during the strong light intensity.

2. Experimental

Molecules are synthesized by Claisen-Schmidt [9-10] condensation procedure. We added acetone in excess (2 mL), benzaldehyde (1mmol), and the reaction is carried at 0°C temperature, then added drop-by-drop sodium hydroxide 10% p/v (5mL) and ethanol (3mL). After 12 hours of agitation a precipitate was obtained, filtered with cold ethanol and water. The butenone crystal obtained was recrystallized with ethanol at 50°C temperature. The molecular structure of (E)-4-(4-dimetilaminofenil) but-3-en-2-ona and (E)-4-(4-nitrophenyl) but-3-en-2-ona is shown in Figure 1.

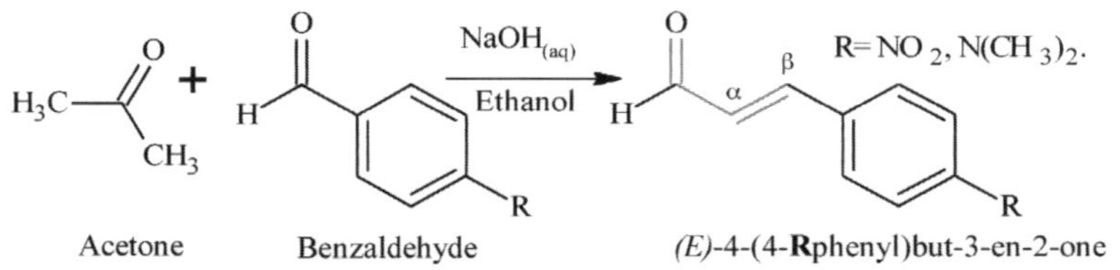

Figure 1. Molecular structure of *(E)*-4-(4-dimetylaminophenyl) but-3-en-2-one and *(E)*-4-(4-nitrophenyl) but-3-en-2-one.

The UV-Vis spectres of the sample were taken in the wavelength range of 200-400nm using an UV-vis EVOLUTION 220 Thermo Scientific spectrophotometer, in Figure 2 high absorbance between 250nm to 425nm is observed for both compounds. In the region of 500nm to 700nm absorbance is practically zero, making possible to use a 532nm laser for the study, this wavelength with high transmittance which allows ensures observe nonlinear effects. From the crystals obtained during the process, compound dilutions were prepared at 10mg/L, taking 1mL aliquots en quartz cells.

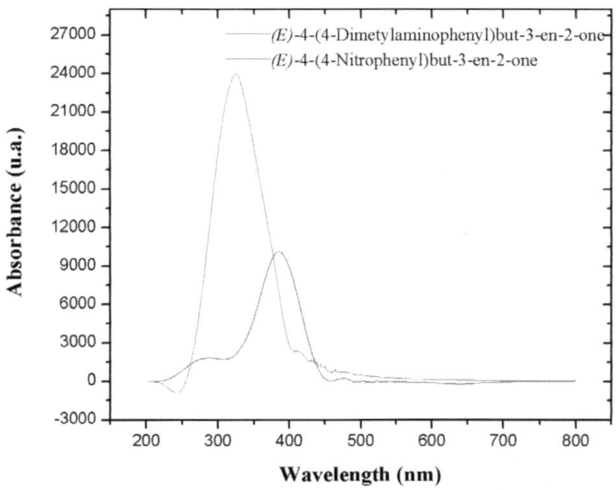

Figure 2. Absorption curves in the UV-Vis of *(E)*-4-(4-dimetylaminophenyl) but-3-en-2-one and *(E)*-4-(4-nitrophenyl) but-3-en-2-one diluted in ethanol at 10mg/L.

Z-Scan measurements were performed using a Nd:YAG laser doubled in frequency emitting at 532 nm, focused by a 10cm focal length lens and with a waist radius of 27 microns; the cell used to contain diluted samples is 1mm thick. The beam transmission obtained in closed and open cell geometry, was determined with a detector (see Figure 3). The Z-Scan curves of the butenones are shown in Figures 4 and 5.

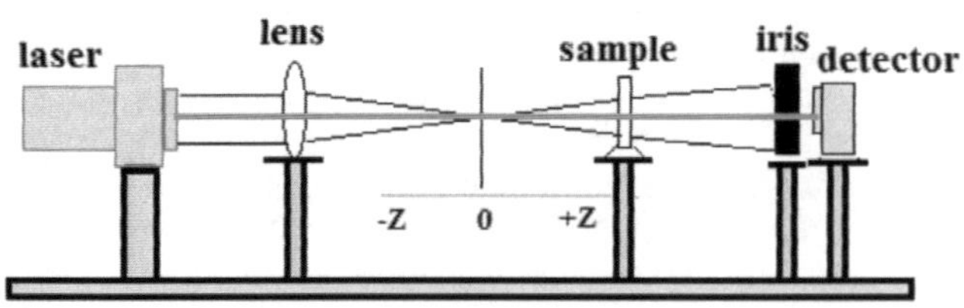

Figura 3. Z-Scan experimental setup.

3. Results and experimental analysis

As seen in the UV-Vis spectrum, the absorption generated by the HOMO-LUMO transition of unsaturated α,β system established for (E)-4-(4-dimetylaminophenyl)but-3-en-2-one and for (E)-4-(4-Nitrophenyl)but-3-en-2-one to contain this system in its structure, exhibit a maximum of absorption of visible light around 375nm and 323nm, respectively. This band of absorption is characteristic of the aromatic ketone, showing that a molecular fragment attracts electrons and other gives electrons (push-pull system) generating a process of electronic intramolecular transfer.

The substitution on the functional group $N(CH_3)_2$ with NO_2 alters conjugation of π system , therefore affects the absorption maximum observed that moves to a shorter wavelength. The spectrum shown in Figure 2 clearly exhibits this behaviour characteristic of the (E)-4-(4-phenyl)but-3-en-2-ones with a yellow coloration due to this displacement bathochromic derived from the change of an electron donor group instead of electron attractors, to realize studies for Z-Scan is recommended to utilize a wavelength of 532nm due to transmittance is high. The compounds demonstrate fluorescence, a phenomenon that is being studied, because of that they can be good candidates to photonic materials.

The Figures 4 and 5 show smooth curves of the experiment result of Z-scan in closed configuration. These materials have a self-defocusing thermal response that can be identified because the curve starts with a peak followed by a valley. The variation in the behaviour of the curves is due to the effect of substituent groups; in the case of the amino group, the nitrogen atom has a free electron pair, conferring an effect delocalized electron donor by interaction with the π system of the aromatic ring. In the case of nitro groups, there is an additional electron density, by the presence of double bonds in the nitrogen, this cause different behaviour and widening nonlinear absorption curve. The Figure 4 presents curves for Z-Scan (E)-4-(4-dimetylaminophenyl)but-3-en-2-one, at a concentration of 0.02M nonlinear effect begins to appear at Z=38.99mm, at a concentration of 0.08 M nonlinear effect appears at Z=21.39mm and at a concentration of 0.23M the effect appears at Z=4.89mm. In Figure 5 the Z-Scan by transmission in closed configuration is observed for (E)-4-(4-nitrophenyl)but-3-en-2-one at a concentration of 0.0047M nonlinear effect occurs in Z=32.43mm, at a concentration of 0.013 M nonlinear effect appears at Z=27.00mm and at a concentration of 0.041M the effect appears at Z=17.17mm. As the concentration increases, $\mathbf{\Delta Z_{p\text{-}v}}$ increases and a broadening in the peak-valley distance, lowest concentration is best because self-_defocusing thermal effect is better.

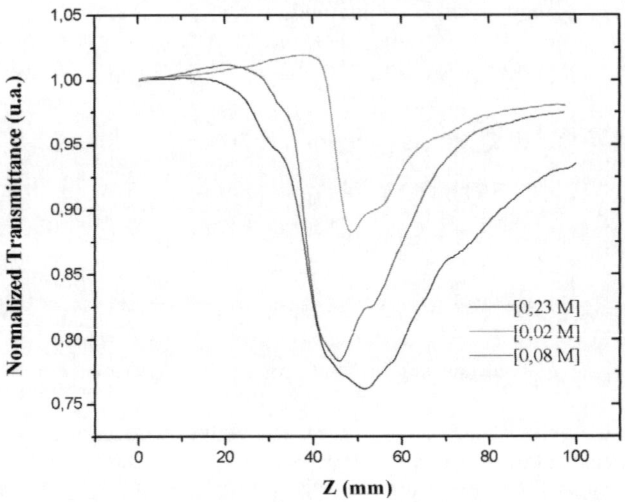

Figure 4. Z-Scan by transmission in closed configuration for (E)-4-(4-dimethylaminophenyl) but-3-en-2-one, Analysing the behaviour at different concentrations, with laser power of 195Mw in ethyl acetate.

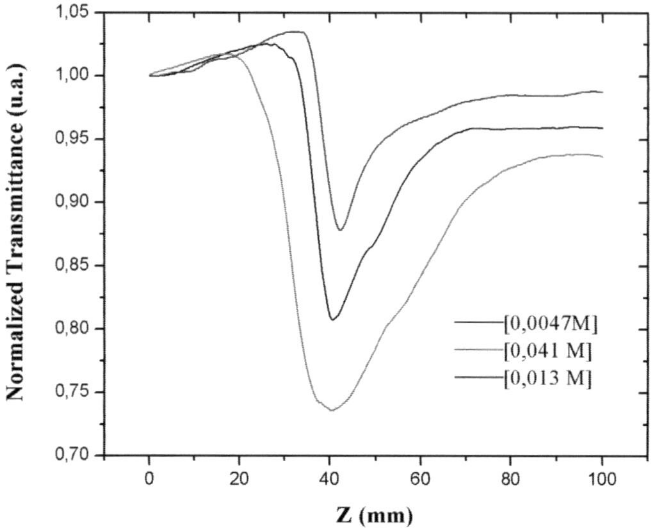

Figure 5. Z-Scan by transmission in closed configuration for (E)-4-(4-nitrophenyl)but-3-en-2-one, Analysing the behaviour at different concentrations with laser power of 195mW in ethyl acetate.

The theory developed by Sheik-Bahae[11] shows that $|\Delta\Phi o| \propto \Delta Tpv$ analogously $|\Delta\Phi o| \propto \eta 2$, Where $\Delta\Phi o$ is the maximum nonlinear phase change induced, it is related to the nonlinear refractive index $\eta 2$, as follows:

$$\Delta T_{pv}=0.406(1\text{-}S)^{0.25}k\ \eta_2 I L_{eff} \quad (1)$$

If clear η_2 de (1),

$$\eta_2=\Delta T_{pv}/(0.406(1\text{-}S)^{0.25}kIL_{eff}) \quad (2)$$

By replacing **k** (number of green laser wave) and **I** (Beam intensity green laser in focus) for $P/\pi\omega_0^2$ an equation that depends on the power and wavelength of laser is obtained.

$$\eta_2=\Delta T_{pv}\omega_0^2\lambda/[1.624(1\text{-}S)^{0.25}PL_{eff}] \quad (3)$$

If used the definition of Rayleigh parameter, we obtained.

$$\eta_2=\Delta T_{pv}\Delta Z_{pv}\lambda^2/[2.7608\pi(1\text{-}S)^{0.25}PL_{eff}] \quad (4)$$

If $S<<1$, (4) takes the following form:

$$\eta_2=\Delta T_{pv}\Delta Z_{pv}\lambda^2/[2.7608\pi PL_{eff}] \quad (5)$$

Equation (5) is used to obtain the value of nonlinear refractive index η_2. $L_{eff} \approx L=0.1cm$ for transparent samples.

The Table 1 contains the values of the nonlinear refractive index calculated according to the concentration of samples, a fixed laser power excitation was set at 195mW, an increase is observed at higher concentrations. Figure 6 It is shown as (E)-4-(4-dimetylaminophenyl) but-3-en-2-one y (E)-4-(4-nitrophenyl)but-3-en-2-one values of η_2 in function of concentration have a linear behaviour to the concentrations tested.

Table 1. Values of refractive index nonlinear obtained from transmittance curve in closed configuration for (E)-4-(4-dimetylaminophenyl) but-3-en-2-one y (E)-4-(4-nitrophenyl) but-3-en-2-one; **P**=0.195 W, L_{eff}=0.1cm.

Molecule	*(E)*-4-(4-dimetylaminophenyl) but-3-en-2-one			(E)-4-(4-nitrophenyl) but-3-en-2-one		
concentration	0.02M	0.08M	0.23M	0.0047M	0.013M	0.041M
ΔT_{p-v}	0.0381	0.2288	0.2186	0.155	0.2171	0.2816
ΔZ_{p-v}(cm)	0.0933	0.2431	0.3625	0.0980	0.1367	0.2328
η_2 (cm²/W)	$5.948x10^{-11}$	$3.424x10^{-10}$	$1.3261x10^{-9}$	$2.55x10^{-10}$	$4.690x10^{-10}$	$1.0970x10^{-9}$

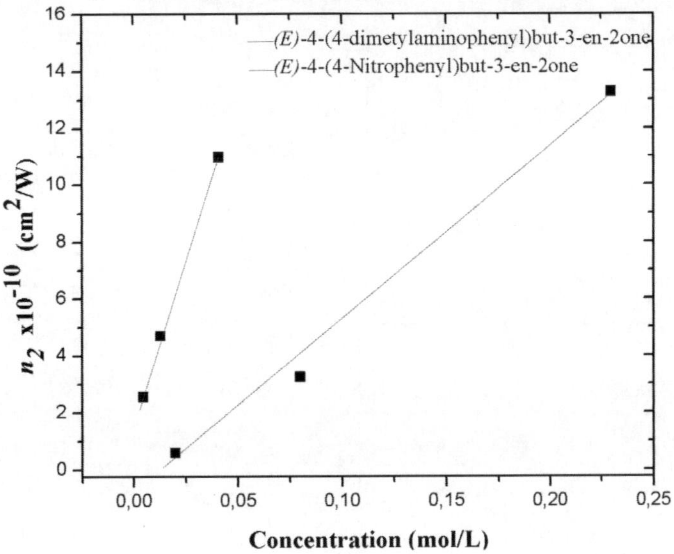

Figure 6. Variation nonlinear refractive index depending on the concentration.

To obtain the nonlinear absorption coefficient β, open cell geometry was used and data of the normalized transmittance for each sample were used as well, from [11] is known that small losses for nonlinear third-order has $\Delta\alpha L=\beta I L_{eff}<1$.

Where L_{eff} =1mm is the sample thickness, $\mathbf{I_0}$=8519W/cm² is the intensity of the laser beam, in Figure 7 shows β Vs Concentration obtained by Z-Scan in open configuration for the studied samples, data can be calculated from.

$$\beta=2(2\Delta T_{vp})^{1/2}/I_0L_{eff} \quad (6)$$

Where ΔT_{vp} is the value of the valley-peak difference in the Z-Scan curve open cell geometry.

To analyse the effect that causes the laser power to influence the sample (E)-4-(4-dimethylaminophenyl)but-3-en-2-one [0.02M], transmittance measurements were performed at different laser powers in the open configuration, in Figure 8, variations were observed for the non-linear effect as the laser power increases to less than 45mW, the nonlinear effect begins to be notice; but for the power between 90-200 range mW increases the effect and it is noted that near the 190mW curves suffer a widening, indicating a local thermal effect, heating product produced by the laser at the point where focuses on the sample begins to be significant. This effect is observed at the beginning and at the end in the Z-Scan curve.

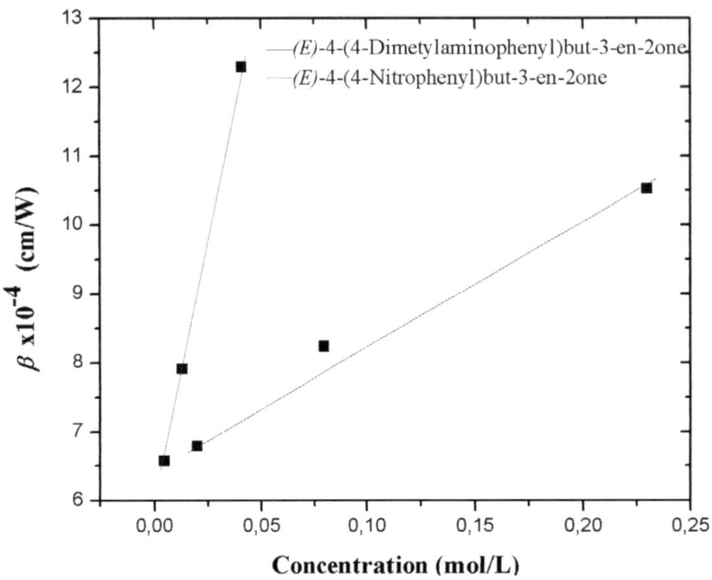

Figure 7. Variation of nonlinear coefficient absorption for samples with different concentrations according to Table 2.

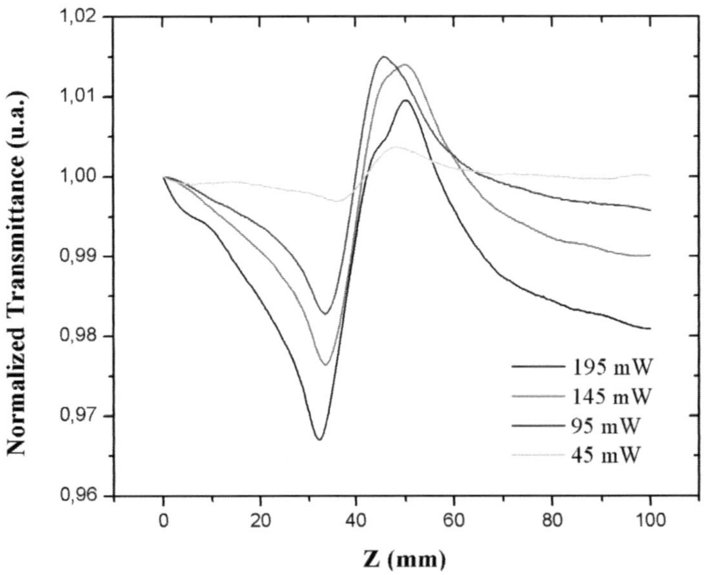

Figure 8. Normalized transmittance curve in open configuration of (E)-4-(4-dimethylaminophenyl)but-3-en-2-one [0.02 M] in ethyl acetate, for each samples a scan of 100 mm was performed at symmetrical lens focus with L_{eff}=1mm.

Determining the refractive index of the samples η_0 became necessary in the parameter to determining third order nonlinear optical susceptibility Re $\chi3$, which it was obtained by applying Snell's law. The obtained values of nonlinear refractive index and absorption coefficient (E)-4-(4-dimetylaminophenyl) but-3-en-2-one and (E)-4-(4-nitrophenyl)but-3-en-2-one are used to determine the actual value of the nonlinear optical susceptibility of the third order according to the following relationship: [12]

$$\mathrm{Re}\chi3\ (esu) = (10^{-4}\xi_0 c^2\ \eta_0^2/\pi)\ \eta_2 \qquad (7)$$

Where ξ_0 is the vacuum electric permittivity ($8.8542x10^{-16}$ C/Nm^2) and **c** ($3x10^{10}$cm/s) is the speed of light.

The results obtained are summarized in Table 2 that shows molecule (E)-4-(4-dimetylaminophenyl)but-3-en-2-one, $\chi 3$ presents values in a range 1.6633×10^{-13} to 3.8511×10^{-13} esu at the concentrations studied, demonstrating an upward linear behaviour, likewise occurs (E)-4-(4-nitrophenyl)but-3-en-2-one with values in a range χ 3 $7.1311\text{x}10^{-13}$ to $3.1857\text{x}10^{-13}$ esu.

Table 2. Values of nonlinear absorption coefficient and nonlinear optical susceptibility of the third order obtained for (E)-4-(4-dimetylaminophenyl)but-3-en-2-one and (E)-4-(4-nitrophenyl)but-3-en-2-one; $\mathbf{I_0}$=8519W/cm^2.

Molecule	(E)-4-(4-dimetylaminophenyl) but-3-en-2-one			(E)-4-(4-nitrophenyl) but-3-en-2-one		
Concentration	0.02M	0.08M	0.23M	0.0047M	0.013M	0.041M
ΔT_{p-v}(cm)	0.0418	0.0761	0.1004	0.0392	0.1391	0.151
β (cm/W)	$6.788\text{x}10^{-4}$	$8.233\text{x}10^{-4}$	$1.0520\text{x}10^{-3}$	$6.5726\text{x}10^{-4}$	$7.912\text{x}10^{-4}$	$1.2901\text{x}10^{-3}$
η_0 (u.a)	1.05	1.06	1.07	1.05	1.06	1.07
Re $\chi 3$(esu)	$1.6633\text{x}10^{-13}$	$9.7587\text{x}10^{-13}$	$38.511\text{x}10^{-13}$	$7.1311\text{x}10^{-13}$	$13.367\text{x}10^{-13}$	$31.857\text{x}10^{-13}$

4. Conclusion

Measurements made with the technique Z-Scan on (E)-4-(4-dimetylaminophenyl)but-3-en-2-one and (E)-4-(4-nitrophenyl)but-3-en-2-one, using a Nd:YAG laser showed high transmittance at 532 nm, they allowed us to obtain experimental values reported for the first time η_0, η_2, β y Re$\chi 3$, The nonlinear refractive index and nonlinear absorption coefficient are linear function of concentration, it shows that they are affected by the ligand group bonded to the molecular structure. The presence of the nitro group in (E)-4-(4-nitrophenyl)but-3-en-2-one shows better linearity than in (E)-4-(4-dimetylaminophenyl)but-3-en-2-one as shown in Figures 6 and 7, for this result the effect of the laser power is highlighted on the samples, when concentration increases there are more particles resulting stirred thermally enhanced nonlinear effect. The values obtained for Re$\chi 3$ in the order of 10^{-13} show that the effect of the interaction of light with molecules samples studied and it make it interesting for application as new optoelectronic materials. Varying the nonlinear refractive index is directly proportional to the third order electric susceptibility and the applied field strength.

References

[1] L E Cuca and O E Bañol 2005 *Revista Colombiana de Química* **34** 35-41
[2] Y Kim, J Kim, H Park and H. Kim 2007 *Biological and Pharmaceutical Bulletin* **30** 1450-1453
[3] J H Isaza, F J Jimenez, J L Galvan and J C Restrepo 2007 *Scientia et Technica* **13** 409-41
[4] P S Patil, S M Dharmapraash, K Ramakrishna, H Fun, R Sai Santosh, R Kumar, D Narayana 2007 Second harmonic generation and crystal growth of chalcone derivates *Journal of crystal growth* **303** 520-524
[5] S L Gomez, F L S Cuppo and A M Figueiredo Neto 2003 *Brazilian Journal of Physics* **33(4)** 813-820
[6] G Tsigaridas, G Fakis, I Polyzos, P Persephonis and V Giannetas 2003 *Appl. Phys. B* **1** 83-86
[7] M G Kuzyk and C W Dirk 1998 *Characterization Techniques and Tabulations for Organic Nonlinear Materials* (New York: Marcel Dkker, Inc.), 655-692.
[8] A Garcia, SValbuena, R Sarmiento, F Racedo 2015 Medida de las propiedades ópticas no lineales de aceite de oliva usando Z-Scan *Opt. Pura Apl.* **48(1)** 55-61
[9] Fuentes A, Marinas J. M, Sinisterra J V 1987 *Tetrahedrons Letters* **28** 4541-4544
[10] Aguilera A, Alcántara R, Marinas J R, Sinisterra J V 1987 *Can. J. Chem.* **65** 1165-1171
[11] Sheik-Bahae M et al. 1990 Sensitive measurement of optical non-linearities using a single beam *IEEE J. Quantum Electron.* **26** 760-769
[12] Sheik-Bahae M, Said A A 1989 Van Stryland EW. High - sensitivity, single - beam η_2 measurements *Opt Lett* **14** 955

Characterization of the nonlinear optical properties of the new 2-amino-4,6-difenilnicotinonitrilo by UV-Vis spectroscopy absorption and Z-Scan

C Cueto[1], A Pérez[1] and F Racedo[1]
[1] Universidad del Atlántico, Barranquilla, Colombia

E-mail: fran@mail.uniatlantico.edu.co

Abstract. The study of the nonlinear optical properties of new organic molecules in solution was performed. Z-Scan technique was used to investigate the nonlinear optical properties of 2-amino-4,6-diphenylnicotinenitrile depending on the solution concentration and the laser power; This compound was diluted in ethyl acetate at fixed concentration of 0.024M. Through this technique, nonlinear parameters such as the nonlinear refractive index (η_2), the nonlinear absorption coefficient (β) and the third-order electric susceptibility (χ^3) were determined. For these measurements, a laser Nd: YAG emitting at 532nm, a lens 10 cm focus, an iris of 1mm and a cell with a thickness of 1mm were used. The study was performed with laser powers of 55mW, 100mW, 145mW and 195mW; All measurements were made by transmission in closed and open configurations. Finally the sample was characterized by absorption spectroscopy UV-Vis. This study allows us to relate the molecular design with the optical properties.

1. Introduction

The aminodiphenylnicotinenitriles are compounds very interesting because they possess a pyridine structure, which is known to have biological activity, they also have an amino group at position 2 and a cyano group in position 3 according to the numeration of pyridine ring IUPAC; allowing cyclocondensation reactions, nucleophilic addition reactions bidentate ligands and therefore are synthesized by conventional and no conventional methods [1].

This compound has a conjugated system or delocalization of the π orbital, which makes it possible participant in charge of transfer processes. Furthermore, they can be synthesized incorporating appropriate substituents on the aromatic rings so that each ring may confer characteristics like charge donor or acceptor [2-4].These charge transfer characteristics and polarizability could make these materials suitable as raw materials for optoelectronic applications [5].

Organic materials with nonlinear optical responses are very interesting in this subject because their time response is fast; they are inexpensive, easy to process and exhibit thermal and chemical stability, among other advantages. The structural requirement to be satisfied by organic materials that can show nonlinear optical phenomena is the presence of a lattice of electron π-conjugated delocalized, which confer a high polarizability and rapid charge redistribution when the conjugated molecule interacts with intense and rapidly variable electromagnetic fields as the ones obtains with laser. The large and ultrafast nonlinearity of these materials can be used in different applications such as optical switches, optical modulators, optical signal processing and as optical limiters [6].

Z-Scan technique [7] is used to monitor the transmittance of the sample according to their position Z on the position of the focal plane of the main lens [8].Scan range required depends on the beam parameters and the sample thickness. The nonlinear optical effects are due to the nonlinear properties of the sample such as changes in the refractive index and absorption coefficient that is taken in the incidence of a beam of high intensity light such as a laser [9,10].

Nowadays, there are not any reports of an optical study done on 2-amino-4,6-diphenylnicotinenitriles. Therefore a detailed analysis of the nonlinear optical properties is performed for a better understanding of the system α, β-unsaturated.

2. Experimental technique

2.1. Synthesis of 2-amino-4,6-diphenylnicotinenitrile

The synthesis of this compound occurs by heating till melting temperature by mixing one mmol (198.04mg) 2-benzylidenemalononitrile, previously synthesized from piperonal with malononitrile through a Knoevena gel condensation; with acetophenone (0.116mL) and ammonium acetate (770mg) throught pyridine synthesys of Hantzsch.The mixture was melted in a 10mL balloon at 200°C for 30 minutes. During reaction, the mixture was controlled by thin layer chromatography (TLC), after that ethanol was added and heated for 10 minutes. After reaction, the mixture is allowed to stand until it reaches room temperature. The resulting precipitate was filtered, washed with cold ethanol, dried and recrystallized. Synthesis reaction is shown in Figure 1.

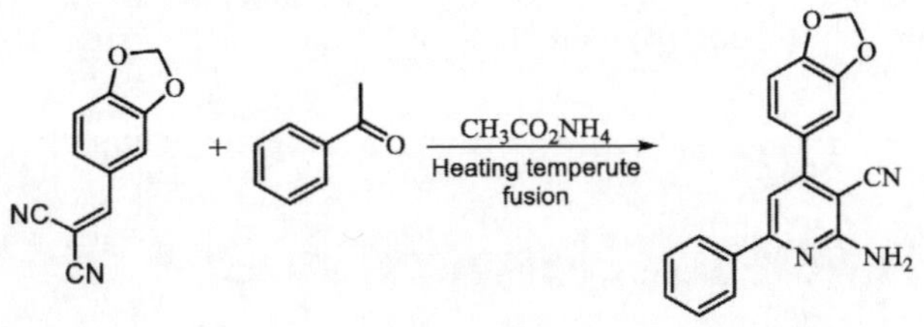

Figure 1. Synthesis of 2-amino-4,6-diphenylnicotinenitrile.

2.2. Characterization of the sample by UV-Vis spectroscopy

The UV-Vis spectrum of the sample was recorded in the wavelength range of 200 to 800nm using a UV-Vis spectrophotometer EVOLUTION60S ThermoScientific, sample preparation was made at a concentration of 20mg/L. The spectrum is shown in Figure 2. It is observed in the spectrum that 2-amino-4,6-diphenylnicotinenitrile absorbs light at a wavelength of 355nm; where it is possible to use a laser emitting at 532nm, since no absorption is observed at this wavelength presenting high transmittance, which is suitable for experiments with Z-Scan.

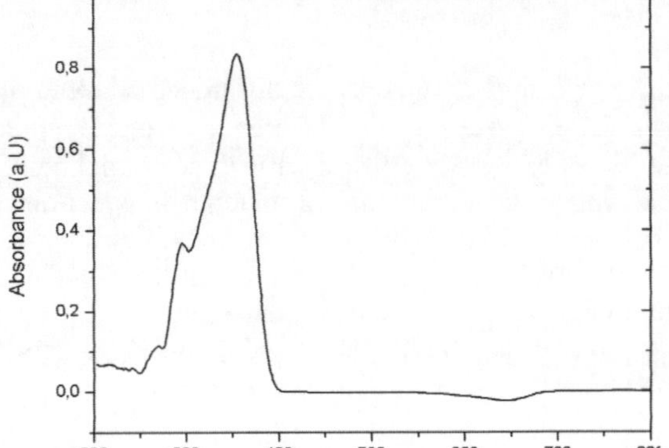

Figure 2. UV-Vis spectrum of 2-amino-4,6-diphenylnicotinenitrile.

3. Results and discussion

Through the Z-Scan technique were carried out measurements using a laser Nd: YAG doubled in frequency emitting at 532nm and focused by a lens 10cm, a radius belt 27 microns at the focal point, the radius of the beam waist of the laser is 1.25mm, as shown in Figure 3; the sample was placed in a glass cell with thickness of 1mm. 2-amino4,6-diphenylenilnicotinenitrile under study was diluted in ethyl acetate at a fixed concentration of 0.024 M, measurements were performed at different laser power: 55mW, 100mW, 145mW and 195mW.

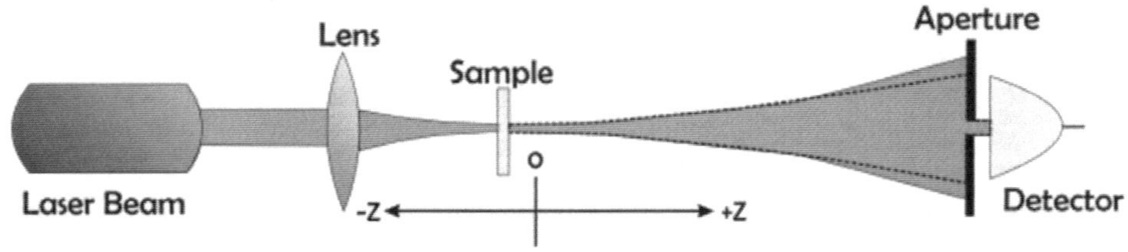

Figure 3. Experimental set of the Z-Scan technique.

The Z-Scan technique is a method used frequently to determine the nonlinear refractive index (η_2), the nonlinear absorption coefficient (β) and the third-order electric susceptibility (χ^3) in certain materials.

As proposed by Sheik-Bahe´s theoretical mathematical model, the nonlinear refractive index (η_2) was calculated by:

$$\Delta Tpv = 0.406\,(1-S)^{0.25}\,|\Delta\Phi| \tag{1}$$

Where he $\Delta\Phi$ is the phase shift in the focus axis and is expressed by $|\Delta\Phi| = kn_2 I_0 L_{eff}$ and $I_0 = P/\pi\omega_0^2$ is the intensity of the laser.

Solving n_2 in Equation 1, and replacing it in the expression above, we obtain:

$$n_2 = \frac{\Delta Tpv\ \omega_0^2\ \lambda}{1.624\ (1-S)^{0.25} P\ L_{eff}} \tag{2}$$

Using the Rayleigh parameter definition is obtained:

$$n_2 = \frac{\Delta Tpv\ \Delta Zpv\ (\lambda)^2}{2.7608\ \pi\ P\ L_{eff}} \tag{3}$$

Equation 3 is used to obtain n_2 where ΔTpv can be defined as the difference between the peak and the valley of the normalized transmittance (T_p-T_v), ΔZpv is the difference of displacement in the Z path, λ is the length of laser wave, P is the power emitted and L_{eff} laser is the cell thickness.

In Figure 4, variations in the nonlinear optical properties of the material in function of the laser power are observed. For displacement between 60mm and 100mm the emergence of an effect of thermal origin primarily to high power, due to local heating of the sample at the point where laser is located . On power of 55mW, thermal and nonlinear effects are less intense.

To obtain β, we proceed to remove the iris in assembling to the Z-Scan. From this normalized transmittance curve in open configuration (Figure 5), we were able to calculate β.

$$\beta = \frac{2(2(\Delta T_{vp}))^{1/2}}{I_0 \, L_{eff}} \tag{4}$$

Where I_0 is the intensity of laser beam, ΔT_{vp} is the difference between valley and peak of the normalized transmittance curve for the open cell configuration, and L_{eff} is cell thickness.

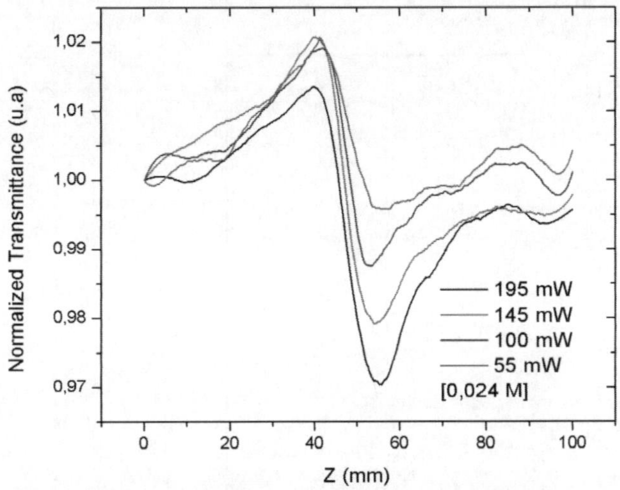

Figure 4. Normalized transmittance for Z displacement of 100mm in closed configuration to different powers of 55mW, 100mW, 145mW and 195mW, and nonlinear refractive index (η_2) positive 2-amino-4,6-diphenylnicotinenitrile.

A comparison Z-Scan scanning in open cell configuration is shown in Figure 5, for different laser powers. The emergence of the thermal effect on the 2-amino-4,6-diphenylnicotinenitrile for a laser power of 55mW is smaller, but has less nonlinearity, opposite occurs in the curve of 195mW of laser power, where the thermal effect is greater compared to the other laser power, but this shows a better nonlinearity. The thermal effect increases when increasing laser power. Furthermore it is observed that close to 195mW power curves suffer a widening indicating that the thermal effect produced by the laser focused on the sample begins to be significant, this effect observed in the curve Z-Scan.

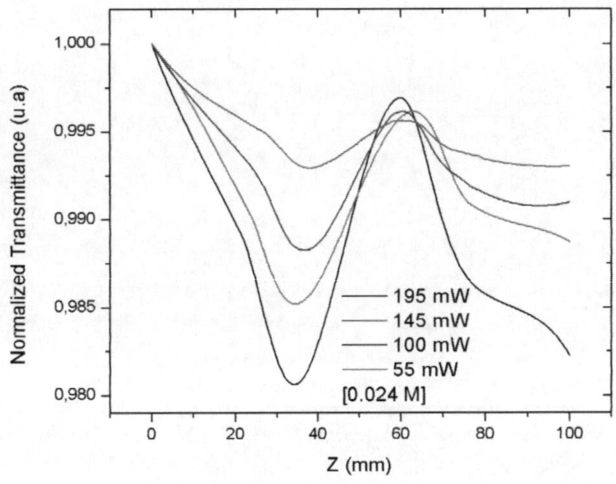

Figure 5. Normalized transmittance for displacement Z of 100mm open cell, laser powers of 55mW, 100mW, 145mW and 195mW, 2-amino-4,6-diphenylnicotinenitrile.

For determining the third-order electric susceptibility (χ^3) the following equation is used:

$$\mathrm{Re}\,\chi^3(\mathrm{esu}) = \left(\frac{10^{-4}\varepsilon_0 c^2 n_0^2}{\pi}\right) n_2 \quad (5)$$

Where ε_0 is the electric permittivity, c is light speed, n_0 is the refractive index of material obtained from Snell's law and n_2 is the nonlinear refractive index.

Figure 6 and 7 shows that the nonlinear refractive index and nonlinear absorption coefficient, depending on the potency show linear behaviour.

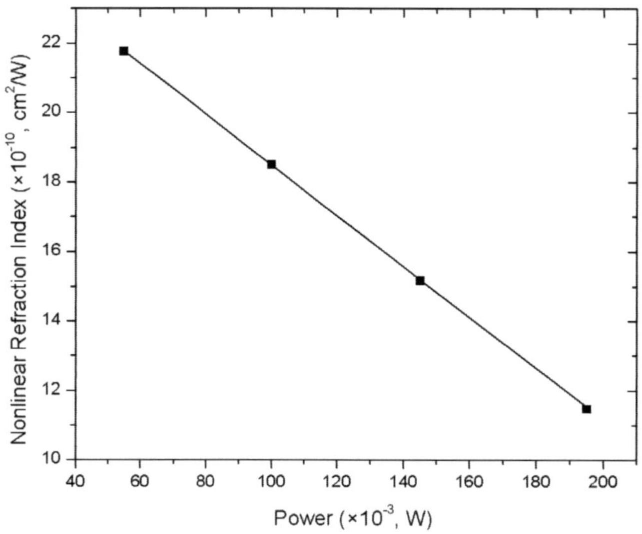

Figure 6. Nonlinear refractive index variation depending on the laser power according to Table 1 for 2-amino-4,6-diphenylnicotinenitrile.

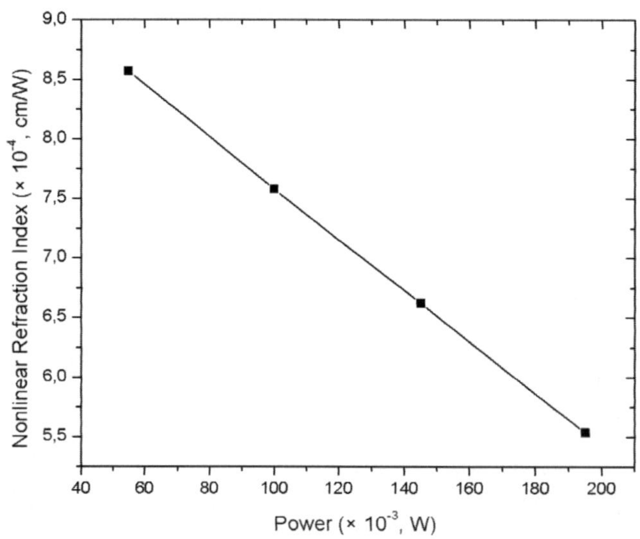

Figure 7. Variation of nonlinear absorption Coefficient based on the laser power according to Table 1, for 2-amino-4,6-diphenylnicotinenitrile.

IOP Conf. Series: Journal of Physics: Conf. Series **786** (2017) 012005 doi:10.1088/1742-6596/786/1/012005

Table 1. Optically nonlinear parameters calculated for the 2-amino-4,6-diphenylnicotinenitrile at a concentration of [0.024M] in different power.

Power (mW)	Refractive index nonlinear			Absorption coefficient nonlinear			Electric susceptibility of third order	
	ΔTpv (u.a)	ΔZpv (cm)	n_2 (cm^2/W)	ΔTvp (u.a)	I_0 (W/cm^2)	β (cm/W)	n_0	Re (χ^3) (esu)
55	0.0248	1.4788	21.7589×10^{-10}	5.3×10^{-3}	2402.7959	8.5697×10^{-4}	1.0707	6.3272×10^{-12}
100	0.0318	1.7826	18.4977×10^{-10}	13.7×10^{-3}	4368.7199	7.5779×10^{-4}		5.3789×10^{-12}
145	0.0415	1.6251	15.1774×10^{-10}	22.0×10^{-3}	6334.6439	6.6227×10^{-4}		4.4134×10^{-12}
195	0.0533	1.2865	11.4746×10^{-10}	27.8×10^{-3}	8519.0039	5.5357×10^{-4}		3.3367×10^{-12}

4. Conclusion

From normalized transmittance curves obtained from technical Z-Scan, it were achieved characterize the various parameters of non-linear optics of 2-amino-4,6-diphenylnicotinenitrile, as shown in Table 1, where it was observed that each parameter decreases as was increased laser power.

Experimental results of these parameters were obtain by first time, in which the study was conducted at different laser power, resulting in a variation on properties of the compound was obtained first, allowing accurate value determined for each power.

References

[1] Yelwande A A, Navgire M E, Tayde D T, Arbad B R and Land M K 2012 *S. Afr. j. chem.* **65** 131-137
[2] Ritesh N, Sethuraman S and Beilstein 2010 *J. Org. Chem* **6** 992-1001
[3] Tao X T, Watanabe T, Kono K, Deguchi T, Nakayama M and Miyata S 1996 *Chem. Mat.* **8** 1326-1332
[4] D'Silva E D, Podagatlapalli G K, Rao S V, Rao D N and Dharmaprakash S M 2011 *Cryst. Growth Des.***11** 5362-5369
[5] Rao Y K, Fang S H and Tzeng Y M 2004 *Bioorg. Med. Chem* **12** 2679-2686
[6] Kirkby C J G, Cush R and Bennion I 1985 *Opt. Commun.* **56** 288
[7] Gomez S L, Cuppo F L S and FigueiredoNeto A M 2003 *Braz. J. Phys.* **33(4)** 813-820
[8] Van Stryland E W and Sheik-Bahae M 1998 Z-scan measurements of optical nonlinearities *Characterization Techniques and Tabulations for Organic Nonlinear Materials* **18(3)** 655-692
[9] Garcia A, Valbuena S, Sarmiento R and Racedo F 2015 Medida de las propiedades ópticas no lineales de aceite de oliva usando Z-Scan *Opt. Pura Apl.* **48(1)** 55-61
[10] Fontalvo M, Garcia A, Valbuena S and Racedo F 2016 Measurement of nonlinear refractive index of organic materials by z-scan *Journal of Physics: Conference Series* **687** 012100

Magneto-electric effect in non-uniform quantum nanotube

L F Garcia[1], I D Mikhailov[1] and J Sierra-Ortega[2]
[1] Universidad Industrial de Santander, Bucaramanga, Colombia
[2] Universidad de Magdalena, Santa Marta, Colombia

E-mail: lgarciar@uis.edu.co

Abstract. A model of non-uniform core-shell type nanostructure with a donor located at a bottleneck point of the core is considered. It is shown that such structure can have a giant polarizability, which, besides can be controlled by applying of the external magnetic field. Presented theoretical analysis reveals a new possibility for the coupling between the polarization and magnetization arising from the quantum-size effect in non-uniform semiconductor nanowires.

1. Introduction

Nanostructures with tubular architecture have exceptional physical properties of great interest for potential applications in electronic devices, solar cells, and sensing. The ring-like geometry of these structures offers an opportunity for controlling their spectral and electrical properties by applying of an external magnetic field. On the other hand, their reduced dimensionality along the axis retains electric carriers within a narrow conducting channel, making them very sensible to the external electric field. One could expect that the combination of these properties offer an additional possibility in device design related to control of the magnetization or polarization by an electric field or magnetic field. Different techniques of crystal growth have been used to fabricate narrow GaAs/AlAs and InAs/GaAs nanotubes (NTs) [1-3], which, as it has been demonstrated, exhibit in their photoluminescence spectrum the Aharonov-Bohm (AB) oscillations in the presence of the external magnetic field [4]. Earlier, it has been also reported on fabrication of non-vertically aligned GaN/GaP core/shell nanowires with variable cross section dimensions [5, 6].

The charge probability distribution along the symmetry axis of the excessive electron inside a NT with variable radius in the ground and excited states is nonhomogeneous, the narrower is the cross section the smaller is the probability for finding the electron close to this point. However, the distribution can be easily changed in the presence of the external and the electric field due to a high electric and magnetic polarizability of such structure. In this paper, it is presented a theoretical analysis of a possibility for designing a tubular architecture with a giant electric polarizability, which besides can be controlled by the external magnetic field.

2. Theoretical model

To this end, below it is considered a model of axially symmetric nanowire with the core/shell type heterojunction, and with a shape of the vertical cross-section along the symmetry axis, schematically represented in Figure1. In this model, the shell region presents a tube-shaped quantum well with variable radius, capable to trap the excessive electron, released from the on-axis donor, located at the bottleneck point. The corresponding geometrical parameters of the structure are the height, H the centreline radius

R_b at the bottleneck, the centreline radius R_t at extreme points of the tube, and the shell thickness w . The external fields, applied parallel to the Z-axis, are magnetic B and electric F. In this model, the shell region presents a tube-shaped quantum well with variable radius, capable to trap the excessive electron, released from the on-axis donor, located at the bottleneck point. The shell layer along the core's lateral frontier regarded below as an infinite-barrier antidot, presents a narrow cylindrical nanotube with the centreline radius depending on the coordinate z as follows:

$$R_c(z) = R_b + (R_t - R_b)\cdot|z/H|^n ;\ n \quad 1, 2 \tag{1}$$

The core/shell frontier presents a hard wall for the electron is located within the quasi-2D region defined in cylindrical coordinates as $\{R_c(z) - w/2 < \rho < R_c(z) + w/2;\ 0 < \varphi < 2\pi;\ 0 < z < H\}$, in which the electron circulates around the core, encompassing the cone's lateral surface. The electrostatic attraction and the external magnetic field B press the electron circular tracks toward the bottleneck point while the external electric field F and the structural confinement push them toward one of two extreme points of the structure. The competition between these two groups of forces can change drastically the electron-donor separation resulting in the inducing of a giant dipole momentum.

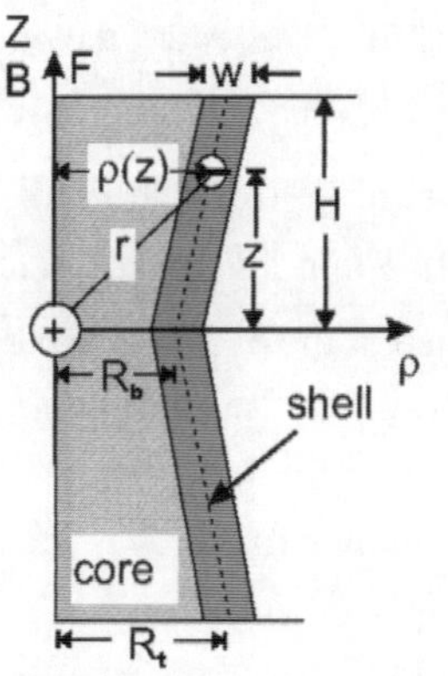

Figure 1. Vertical cross-section of the nanowire with core/shell type junction and with on-axis donor.

In our calculations there are used effective Bohr radius $a_0^* = {}^2\varepsilon/m^* e^2$ and the effective Rydberg, $R_y^* = e^2/2\varepsilon a_0^*$ as units of the length and energy, while the values $\xi = eF a_0^*/R_y^*;\ \gamma = e\ B/2m^* R_y^*$ as dimensionless parameters of electric field F and magnetic field B, respectively, being m^* the electron effective mass and ε dielectric constant of the shell material parameters. By using the adiabatic approximation [7,8] and considering the ratio $w/R_c(z)$ as a small parameter one can represent wave functions for low-lying energy levels as follows: $\Psi_m(\rho,\varphi,z) = e^{im\varphi}\phi(\rho) f_m(z);\ m = 0, \pm 1, \pm 2,$. Here m is the Z-projection of the angular momentum, $\phi(\rho)$ is the radial eigenfunction corresponding to the lowest energy in the circular 2D ring of the width w and the radius $R_c(z)$, while the axial part of the wave function $f_m(z)$ in the limit $w/\rho_c(z) \to 0$ is the solution of the one-dimensional boundary problem with the adiabatic potential $V_{eff}(z)$:

$$-f_m''(z) + V_{eff}(z) f_m(z) = E_m \cdot f_m(z);\ -H < z < H;\ f_m(\pm H) = 0 \tag{2a}$$

$$V_{eff}^{(m)}(z) = \xi \cdot z + \left(m^2 + 1/2\right) \Big/ R_c^2(z) + \gamma^2 \cdot R_c^2(z)/4 - 2\Big/\sqrt{z^2 + R_c^2(z)} - \gamma \cdot m \quad (2b)$$

3. Results and discussion

Eigenvalues of Equation (2) are found by means of the numerical trigonometric sweep method [9,10]. Results, presented below are given for the structure with parameters, $R_b = 1a_0*$, $R_t = 4a_0*$ and $H = 20a_0*$. The type of the tube's profile no homogeneity in our model (1) is given by the parameter n. Below, the profile is called as linear if $n = 1$ and parabolic if $n = 2$. In presented below calculations dimensionless units typical for GaAs material $a_0* \approx 1nm$, $Ry* \approx 6meV$ are used.

Classical electron paths present circular horizontal tracks whose radii $R_c(z)$ depend on the distance $|z|$ from the bottleneck point, given by the minimum position $z_{\min}$ of the adiabatic potential (2b). The stationary point position $z_{\min}$ in accordance with Equation (2b) is defined by interplay between two groups of forces. One of them is induced by the external electric field and the centrifugal force, both pushing the electron toward one of two extremes, and other induced by the magnetic confinement and the Coulomb attraction to the donor that drive the electron toward the bottleneck point $z = 0$. Analyzing the displacement of $z_{\min}$ provided by external fields one can give a simple interpretation of the spectral, electric and magnetic properties of this structure. In Figure2 it is presented the curves of the adiabatic potentials $V_{eff}^{(0)}(z)$ for states with zero angular momentum $(m = 0)$, for two different magnetic fields $\gamma = 0$ and $\gamma = 2$, and for three different values of the electric field. At the zero magnetic field case $(\gamma = 0)$ the potential curves presented in Figure 2(a) have two minima, one deep main minimum, close to the donor $(z = 0)$ and other, the shallow minimum located at the extreme of the NT $(z = H)$. As the electric field increasing surpasses the value $F = 0.5kV/cm$, the second minimum becomes more profound, while the electron in its ground state is jumped from the bottleneck point $(z = 0)$ up to the tube extreme $(z = H)$, generating in this way a giant dipole momentum.

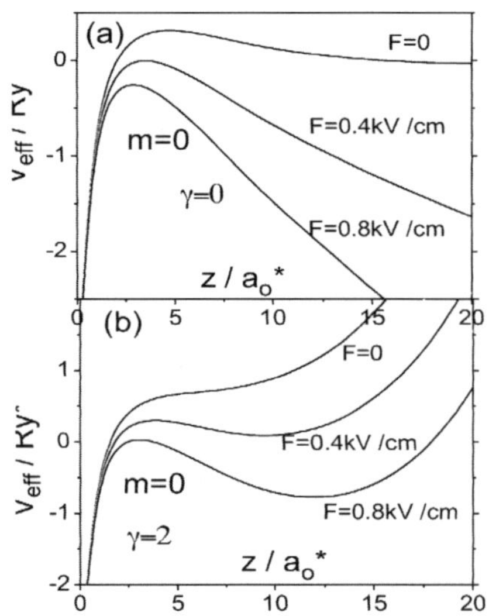

Figure 2. Adiabatic potential along symmetry axis of the nanotube with linear profile in the presence of external fields.

The shape of the potential curves is changed essentially in the presence of the magnetic field as it can be seen from Figure 2(b), where similar results for $\gamma = 2$ are shown. If the electric field pushes the electron toward the tube's extremes, the diamagnetic confinement (the third term in Equation 2(b) drives it toward the bottleneck point and therefore under increasing external magnetic field the second minimum in the adiabatic potential in Figure 2(b) is successively displaced in the direction of the bottleneck point. Such modification of the adiabatic potential under external field affected also the charge distribution in the ground state of the structure. In Figure 3(a) are presented results of calculation of the electron ground state energy E_0 as function of the electric field, found by solving eigenvalue problem (2). For small electric field the ground state energy is almost constant about 1Ry*due to the fact that the electron-donor separation remains unchanged until F increasing reaches a critical value. As F further increases, the electron-donor separation is changed abruptly and the energies begin to grow linearly with a sharply increased slopes. In Figure 3 the curves are presented for three different values of the external magnetic field applied at the same direction. The larger applied magnetic field, the greater is the critical value of the electric field at which the charge distribution and the slopes of curves are changed significantly.

In Figure 3(b) it is presented the electric moment d of the structure at zero temperature as function of the electric field calculated by means of the numerical derivation of the ground state energy E_0 accordingly to the relation:

$$d = -ea_0 * \cdot \partial E_0 / \partial \xi \tag{3}$$

It is seen that the value of the dipole moment of the structure for small electric field corresponds to the electron-donor separation about $1a_0*$ and rises abruptly up to a value corresponding to a giant electron-donor separation close to $20a_0*$, as the electric field exceeds its critical value. The value of the critical electric field can be changed by applying the magnetic field. As the magnetic field increased, the jump of the curves of the dipole moment dependency becomes smoother. The curves presented in Figure 3 point out a possibility of the existence of the magnetoelectric effect in nanostructures with tubular architecture.

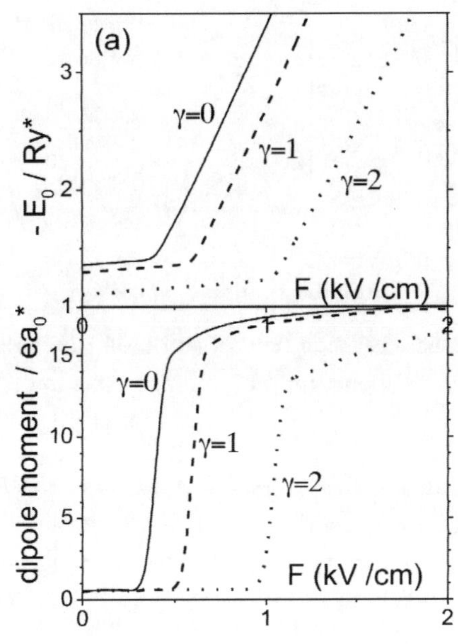

Figure 3. (a) Ground state energy and (b) Dipole moment as functions of the external electric field, given for three different values of external magnetic field.

Other manifestation of the quantum-size magnetoelectric effect in tubular nanostructures one could reveal by analyzing AB oscillations of the energy levels in the presence of the external magnetic field. AB oscillations of energy levels in homogeneous tubes are similar to those in one-dimensional quantum, where the period of the oscillation of the ground state energies $\Delta\gamma$ depends on the radius R as $\Delta\gamma = 2/R^2$. One can observe in Figure 4 AB oscillations of the eigenvalues E_m of the Equation 2 for some lower levels $m = 0, \pm 1, \pm 2,$ in uniform nanotubes with radii $R = 1a_0{}^*$ and $R = 4a_0{}^*$, where periods of oscillations are equal to $\Delta\gamma = 2$ and $\Delta\gamma = 1/8$, respectively.

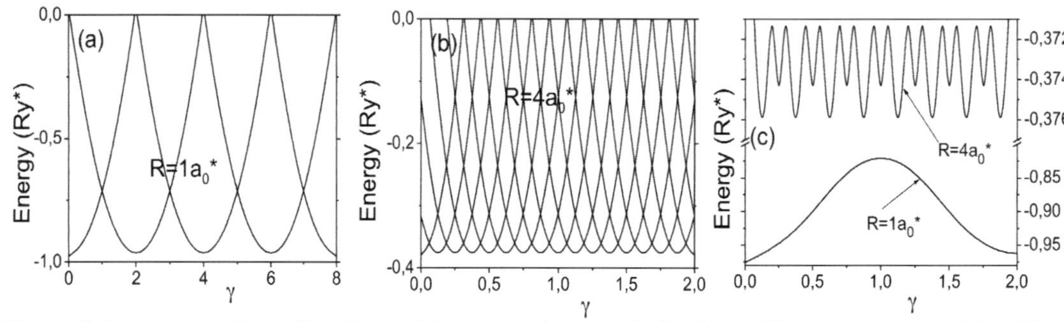

Figure 4. Lower energies as functions of the external magnetic field in uniform nanotubes with radii: (a) $R_c = 1a_0{}^*$ and (b) $R_c = 4a_0{}^*$. In (c) the comparison of the oscillations of the ground state energy in nanotubes with different radii are shown.

In order to demonstrate that the shape of the AB oscillation in tubes with non-uniform profiles are changed essentially, in Figure 5 are shown the lower eigenvalues of the problem (2) as functions of the external magnetic field for the zero-electric field case and for two type of profiles, linear $(n = 1)$ and parabolic $(n = 2)$ with parameters $R_b = 1a_0{}^*$, $R_t = 4a_0{}^*$ and $H = 20a_0{}^*$.

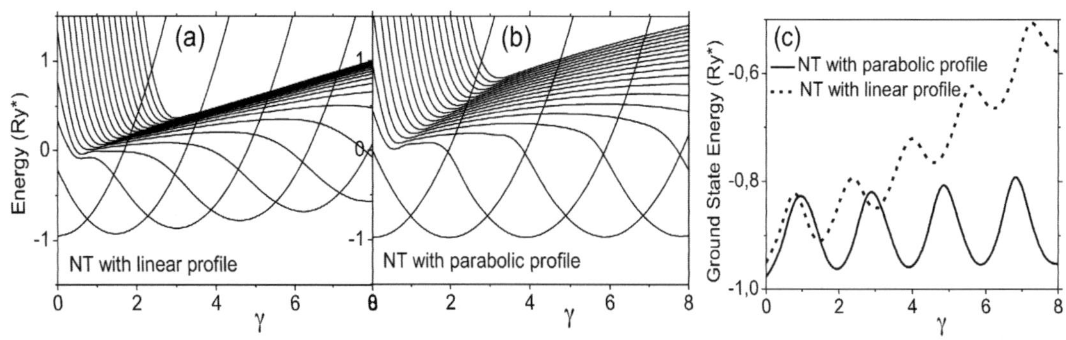

Figure 5. Lower energies as functions of the external magnetic field in non-uniform nanotubes with parameters $R_b = 1a_0{}^*$, $R_t = 4a_0{}^*$ and $H = 20a_0{}^*$, for (a) linear profile, (b) parabolic profile and (c) comparison of the oscillations of the ground state energy in tubes with different profiles.

It is seen that in both cases the period of oscillation of the lowest energy level is equal to $\Delta\gamma = 2$, coinciding with the corresponding value for the 1D quantum ring of the radius $R_b = 1a^*$. This result demonstrates that in the zero electric field case the electron is mainly localized close to the bottleneck point $z = 0$. In spite of a general similarity of the curves of the AB oscillations for both types of profiles one can see particular differences on their magnetic field dependencies. The comparison of the curves for the excited states in Figures 5(a) and 5(b) reveals that unlike the linear profile for which all excited states in the high magnetic field becomes almost degenerated, the splitting between excited states in nanotubes with parabolic profile remains meaningful for all magnetic field values. There is also a significant difference in the magnetic field dependence of the ground state energies of these two types of non-uniform profiles, as it can be seen from Figure 5(c). One can observe a similarity of the oscillations of the ground state energy in Figure 4(a) for the uniform NT of radius $R = 1a_0{}^*$ and for non -uniform NT with the parabolic profile in Figure 5(c), the periods and amplitudes of the NT oscillations in both cases are practically equals. Unlike, for non-uniform NT with linear profile the AB oscillation of the ground state energy level in Figure 5(c) exhibits a weakly reduced amplitude and increased period and a linear trend with slope different of zero. These differences are associated with a significantly larger separation, existing between the donor and electron in the excited states in the non-uniform structure with linear profile, in comparison with one in the parabolic profile structure.

One can expect that the curves of the AB oscillations should be very sensible to applying the external electric field, which by forcing the electron to move away the donor also makes to rise its rotation radius. In Figure 6, there are presented lower energies as functions of the external magnetic field in non-uniform NT with linear profile for four different values of the external electric field.

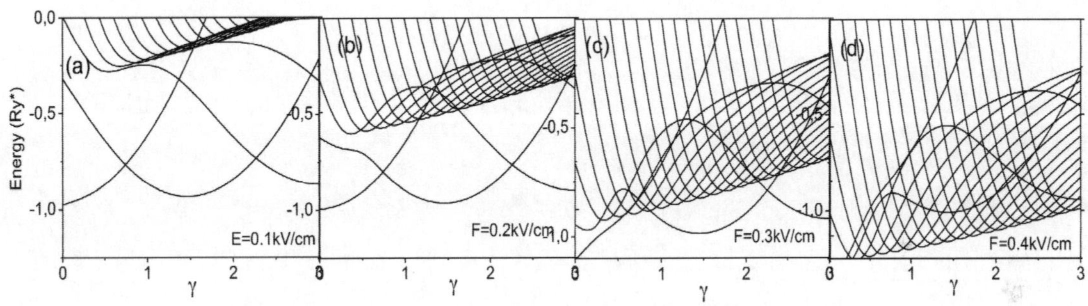

Figure 6. Successive transformation of curves of the Aharonov-Bohm oscillations under increasing electric field in non-uniform nanotube with linear profile.

It is seen that there is two group of the curves of the energies dependencies with different periods of the AB oscillations, equal approximately to $\Delta\gamma = 2$ and $\Delta\gamma = 1/8$, respectively. The first group with the large period of the oscillation can be attributed to the states with small dipole momenta when the electron is located close to the bottleneck point $z = 0$, while the second group corresponds to states with large dipole momenta as the electron is located close to the end of the NT. One can observe in Figure 6 a successive convergence of these two groups of levels with increase of the electric field when the electric field is grown from zero to 0.3kV/cm. As the electric field further increases and exceeds a critical value, the period of the AB oscillations is dropped abruptly, from $\Delta\gamma = 2$ to $\Delta\gamma = 1/8$. In this way, by applying the external electric field it is possible to change also magnetic properties of the structure.

4. Conclusions

A simple model, in which on-axis donor is located at the bottleneck point of structure, is considered in order to analyze the effect of the electric and magnetic fields applied along the symmetry axis of non-uniform nanotubes on their spectral, electric and magnetic properties. It is revealed that such structure can exhibit a giant polarizability under external electric field, which besides can be controlled by applying additional magnetic field. The analysis of the Aharonov-Bohm oscillations under external electric field reveals their significant sensibility to applied external electric field that can change magnetic properties in this structure abruptly. Our theoretical analysis predicts a possibility of the existence of the magnetoelectric effect, arising from the quantum-size effect in non-uniform nanotubes.

Acknowledgments
This work was financed by the Universidad del Magdalena through Vicerrectoría de Investigación and the Universidad Industrial de Santander (UIS) through Vicerrectoría de Investigación y Extensión, DIEF de Ciencias (Cod. 1350).

References

[1] K Kubota, P O Vaccaro, N Ohtani, Y Hirose, M Hosoda, T Aida 2002 *Physica E* **13** 313
[2] Premila Mohan, Junichi Motohisa and Takashi Fukui 2006 *Appl. Phys. Letts.* **88** 013110
[3] Ik Su Chun and Xiuling Li 2008 *IEEE Transactions on Nanotechnology* **7** 493
[4] K Tsumura, Sh Nomura, P Mohan, J Motohisa and T Fukui 2007 *Japan J Appl Phys* **46** L440
[5] H M Lin, Y L Chen, J Yang, Y C Liu, K M Yin, J J Kai, F RChen, L C Chen, Y F Chen, C C Chen 2003 *Nano Lett.* **3** 5372003
[6] Kai Wang, Jiajun Chen, Weilie Zhou, Yong Zhang, Yanfa Yan, John Pern and Angelo Mascarenhas 2008 *Adv. Mater* **20** 3248
[7] K L Janssens, F F M Peeters and V A Schweigert 2001 *Phys. Rev. B* **63** 205311
[8] L C Porras, I D Mikhailov 2013 *Physica E* **53** 41
[9] F J Betancur, I D Mikhailov and L E Oliveira 1998 *J. Appl. Phys. D* **31** 33919
[10] W. Gutiérrez, L. F. García, and I. D. Mikhailov 2010 *Physica E* 2010 **43** 559

Effect of variation of the central-hole depth and the axial anisotropy on the AB oscillations in a wide nanoring

F A Rodríguez Prada[1,2], W Gutiérrez[1] and I D Mikhailov[1]
[1] Universidad de Santander (UDES), Bucaramanga, Colombia
[2] Universidad Industrial de Santander (UIS), Bucaramanga, Colombia

E-mail: inlogin@udes.edu.co

Abstract. The effect of the external magnetic field on the spectral properties of one-electron non-uniform quantum ring with radially directed hills is analysed. The corresponding one-particle wave equation is separable in the adiabatic limit, when the layer thickness is essentially smaller than its lateral dimension. Our calculations show that the presence of a single axially directed hill as well as a rise of the central hole thickness produce a quenching of the Aharonov-Bohm (AB) oscillations of the lower energy levels and of the magnetic momentum. However, as the number of radially directed hills is increased, the system exhibits again oscillations, resulted from an enhancement of tunnelling circular currents.

1. Introduction

The development of new semiconductor growth techniques has made possible the fabrication of the self-assembled quantum rings (QRs) which in the presence of the magnetic field exhibit an interesting quantum-interference phenomenon, denominated AB effect [1]. Theoretical analysis of this phenomenon reveals its delicate nature related to a destructive interference induced by a possibility to appearance of multiple different spatially separated paths in QR under a diminishing of the central-hole dimension or any non-uniformity that breaks the axial symmetry [2,7]. This destructive interference disappears if the ring is sufficiently narrow, and uniform. However, further numerical analysis of more realistic 3D model of non-uniform crater-like QRs shows that, although the real QR shape differs strongly from an idealized circular-symmetric narrow ring structure, AB-type oscillations in the magnetization survive [8,10].

A 3D exactly solvable model of the crater-like one-electron QR, proposed recently in reference [11,12] provides an explanation of the stability of the AB oscillations with respect to structural non-uniformities. In the presence of the magnetic field applied along the symmetry axis, more probable electron paths in this model are clustered close to the crater's rim being similar to those in quasi-one-dimensional quantum rings independently on the crater width. However, any slight non-uniformity produced by a single radially directed valley or single hill suppresses the oscillations of several lower levels due to the localization of the corresponding rotational states. Nevertheless, when the non-uniformity becomes substantial due to the presence of multiple valleys and hills, the AB oscillations generated by the external magnetic field becomes again possible owing to the electron tunnelling through thin potential barriers.

It is possible to extend the model considered in reference [11,12] in order to analyse additionally the effect of the variation of the width and the depth of the central-hole on the AB oscillations. In this extended model, proposed in this paper, the crater's height between the central hole and the outer border grows linearly with different slopes at different radial directions in the case of non-uniform structure, while the corresponding slopes in the central-hole at all radial directions are zero. By varying the depth and the radius of the central hole in this model, one can analyse a successive change of the spectral and magnetic properties during a transformation of the nanostructure morphology, from disk-like to ring-like.

2. Theoretical model

It is considered a model of a crater-like non-uniform QD as a thin layer whose thickness d depends on polar coordinates (ρ,φ) as:

$$d(\rho,\varphi)=\left[h_a\vartheta(\rho_a-\rho)+h_b\,\rho/\rho_b\,\vartheta(\rho-\rho_a)\right]\vartheta(\rho_b-\rho)\Big/\sqrt{1+\sigma(\rho/\rho_b)^2 f^2(\varphi)};\ \ h_a=h_b\,\rho_a/\rho_b; \qquad (1)$$

Here $f(\varphi)=\sin p\varphi;\ p=1/2,1,2,...$ and $\vartheta(x)$ is Heaviside step function, equal to zero, for $x<0$ and to one, for $x>0$; h_a is the layer thickness at the bottom of the crater and ρ_a is the radius of the central hole and h_b, ρ_a are corresponding values for height and radius of outer border of the isotropic crater. In the relation (1), the parameter σ and the function $f(\varphi)$ describe the degree and the type of the anisotropy of the crater, respectively. As $p=1/2$, the structure has a single radially directed hill, as $p=1$ two radially directed hills, separated by two valleys, as $p=2$ four hills, etc. In Figure 1 can be observed examples of 3D images of corresponding models of layers with profiles given by the relations (1) for $\sigma=1$ and p=1, 2, or 4.

In order to simplify our numerical analysis in what follows, it is considered a model with the infinite-barrier confinement potential $V(\mathbf{r})$, which is supposed to be equal to zero inside the crater and to infinity otherwise. The external homogeneous magnetic field $\mathbf{B}=B\hat{z}$ is applied along the Z-axis. The effective Bohr radius $a_0^*=\ {}^2\varepsilon/m^*e^2$, the effective Rydberg $R_y^*=e^2/2\varepsilon a_0^*$ and $\gamma=e\ \ B/2m^*cR_y^*$ as units of length, energy and the dimensionless magnetic field strength are used below, being m^* the electron effective mass and ε the dielectric constant.

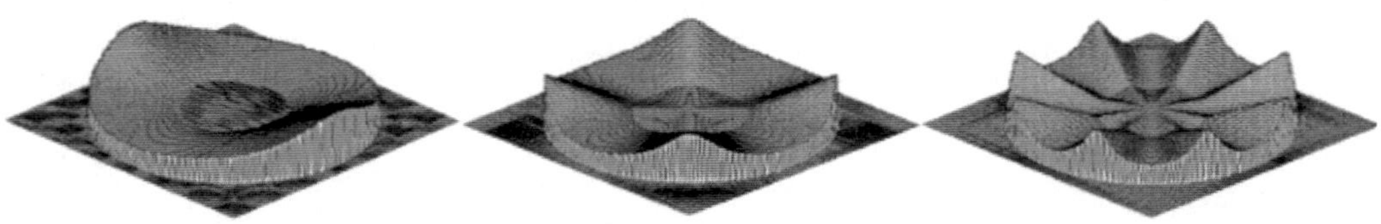

Figure 1. 3D images of non-uniform quantum ring with morphology given by the relation (1).

As the thicknesses of actual self-assembled QDs manufactured up to now are much smaller than their lateral dimensions one can take advantage of the adiabatic approximation assuming a model with infinite-barrier confinement. In framework of this approximation, the fast movement of the electron along the Z-axis inside the layer is supposed to be independent on its in-plane slow displacements [13,15]. Therefore, the ground state energy for each the electron in-plane position given by polar coordinates (ρ,φ), given by the well-known expression corresponding to infinite-barrier quantum well of width $d(\rho,\varphi)$, in dimensionless units is:

$$E_z(\rho,\varphi)=\pi^2/d^2(\rho,\varphi) \qquad (2)$$

Following the adiabatic approximation procedure, one can consider afterwards the in-plane electron motion as 2D problem with additional adiabatic potential given by this function. The renormalized 2D Hamiltonian describing in the effective-mass approximation the in-plane electron slow motion inside the layer with profile given by Equation (1) in the presence of the adiabatic potential (2) and the magnetic field applied along Z- axis has the following form:

$$H = -\Delta_{\rho,\varphi} - i\gamma\frac{\partial}{\partial\varphi} + \frac{\gamma^2\rho^2}{4} + \frac{\pi^2}{h_a^2}\left(1+\left(\frac{\rho}{\rho_a}\right)^2\right)\vartheta(\rho_a-\rho) + \sigma\frac{\pi^2 f^2(\varphi)}{h_a^2}\left[\left(\frac{\rho}{\rho_a}\right)^2 \vartheta(\rho_a-\rho)+\vartheta(\rho \quad \rho_a)\right] \quad (3)$$

For the uniform crater ($\sigma = 0$) eigenfunctions of the Hamiltonian (3) depends on two quantum numbers, radial $n = 1,2,3,...$ and angular $m = \pm1,\pm2,\pm3,...$ and they can be found exactly in a form of the linear combination of Hypergeometric confluent functions. The energies $E^{(0)}(n,m)$ and the wave functions $\phi_{n,m}(\rho)$ of the uniform crater, found by solving the transcendental equation, can be used further to calculate the matrix elements $\langle\phi_{n,m}|U|\phi_{n',m'}\rangle$ of the perturbation U induced by the non-uniformity and given by the relation (3). Resulting matrix elements of the Hamiltonian for non-uniform crater are:

$$\begin{gathered}\langle\phi_{n,m}|H|\phi_{n',m'}\rangle = E^{(0)}(n,m)\delta_{n,n'}\delta_{m,m'} + \langle nm|\eta|n'm'\rangle\left(2\delta_{m,m'} - \delta_{m,m'+2p} - \delta_{m,m'-2p}\right) \\ \langle n|\eta|n'\rangle = \frac{\pi^2\sigma}{4h_a^2}\left[\int_0^{\rho_a}\phi_{n,m}(\rho)\phi_{n',m'}(\rho)\left(\frac{\rho}{\rho_a}\right)^2\rho d\rho + \int_{\rho_a}^{\rho_b}\phi_{n,m}(\rho)\phi_{n',m'}(\rho)\rho d\rho\right]\end{gathered} \quad (4)$$

In what follows, the results of calculation of some lower energies $E_m(\gamma)$ corresponding to radial quantum number $n = 1$ and of the magnetic moment $p_m(\gamma)$ at zero temperature, as functions of the magnetic field γ, are presented. The magnetic moment $p_m(\gamma)$ at zero temperature, one can calculate according to the Hellmann-Feynman theorem, as follows:

$$p_m = -\mu_B\frac{\partial E_1}{\partial\gamma} \quad (5)$$

3. Results and discussion

The geometrical parameters used below are: a typical value for the InAs/GaAs material effective Bohr radius $a_0{}^* \approx 10nm$, the outer radius $\rho_b = 20nm$ and the height of crater at the rim $h_b = 4nm$, dimensions close to those of crater like QDs fabricated earlier by means of the droplet epitaxial technique [16,18]. The layer thickness at the bottom of the crater is considered as variable between $h_a = 1nm$ and $h_a = 3.5nm$.

In order to facilitate the interpretation of numerical results, it has been found the electron density distribution in a simple model with infinite-barrier confinement and in crater-like uniform InAs/GaAs structures with profile given by the relation (1) for $\sigma = 0$, by using the software of COMSOL MULTIPHYSICS (v.5.1) [19]. At first row of Figure 2, it is presented contour plots of the radial charge distribution in a plane through the axis of symmetry of uniform craters with central-hole thicknesses $h_a = 1$, 2, and $3.5nm$, calculated for infinite-barrier model, while at the second row are shown a similar result, obtained for the finite barrier model, using material parameters of InAs and GaAs.

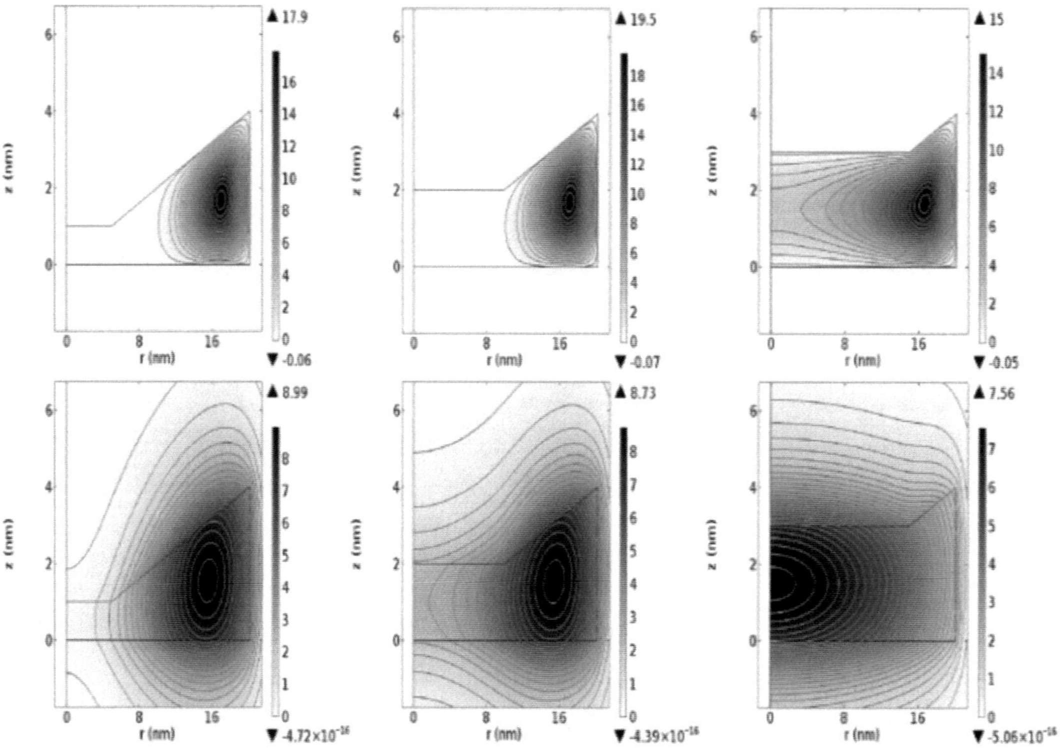

Figure 2. Contour plots of the density of the radial charge distribution in a plane through the axis of symmetry of uniform craters with three different central-hole depths in the ground state calculated for models with infinite (first row) and finite (second row) barrier heights.

For h_a less than 3.5nm, one can observe AB oscillations of energy levels, typical for a 1D ring-like structure, which generate multiple crossovers between the curves with different magnetic quantum numbers m. As it is well known, the period of AB oscillations, $\Delta\gamma$ in 1D QR of the radius R is equal in dimensionless units to. $\Delta\gamma = 2/R^2$. According to this estimation, AB oscillations of energy levels at the first column of Figure 3 at the crater with the layer thickness in central hole region $h_a = 2nm$ are similar to one-dimensional QR of radius $R = 18.3nm$, those at the second column for $h_a = 3nm$ to a QR with radius $R = 17.3nm$, while at the third column for $h_a = 3.5nm$ AB oscillations are absent. It is largely consistent with results of calculation of the charge distribution presented in Figure 2, where one can see that for $h_a < 3.5nm$, 80-90% of the electric charge is localized inside a narrow region close to the outer border, while for $h_a \geq 3.5nm$ the electric charge leaks inside the central-hole region. It results in transformation of the ring-like charge distribution in the disk-like one. When initially h_a grows from 2nm to 3nm the maximum of the density is displaced slightly toward the border, the confinement is increased and the averaged value of the electron rotation radius is enlarged.

A rise of the energy band bottom and an increase of the period of the AB oscillation in Figure 3 is associated with such change of the charge distribution. For $h_a \geq 3.5nm$ the electron configuration becomes rather similar to one of the disc-like structure for which there are no the AB oscillation of the energy levels.

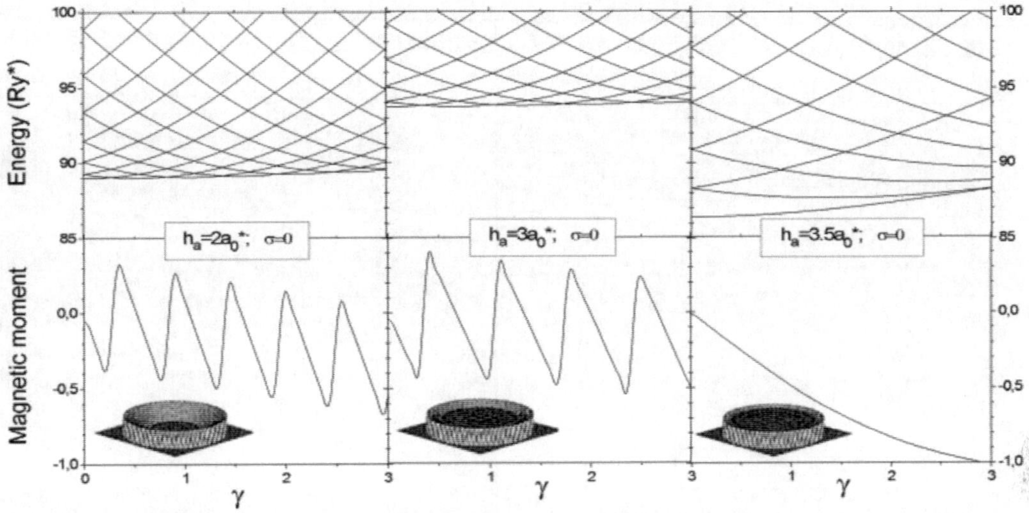

Figure 3. Lower energies (upper row) and induced magnetic momenta (lower row) in one-electron axially symmetrical crater-like QDs as functions of the external magnetic field for three different depths of the central-hole. Corresponding images are shown at the bottom of the Figure.

In Figure 4 similar curves are presented for wide axially non-uniform QRs, whose morphology is given by the layer thickness dependency specified by the relation (1) with parameters, inner and outer radii and thicknesses $h_a = 1nm,\ \rho_a = 5nm$ and $h_b = 4nm,\ \rho_b = 20nm$, respectively. In the first row is displayed the energies dependencies on the external magnetic field in the uniform crater $(\sigma = 0)$ and in non-uniform structures with one, two and four radially directed hills, corresponding to parameters $p = 1/2$, $p = 1$, and $p = 2$, respectively, for $\sigma = 0.01$ in the relation (1). Similar dependencies for induced magnetic momenta are observed in the second row of Figure 4.

In Figure 4 is shown a similarity of the curves of the energies dependencies on the magnetic field for upper energy levels, which are due to rotational states. The electron density distribution in these states is practically insensible to the structural non-uniformity due to the fact, that the energy of the electron in these states are superior than the heights of barriers generated by the axial non-uniformity. On the contrary, the magnetic field dependencies of five lower levels and the magnetization in the single-hill structure ($p = 1/2$) in Figure 4 do not exhibit any oscillation, such as observed in the second column of Figure 4.

This effect is attributed to a localization of the lower rotational states, induced by single hill-type non-uniformity, and to relate to this localization insensibility to the external magnetic field.

It is seen also, that these lower energies in the third and fourth columns, for cases of non-uniform craters with two ($p = 1$), and four ($p = 2$) hills, are clustered inside separated bands with two and four levels in each one, respectively. The presence of a single radially directed hill in the crater produces an effective adiabatic potential with a wide barrier that impedes a cyclic displacement of the electron with a low energy along the rotational path. With a growth of the number of the hills, the potential barriers along circular paths between adjacent hills become narrower.

The bigger the number of the hills, the narrower are the barriers for circular paths and the larger is the tunnelling current generated by the external magnetic field. An increase of the tunnelling current provides a growth of the amplitude of the AB oscillations and a clustering of the vibrational levels in separated non-crossing bands, while inside each band one can see in Figure 4 multiple crossovers and reordering of the levels, typical for AB oscillations.

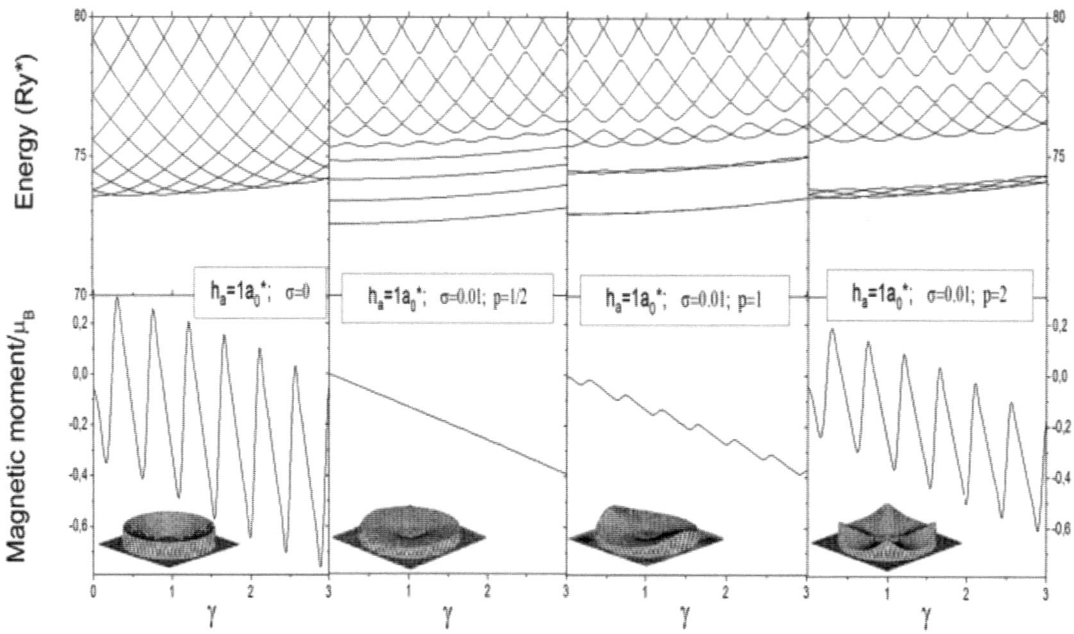

Figure 4. Lower energies (upper row) and induced magnetic momenta (lower row) in one-electron non-uniform crater-like QDs with different number of radially directed hills as functions of the external magnetic field. Corresponding images are shown at the bottom of the Figure.

The period of AB oscillations in the first column of Figure 4 for uniform crater-like wide QR with inner and outer radii 5nm and 20nm, respectively coincides with the corresponding value in 1D QR of the radius 20nm. Its means that lines of the circular current induced by the magnetic field in uniform crater goes along the external border. It is interesting that AB oscillations in a non-uniform crater with multiple hills and valleys restored due to induced tunnelling current have the same period of oscillation as it one can observe at the last column of Figure 4 It means that lines of the circular tunnelling current induced by the magnetic field in non-uniform crater also goes along the outer border.

4. Conclusions

In this work is presented a theoretical analysis of the effect of the variation of the depth of the central hole in uniform and in non-uniform crater-like quantum dots on their energies and the induced magnetic momenta dependencies on the external magnetic field applied in the Z-axis direction. To this end is used a simple model of one-electron crater-like quantum dot with ideally circulate inner and outer radii but with non-uniform thickness due to the presence of radially directed hills. It is shown that in the adiabatic limit, when the thickness of the crater grows linearly in all radial directions keeping its value significantly smaller than the outer radius, the energies and wave functions of the electron can be found analytically.

The dependencies of the energies and induced magnetic momentum on the magnetic field in uniform craters reveal the presence of the AB oscillations typical for 1D quantum ring independently of the size of the central hole of the crater until the layer thickness of the central-hole region becomes almost equal to the thickness of the outer border. This result is attributed to a strong confinement that retains the electron in the crater-like structure inside a very narrow circulate region close to the exterior frontier, independently on width of the crater and the value of the external magnetic field, in contrast to the case of nanorings with rectangular cross-sections.

Similar analysis for crater-like non-uniform nanostructures with radially directed hills show that the barriers in the effective adiabatic potential in the regions of the valleys between adjacent hills along the circular paths, provides a localization of the electron in the vicinities of tops of the hills if the electron energy is smaller than barrier height. It has been shown that in the case of the crater-like structure with a single hill, a small non-uniformity of the crater thickness might suppress the electron rotation induced by the external magnetic field in various lower energy states, producing a quenching of the AB oscillations of the corresponding energy levels. The higher the non-uniformity, the bigger is the number of the separated levels whose energies does not depend on the external magnetic field. Its consider that these states (whose energies does not exhibit AB oscillations) as vibrations, localized close to the hill.

It was also found that an increase of the number of hills in a non-uniform crater produces the assembling of the localized states in independent sub-bands. The number of levels inside of each of them coincides with the number of hills in nanostructures, while the dependencies of the energies inside sub-bands on the magnetic field exhibit AB oscillations with crossovers between them, similar to those for extended states. The period of these oscillations, coincide with those for the extended states. These oscillations are attributed to the tunnelling currents generated by the external magnetic field. The bigger the number of the valleys, the larger are tunnelling currents, the wider are these bands, and the higher are the amplitudes of the corresponding AB oscillations. It has been shown that it is due the fact that the increase of the number of the valleys produces a reduction of the wide of the potential barriers between adjacent hills, making easier the tunnelling through narrower barriers along circular paths.

Acknowledgments

«Universidad de Santander (UDES)» through the «vicerrectoría de Investigaciones»; «Universidad Industrial de Santander (UIS)» through the «vicerrectoría de Investigación y Extensión (VIE), DIEF de Ciencias; «Centro de Investigación Científica y Tecnológica en Materiales y Nanociencias (CMN)» of «Universidad Industrial de Santander (UIS)».

References

[1] Aharonov D Y and Bohm D 1959 *Phys. Rev.* **115** 485
[2] Lorke A, Luyken R J, Govorov A O, Kotthaus J P, García J M and Petroff P M 2000 *Phys. Rev. Lett.* **84** 2223
[3] Pietiläinen P, Halonen V and Chakraborty T 1995 *Physica B* **212** 256
[4] Jacak L, Hawrylak P and Wojs A 2012 *Quantum Dots* (London: Springer)
[5] Fomin V M 2014 *Physics of Quantum Rings* (Berlin: Springer)
[6] Pershin Y V and Piermarocchi C 2005 *Phys. Rev. B* **72** 195340
[7] Bruno-Alfonso A and Latgé A 2008 *Phys. Rev. B* **77** 205303
[8] Offermans P, Koenraad P M, Wolter J H, Granados D, García J M, Fomin V M, Gladilin V N and Devreese J T 2006 *Physica E* **32** 41
[9] Fomin M, Gladilin V N, Klimin S N, Devreese J T, Kleemans N A J M and Koenraad P M 2007 *Phys. Rev. B* **76** 235320
[10] Kleemans N A J M, Bominaar-Silkens I M A, Fomin V M, Gladilin V N, Granados D, Taboada A G, García J M, Offermans P, Zeitler U, Christianen P C M, Maan J C, Devreese J T, and Koenraad P M 2007 *Phys. Rev. Lett.* **99** 146808
[11] F Rodríguez-Prada, L F García, I D Mikhailov 2014 *Physica E* **56** 393
[12] F Rodríguez-Prada, L F García, I D Mikhailov 2014 *Physica E* **60** 23
[13] Peeters F M and Schweigert V A 1996 *Phys. Rev. B* **53** 1468
[14] Szafran B, Adamowski J and Bednarek S 2000 *Physica E* **5** 185
[15] Mikhailov I D, Marín J H, and García F 2005 *Phys. Stat. Sol. b* **242(8)** 1636
[16] Boonpeng P, Jevasuwan W, Suraprapapich S, S and Panyakeow S 2009 *Electron. Eng.* **86** 853
[17] Lee J H, Wang Z M, Ware M E, Wijesundara K C, Garrido M, Stinaff E A and Salamo G J 2008 *Cryst. Growth Des.* **8(6)** 1951
[18] Strom N W, Wang Z M, Lee J H, Waar Z Y A, Mazur Y I and Salamo G J 2007 *Nanoscale Res. Lett.* **2** 112
[19] http://www.comsol.com

Identification of metal elements by time-resolved LIBS technique in sediments lake the "Cisne"

P Pacheco[1], E Arregui[1], J Álvarez[1], N Rangel[1] and R Sarmiento[1]
[1] Universidad del Atlántico, Barranquilla, Colombia

E-mail: paolapacheco@mail.uniatlantico.edu.co

Abstract. Laser induced breakdown spectroscopy (LIBS), is a kind of spectral method of atomic emission that uses pulses of radiation high energy laser as excitation source. One of the advantages of technical LIBS lies in the possibility of analyse the substances in any State of aggregation, already is solid, liquid or gaseous, even in colloids as aerosols, gels and others. Another advantage over other conventional techniques is the simultaneous analysis of elements present in a sample of multielement. This work is made in the use of this technique for the identification of metal pollutants in the Swan Lake sediment samples, collected by drilling cores. Plasmas were generated by focusing the radiation of Nd: YAG laser with an energy per pulse 13mJ and 4ns duration, wavelength of 532nm. The spectra of radiation from the plasmas of sediment were recorded with an Echelle spectrograph type coupled to an ICCD camera. The delay times were between $0.5\mu s$ and $7\mu s$, while the gate width was of $2\mu s$. To ensure the homogeneity of the plasmas, the sediment sample was placed in a positioning system of linear and rotary adjustment of smooth step synchronized with the trigger of the laser pulse. The registration of the spectra of the sediment to different times of delay, allowed to identify the lines prominent of the different elements present in the sample. The analysis of the Spectra allowed the identification of some elements in the sample as if, Ca, Na, Mg, and Al through the measurement of wavelengths of the prominent peaks.

1. Introduction
Laser Produced Plasma by Spectroscopy (LPPS), also known as LIBS (Laser Induced Breakdown Spectroscopy) [1,2], is a method of Atomic Spectroscopy that has become a very effective technique for the analysis of sediments from lagoon origin [3]. Works like Harmon R, et. al. [4]; B Salle [5] used this technique to detect metals in sediments that only requires a small amount of the sample to spectrally analyse its components and in many cases do not need advance preparation. With the appropriate instrumentation, the recording and the analysis of the spectra can be spatial or temporarily according to the detection to be studied [1].

In LIBS technique, a plasma is generated by the interaction of the laser pulses with the material, what allows the detection of spectral lines of elements present in the excited sample. The spectra obtained reveal the basic characteristics of the composition of the sample. Measuring the wavelength of each peak intensity of the spectra it is compared with the spectral database of National Institute Standard and Technology (NIST) [6].

The implementation of the LIBS technique resolved in time to analyse the toxic components of the soil, this reduces the time of analysis of the samples and the results associated facilitate the design of strategies to combat environmental pollution in targeted sites. In this paper this technique is used for the identification of polluting metallic elements in samples lagoon sediments, which were collected in the lake the "Cisne" by core drilling is done.

2. Sample preparation and experimental description

The proper choice of instruments is essential for a good sampling and laboratory analysis. In the case of metal contaminated soils, elements for sampling must be plastic, Teflon or stainless steel; for this work nucleators pvc were used [7].

The samples were collected on the lakeshore by introducing into the sediment the corer, so that a portion thereof to a depth of 25cm was mined, then the pellets were homogenized in a plastic tray for storage in airtight bags until Analysis.

Once in the laboratory, samples were dried in an oven at a temperature of 110°C for three hours to extract the moisture content. With the dry sediment, uniform pellets of 10mm in diameter and 8mm thick with a press were developed.

Plasmas sediment samples were generated in air atmosphere and argon atmosphere, focusing the radiation from a Nd: YAG (Tempest 10, New Wave Research) operating at a wavelength of 532nm, with an average energy per pulse 13mJ and duration of 4ns. The firing frequency was 2Hz. The sample was mounted in high precision motor stepper stage as possible plasmas in the same plane and under the same conditions engine, because in each point of interaction of the laser pulse with the material is causing erosion and leaving a small crater. Figure 1 shows a basic diagram of the experimental set LIBS time resolved technique. The radiation emitted by the plasma was collected with an optical fibre. The collimated light was focused to the entrance slit of an Echelle spectrograph (Mechell 5000, Andor Technology) equipped with a ICCD camera (iSTAR 334T, Andor Technology), giving a broadband coverage from 200nm to 900nm with resolution of 0.1nm. The acquisition and recording of the spectra was set in Solis T operating software, the parameters of delay time was set between 0.5μs and 7μs, with steps of 0.5μs each record with 20 accumulations, and temperature cooling to -10°C [8].

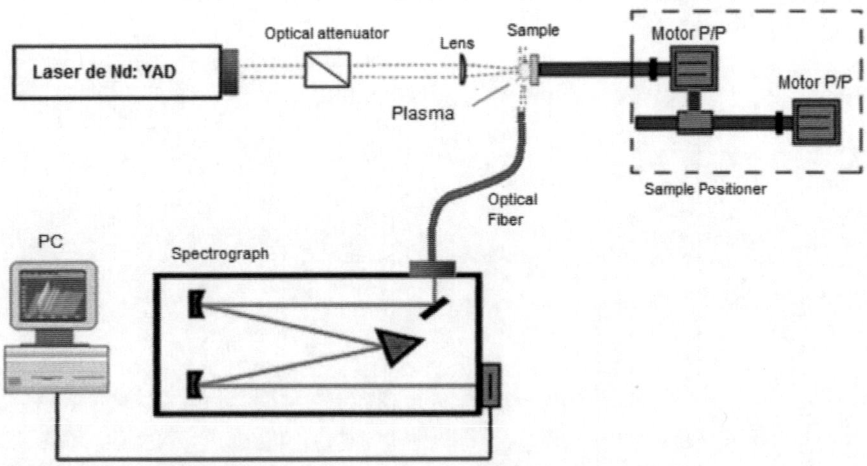

Figure 1. Experimental setup of the LIBS technique resolved in time.

3. Discussion of results

For the analysis of the spectra of laser plasmas, sediment samples variation of the various parameters of acquisition of the spectra were taken into account delay time to select the most appropriate spectra; the type of atmosphere, humidity conditions of the sample which affects the spectra obtained. The spectra are affected by humidity and a significant drop in the relative intensity of the spectral lines compared to the spectra of the dry sample for both atmospheres as shown in Figure 2. To a higher humidity, plasma is cooled by the water vapour in full accordance with reference [9]. It is also observed that the relative intensity of the lines and appears more intense spectrum under argon in air atmosphere for both dry and wet samples.

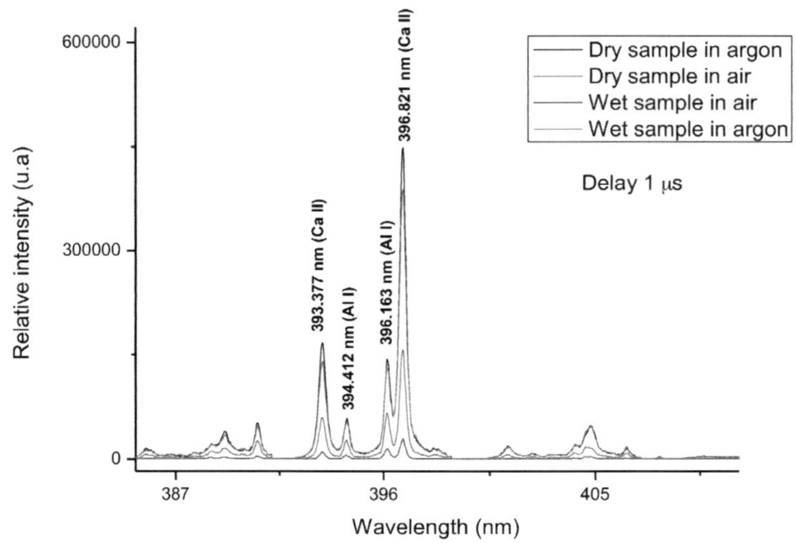

Figure 2. Overlapping spectra of laser plasma sediment wet and dry in air and argon atmosphere with delay time t_d=1μs.

For delay time of 1μs, the spectra was display with a continued strong and ionic lines broadened Ca II and Mg II as shown in Figures 2 and 3. For longer delays to 1μs persistent atomic lines of neutral atoms Ca I, Si I, Al I, Mg I, K I, Na I.

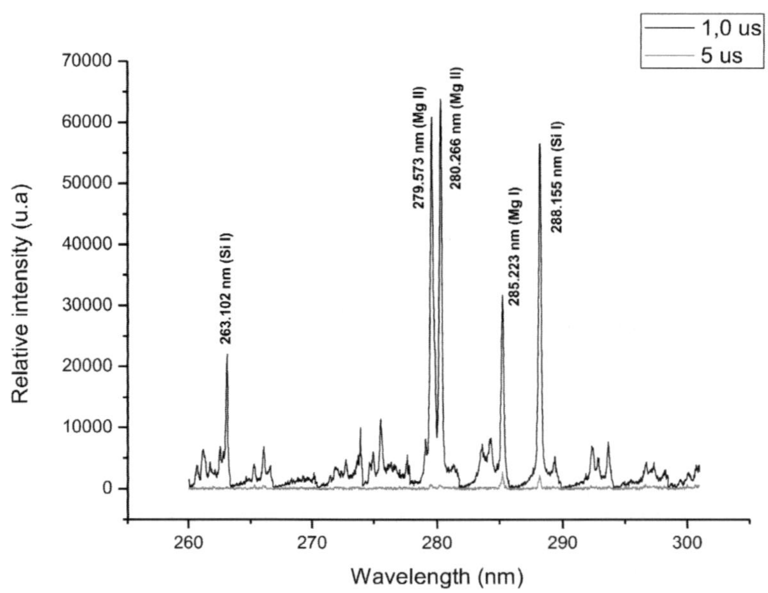

Figure. 3 Spectra overlapping laser plasma sediment in the region between 260nm-300nm for delay times td=1μs (black) and td=5μs (red).

Figure 4 shows the behaviour of LIBS spectra for different delay time in the region between 300 and 450nm. The analysis of temporal records allows better discrimination and allocation of atomic and ionic lines simultaneously presented in the multielement spectra.

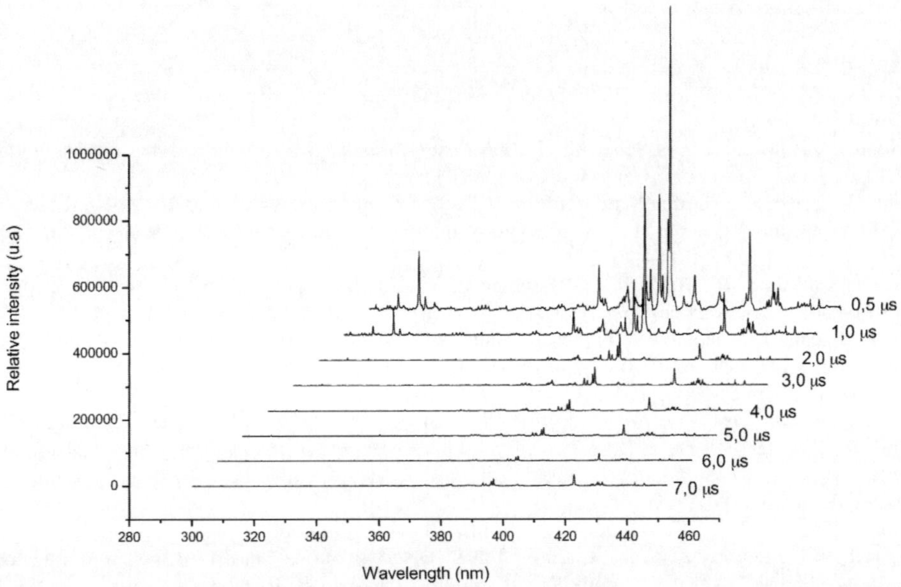

Figure 4. Plasma-laser spectra sediment taken with different delay times t_d.

Table 1 shows some of the observed and measured in the LIBS spectra sediments lake the "Cisne" located in the north of Barranquilla, Colombia.

Table 1. Spectral lines of the species presented in the plasma-laser sediment.

Element	Wavelength (nm)	Element	Wavelength (nm)
Si I	243.515	Ca I	422.705
Si I	251.647	Ca I	442.454
Si I	251.902	Ca I	445.456
Si I	252.369	Ca I	558.873
Si I	252.843	Ca I	559.492
Si I	263.102	Na I	588.968
Mg II	279.573	Na I	589.572
Mg II	280.266	Ca I	610.298
Mg I	285.223	Ca I	612.209
Si I	288.155	Ca I	616.206
Al I	309.238	Ca I	616.975
Ca II	315.906	Ca I	643.902
Ca II	373.621	Ca I	646.247
Si I	390.554	Ca I	649.365
Ca II	393.377	Ca I	671.796
Al I	394.412	Ca I	714.803
Al I	396.163	Ca I	720.283
Ca II	396.821	K I	766.475
K I	404.730	K I	769.873

4. Conclusions

The LIBS technique was applied for analysis of time resolved spectra of the sediment lake the "Cisne", which allowed us to detect some elements present in the samples collected. The main elements were detected Ca, Si, Al, Mg, K and Na. Other lines present in the spectra are being validated for confirmation.

References

[1] L Radziemski and D Cremers 2006 *Handbook Laser-Induced Breakdown Spectroscopy* (Chicherster: John Wiley & Sons)

[2] Miziolek A, Palleschi V and Schechter I 2006 *Laser–Induced Breakdown Spectroscopy: Fundamentals and Applications* (New York: Cambrige University Press)

[3] Arroyo L, Trejos T, Gardinali P, Almirall J 2009 Optimization and validation of a Laser Ablation Inductively Coupled Plasma Mass Spectrometry method for the routine analysis of soils and sediments *Spectrochimica Acta Part B* **64** 16–25

[4] Harmon R, Russo R, Hark R 2013 Applications of laser-induced breakdown spectroscopy for geochemical and environmental analysis *Spectrochimica Acta Part B* **64** 16–26

[5] Sallé B, Lacour J, Mauchien P, Fichet P, Maurice S, Manhès G 2006 Comparative study of different methodologies for quantitative rock analysis by Laser-Induced Breakdown Spectroscopy in a simulated Martian atmosphere *Spectrochimica Acta Part B* **61** 301-313

[6] NIST *Atomic data base Lines Form* http://physics.nist.gov/PhysRefData/ASD/lines_form.html

[7] Sepúlveda T, Velasco J, De la Rosa P 2009 *Suelos contaminados por metales y metaloides: muestreo y alternativas para su remediación* (México: Secretaria de Medio Ambiente y Recursos Naturales -Instituto Nacional de Ecología)

[8] Andor Technology 2011 *User Guide Mechelle* http://www.andor.com

[9] Paris P, Piip K, Leep A, Lissovski A, Aints M 2015 Discrimination of moist oil shale and limestone using laser induced breakdown spectroscopy *Spectrochimica Acta Part B* **107** 61-66

Global characterization of a nematic liquid crystal display LCX038ARA using the retarder-rotor model in the modulation amplitude regime-coupled without applied voltage

C J Cuevas Cely[1], C H Acevedo[1] and Y Torres Moreno[1]
[1] Universidad Industrial de Santander, Bucaramanga, Colombia

E-mail: cuevasjennifer07@gmail.com

Abstract. This work shows experimental and theoretical results of the characterization of a nematic liquid-crystal spatial light modulator Sony model LCX038ARA for the parameters angle of molecular rotation, the birefringence and angle of the molecular axis, using the retarder-rotor model without electric field applied in the amplitude regime-coupled.

1. Introduction

Liquid crystal displays (LCDs), are electronic devices contains a material that has liquid/solid properties sandwiches between two transparent electrodes. The unique properties of the LCDs make them attractive for applications such as spatial light modulators (SLM), which are widely useful in single molecular imaging [1-2] and optical tweezers [3-4]. Most SLM are nematic type and its properties can be explained using optics, electronics and materials science. According to the particular helical structure of each nematic liquid-crystal spatial light modulator can find some slight differences for the parameters: molecular twist angle, birefringence and angle of the molecular axis in an equal model of nematic liquid-crystal spatial light modulator. The main characteristic of SLM nematic type is that molecules of each layer tend to be parallel but their positions are random. If the angular orientation increases as going in a direction from the first layer to the last layer, then the LCD is called Twisted Nematic Liquid Crystal Display (TNLCD) [5-7].

To describe the behavior of a TNLCD, it is necessary to know the value of the parameters: the ΨD), and the birefringence (β). The birefringence parameter (β) is defined as [8-10],

$$\beta = \pi\lambda n_e - n_0 d \tag{1}$$

Where λ is the wavelength, d is the thickness of the LCD layer, while n_e and n_0 are the extraordinary and the ordinary indices of refraction, respectively.

In this work present the model by Soutar-Lu (retarder-rotor model) combined with the Stokes parameters in order to experimentally find the values of the parameters of the LCD without applied voltage.

2. Methodology

Soutar-Lu method can be used to characterize a SLM in the amplitude coupled mode, i.e. maximize intensity response and minimize the response phase of the LCD to find the values of the twister angle ΨD), and the birefringence(β).

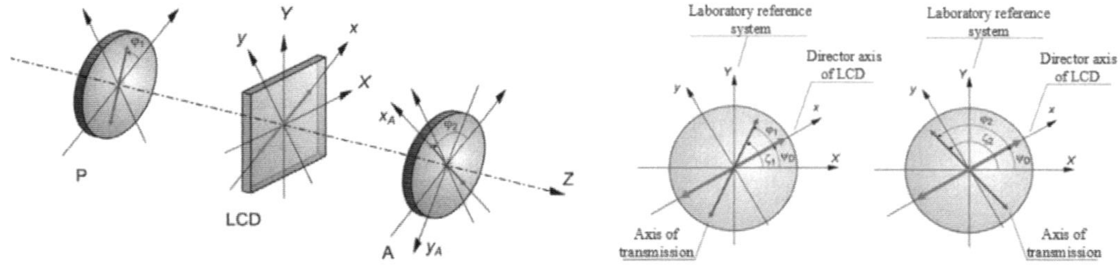

(a) (b)

Figure 1. (a)-(b) Diagrams of reference and angles involved in the Soutar&Lu technique. (b) Detailed diagram for the polarizer and analyzer angles. Source [11].

Figure 1 shows the experimental setup used in Soutar-Lu method. In this figure can observe that the LCD is sandwiching between two lineal polarizer with of purpose to obtain the intensity of the system (Polarizer-LCD-Polarizer) in two configurations, which the polarizers are cross and the polarizers are parallel [11,12]. Using the model based on the Jones matrix theory it possible to find the intensity of the optical system given by:

Parallel configuration:

$$Tp=X\cos\varnothing+Z\sin\varnothing 2+Y\cos 2\zeta 1-\varnothing-2\Psi D\ 2 \quad (2)$$

$$Tp=\cos\gamma\cos\varnothing+\varnothing\gamma\sin\gamma\cos\varnothing 2+\beta\gamma\sin\gamma\cos(2\zeta 1-\varnothing-2\Psi D)2 \quad (3)$$

Cross configuration:

$$Tc=-X\sin\varnothing+Z\cos\varnothing 2+Y\sin(2\zeta 1-\varnothing-2\Psi D)2 \quad (4)$$

$$Tc=-\cos\gamma\sin\varnothing+\varnothing\gamma\sin\gamma\cos\varnothing 2+\beta\gamma\sin\gamma\sin(2\zeta 1-\varnothing-2\Psi D)2 \quad (5)$$

With:

$$X = \cos\gamma \quad (6)$$

$$Y = \left(\frac{\beta}{\gamma}\right)\sin\gamma \quad (7)$$

$$Z = \left(\frac{\varnothing}{\gamma}\right)\sin\gamma \quad (8)$$

$$\gamma = \sqrt{\varnothing^2 + \beta^2} \quad (9)$$

2.1. Determination of the twist angle (ϕ) and the angle of the molecular axis of the TNLCD (ΨD)

With the Soutar-Lu method is possible to find the magnitude of the twist angle (ϕ) and (ΨD) director axis orientation To at the input surface of the TNLCD, however the signs of these parameters cannot be determined. The use of Stokes parameters is proposed to find the sign of ϕ and ΨD. The Stokes parameters can be interpreted as a representation in the Poincaré sphere, of the action of a liquid crystal device and is the most general method for geometrically represent the polarization states of a light beam and its transformations undergone by the action of a polarization system [11]. In the representation of the Poincare sphere, there is a univocal correspondence between the pure polarization states and points of the surface of the sphere surface as shown in Figure 2.

The Cartesian coordinates of each one of these points are the Stokes parameters S_1, S_2, S_3 of the ΨD, using the unique relationship between a state of polarization incident on the LCD and a point on the Poincare sphere [14]. Such analysis can be performed by determining the value of the Stokes parameters of the emerging light from the device using the following expressions [11]:

$$S_0 = X^2 + Y^2 + Z^2 = 1 \quad (10)$$

$$S1=(X2-Z2)\cos[2(\emptyset+\zeta1)]+2XZ\sin[2(\emptyset+\zeta1)]+Y2\cos[2\emptyset+2\psi D-\zeta1 \quad (11)$$

$$S2=(X2-Z2)\sin[2(\emptyset+\zeta1)]-2XZ\cos[2(\emptyset+\zeta1)]+Y2\sin[2\emptyset+2\psi D-\zeta1) \quad (12)$$

$$S3=-2Y(Z\cos[2(\psi D-\zeta1)]+X\sin[22\psi D-\zeta1) \quad (13)$$

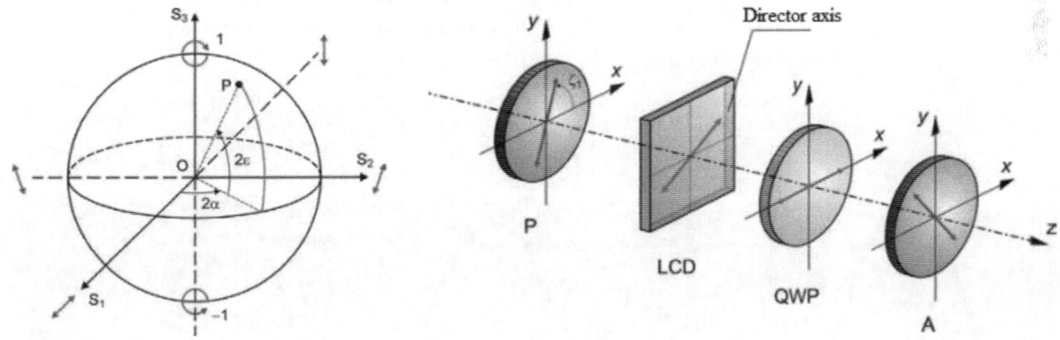

(a) (b)

Figure 2. (a) Poincaré sphere representation of a polarization state. (b) Experimental setup for the measurement of the Stokes parameters. QWP: Quarter wave plate, it is used for measuring the parameter S3. From reference [11].

3. Results

The experimental and theoretical results of the intensity obtained with Soutar-Lu technique are shown in Figure 3. In these results can observe a good correspondence between the theory and the experimental procedure. Finally, using a nonlinear regression ΨD have the values shown in Table 1.

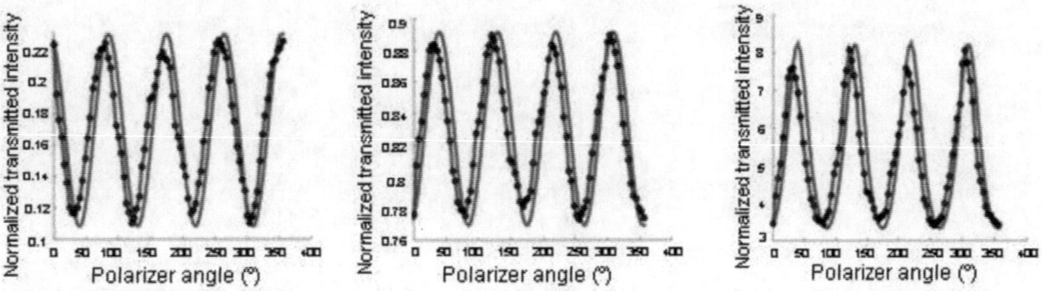

(a) (b) (c)

Figure 3. Experimental (red line) and theoretical (blueplot) curves obtained for (a) the parallel configuration, (b) cross configuration and (c) setting the ratio of the cross-configuration and the parallel configuration.

Table 1. Results for the TNLCD parameters using the Soutar-Lu method.

Parameter	Value
Twist angle (ϕ)	+93.8° or -93.8°
The angle of the molecular axis (ΨD)	+75.1° or -14.9°
The birefringence (β)	2.126 rad

3.1. Determination the sing of the twist angle (ϕ) and the angle of the molecular axis (ΨD)

Using the stokes parameters can find the sign of ϕ and β [10]. The results obtained are presented theoretically and experimentally for Stokes parameters. The results obtained theoretically and experimentally for Stokes parameters are presented in the Figures 4 to 7. In these results can see that value of S_3 changes sign when ΨD is replaced by 2π+ΨD, while S_1 and S_2 remain invariant. Then the ΨD was found from S3 parameter. While the change sign of ϕ causes a modification in the value of the three parameters S_1, S_2 and S_3. For this reason using a combination of the value found of ΨD, the transmission curves and the stokes parameters is possible the sign of ϕ.

−14.9°.

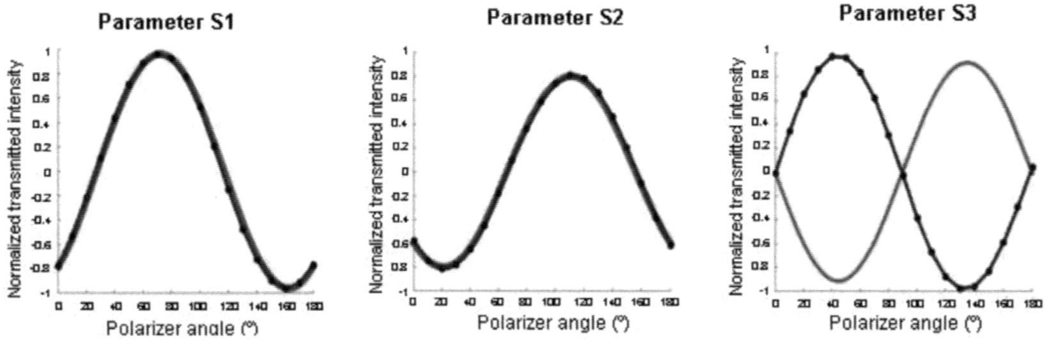

(a) (b) (c)

Figure 4. Experimental (blue plot) and theoretical (red plot) curves obtained for the Stokes parameters by adding π/2 to -14.9°. (a) S_1. (b) S_2. (c) S_3.

+75.1°.

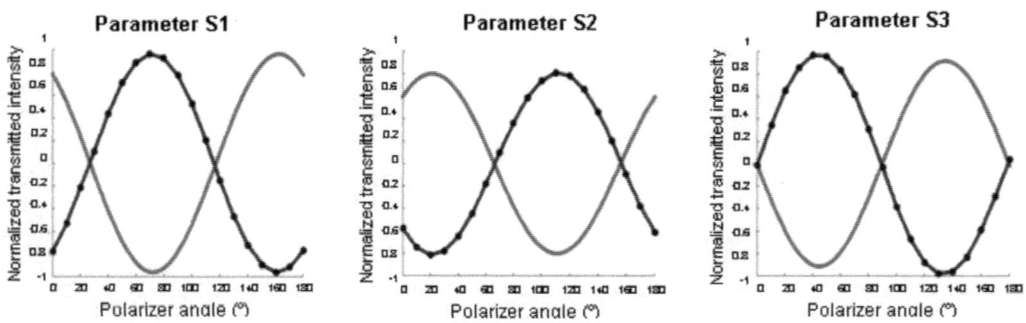

(a) (b) (c)

Figure 5. Experimental (blue plot) and theoretical (red plot) curves obtained for the Stokes parameters by adding π/2 to +75.1°. (a) S_1. (b) S_2. (c) S_3.

For the twist angle (ϕ)+93.8°.

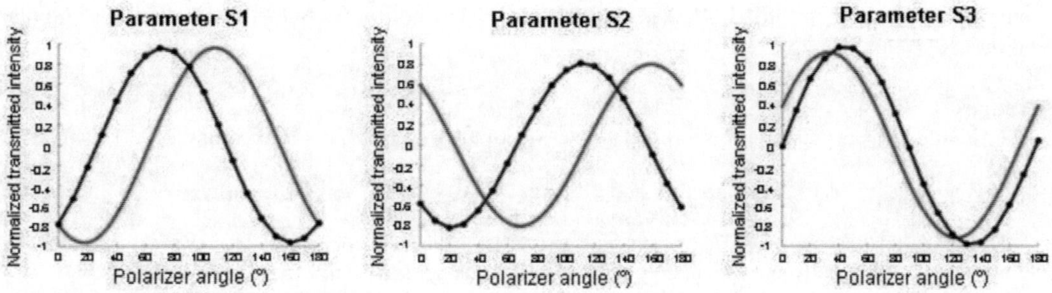

(a) (b) (c)

Figure 6. Experimental (blue plot) and theoretical (red plot) curves obtained for the Stokes parameters whit twist angle +98.3°. (a) S_1. (b) S_2. (c) S_3.

−98.3°.

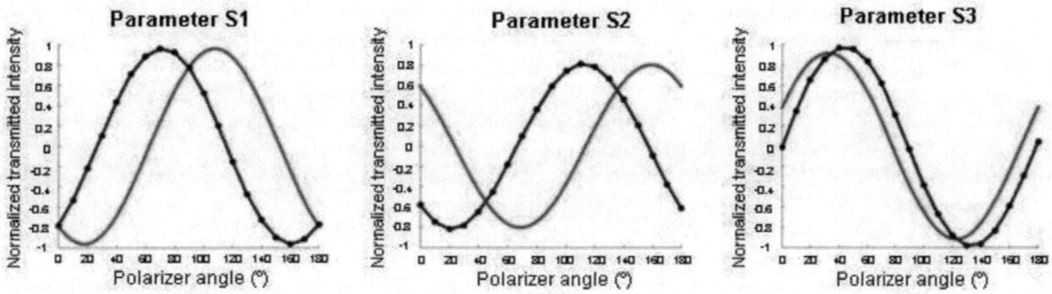

(a) (b) (c)

Figure 7. Experimental (blue plot) and theoretical (red plot) curves obtained for the Stokes parameters whit twist angle -98.3°. (a) S_1. (b) S_2. (c) S_3.

Finally, the Table 2 show the final results obtained for the twist angle (ϕ), the angle of the molecular axis (ΨD), and the birefringence(β) when the LCD is off.

Table 2. Finals Results of the TNLCD parameters.

Parameter	Value
Twist angle (ϕ)	+93,8°
The angle of the molecular axis (ΨD)	-14,9°
The birefringence (β)	2,126 rad

4. Conclusions

retarder-rotor model using the Stokes parameters. Therefore, a unique and unambiguous response without the application of an external voltage for the liquid crystal matrix LCX038ARA by transmission is obtained.

Acknowledgments
Research partially supported by Universidad Industrial de Santander with the projects 5708 and 1788. Thanks to Colciencias for financial support to project 110256934957: "Dispositivos ópticos para sistemas de alta dimensionalidad QKD (Quantum Key Distribution) basados en el momento angular orbital de la luz".

References

[1] T Gould, J Myers and J Bewersdorf 2011 Total internal reflection STED microscopy *Optics Express* **19** 13351-13357
[2] M Zhongsheng, M Changjun, S Zhu and X Yuan 2014 Tight focusing of quasi-cilyndrically polarized beams *Journal of the Optical Society of America A* **31(2)** 373-378
[3] T Ehmke, T Heiko, A Knebl and A Heisterkamp 2014 Molecular orientation sensitive second harmonic microscopy by radially and azimuthally polarized light *Biomedical Optics Express* **5(7)** 2232-2246
[4] H Wang, X Zhenwei, Z Mile, C Haili, H Jingsuo, F Shengfei, W Xinke, S Wenfeng, Y Jiasheng, H Peng and Z Yan 2015 A miniaturized optical fiber microphone with concentric nanorings grating and microsprings structured diaphragm *Optics and Laser Technology* **21(1)** 102-111
[5] Fujitsu 2006 Fundamentals of Liquid Crystal Displays – How They Work and What They Do (California: Fujitsu Microelectronics America, Inc.) pp 1-14
[6] C Zafra 2009 *Caracterización,Modelado Eléctrico y Desarrollo de Nuevas Aplicaciones de Dispositivos Basados en Cristales Líquidos* (Madrid: Universidad Carlos III de Madrid)
[7] E Martín B 1998 *Correlador óptico para el reconocimiento de objetos basado en las propiedades de modulación de los dispositivos de cristal liquido* (Barcelona: Universitad de Barcelona)
[8] J M Vilardy, M S Millán and E Pérez-Cabré 2013 Static and dynamic amplitude modulation of light in a twisted nematic liquid crystal display *Proc. SPIE - Int. Soc. Opt. Eng.* **8785** 10
[9] H Fernández V 2008 *Estudio y optimización del almacenamiento de información en una memoria holográfica* (Alicante: Universidad de Alicante)
[10] C Soutar and K Lu 1994 Determination of the physical properties of an arbitrary twisted-nematic liquid crystal cell *Opt. Eng.* **33(8)** 2704
[11] M Mora-gonzález and F J Casillas Rodríguez 2016 Comparación de Métodos de Calibración para Cristales Líquidos (Nemáticos) *Investigación y Ciencia* **14(36)** 4-9
[12] M M González 2002 *Aplicaciones de cristales líquidos a pruebas ópticas no destructivas* (Guanajuato: Universidad de Guanajuato)
[13] V B Durán 2015 *Optimización del Funcionamiento de un Modulador Espacial de Luz de Cristal Líquido Mediante el Modelo Retardador-Rotor. Aplicaciones en Óptica Adaptativa* (Valencia: Universidad de Valencia)
[14] E Hecht 2002 *Optics* 4th edition (San Francisco: Pearson International Edition)

HOS cell adhesion on TiO_2 nanotubes texturized by laser engraving

A Sandoval Amador[1], N D Montañez Supelano[1], A M Vera Arias[1], P Escobar Rivero[1] and D Y Peña Ballesteros[1]
[1] Universidad Industrial de Santander, Bucaramanga, Colombia

E-mail: anderson84f@outlook.com

Abstract. Due to its outstanding properties, the titanium and its alloys have been widely used in the dental and orthopaedic fields as biomaterials. The TiO_2 nanotubes surface and the texturized process by laser engraving enables significantly accelerated osteoblast adhesion on the biomaterial. For this reason in this paper, the HOS cell responses on TiO_2 nanotubes fabricated on Ti6Al4V alloy and texturized by laser engraving were evaluated. The test surfaces were carried out on smooth Ti6Al4V as control, TiO_2 nanotubes (NT) and surfaces with micropoints obtained by laser engraving, with 1mm spacing (NTP1) and 0.5mm (NTP2). The results show that the texturized process enables decreases the contact angle thus improving wettability of the TiO_2 nanotubes surface. The NTP1 and NTP2 surfaces show excellent cell adhesion and spreading on the surface, which is evident in epifluorescence microscopy images. Furthermore, the NTP1 and NTP2 surfaces improved the cell proliferation at 18% and 16% respectively in relation with NT surface, showing that the laser texturing improves cell response of TiO_2 nanotubes.

1. Introduction

The biomaterials were developed in order to fulfill needs that could not be met naturally [1], for this reason this materials are used in a large number of applications, bone implants, joints, vascular stents, skin substitutes, plastic surgery, cranial and dental implants among many others [2-4]. Bioactive materials used in orthopaedic implants to improve the integration of the implant to the bone and enhanced osseointegration process [5, 6]. One of the most widely used materials for this purpose is titanium and its alloys, Ti6Al4V alloy mainly because of their excellent properties.

Currently different methods are being studied to improve the surface properties of titanium and its alloys in order to achieve and accelerate the osseointegration process in the surface of these materials. One of the most widely used methods has been the electrochemical anodization process [7-9]. This method allows grow controlled the titanium oxide layer, which is formed naturally in the surface of this material. Modifying anodizing parameters, it can be generated different morphologic surfaces and properties that can improve the material behaviour to be implanted in the human body. Another method for modifying the surfaces of orthopaedic implant materials is the textured by laser engraving [10-14]. This method has improved the cell adhesion of Ti and Ti6Al4V surfaces by creating patterns on surfaces. These patterns become places where a preferential cell adhesion occurs, which increase the adhesion of osteoblast cells.

Therefore, in this work we study the effect of laser-textured surface in the cellular response of TiO2 nanotubes grown on Ti6Al4V.

2. Experimental

2.1. Sample preparation

Ti6Al4V cylindrical disc with 14mm diameter and 3mm thick was grinding with silicon carbide paper in successive grades from 220 to 1200 grit and subsequently the disc were cleaned by sonication in ethanol and deionized water for 15 minutes. Some these substrates were anodized to get TiO_2 nanotubes, theses substrates were immersed in a dilute acid mixture of nitric acid (HNO3) and hydrofluoric acid (HF) for 1 minute in order to remove the thin oxide layer that spontaneously forms on the Ti6Al4V surface in presence of the air. The anodization process was carried out in a conventional two electrodes cell, using a solution of 0.5wt% HF, applying 10V during 1 hour at room temperature. Fourth different surfaces were used in this study. They include untreated surface (Ti6Al4V), TiO_2 nanotubes (NT TiO_2), and laser engraving with two different circular patterns (NT TiO2 P1 and NT TiO_2 P2). The patterning surfaces were obtained by laser engraving using a methodology previously described in [15] using a sealed CO_2 laser with maximum power of 40W. The circular patterns generated on the surface can be seen in the Figure 1. All samples were sterilized in autoclaved at 120°C for 20 minutes and then by UV radiation for 30 minutes on each side.

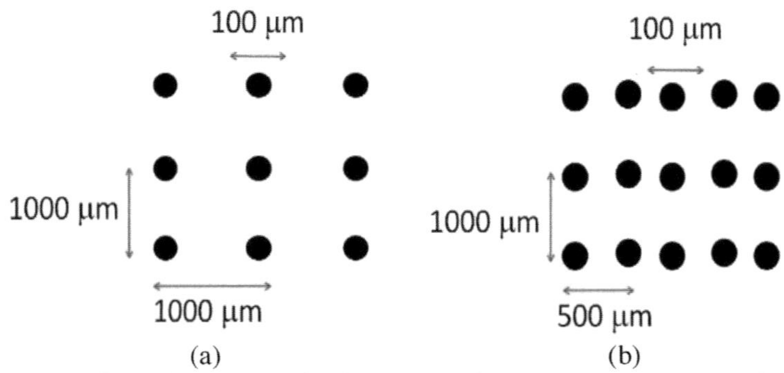

Figure 1. Design pattern for laser engraving process, (a) Pattern 1 and (b) Pattern 2.

2.2. Biological test

These test were performed using a methodology previously described in [15]. For this, HOS cells was cultured in $25cm^2$ Falcon culture plates and maintained in an incubator at a temperature of 37°C regulated with 5% CO_2, 95% air, and saturated humidity. RPMI-1640 medium supplemented with 10% fetal bovine serum and 1% penicillin/streptomycin was used as the cell-cultured medium. For adhesion assay, 2.5×10^4cells/ml in cultured medium were placed on each material and incubated at 37°C. After 1, 3 and 5 days the cells were detached using 0.5ml Trypsin-EDTA for 5 minutes and counted microscopically in a Neubauer chamber. In addition a fluorescence staining were made directly in each sample using Hoechst 33342 fluorescent stain and epifluorescence microscopy (Nikon Eclipse E4000), B-2A filter (Ex=450-490, DM=500, Ba=515). Images were obtained using a camera Nikon Coolpix 5000. UV2A. Each time experiments were performed in duplicate and the results were calculated as number of cells per surface area.

3. Results and discussion

The TiO_2 nanotubes morphology grown on Ti6Al4V alloy by applying 10V in bath containing 0,5% HF concentration is shown in Figure 2. The nanotube inner diameter in the mouth is approximately 30nm. The formation of nanotubes preferentially grown in the alfa phase along with beta phase, besides dissolution of the oxides was observed. Such morphology has been widely described for TiO_2 nanotubes grown in acid electrolytes with F containing [16,17].

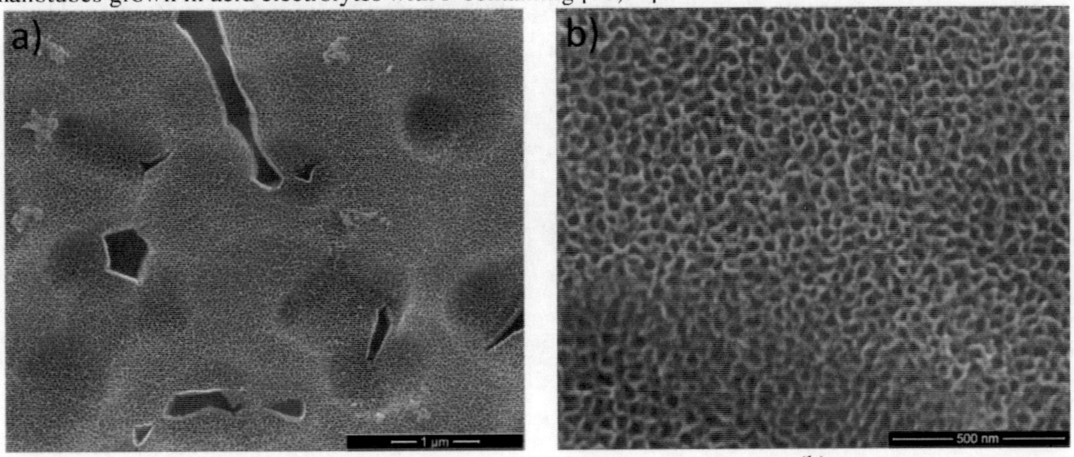

(a) (b)

Figure 2. Ti6Al4V surface morphology (a) microstructure with nanotubular layer (b) TiO_2 nanotubes morphology surface.

As shown in the Figure 3, the patterns has a semicircular shape with diameter of 150μm, the separation spots on the vertical axes is about 1250μm and the horizontal axes is 1295μm for NT TiO2 P1 and 630um for NT TiO_2 P2 respectively. This variation in the dimensions of the patterns was previously described in [15] and the movement performed by the laser engraving equipment to achieve surface texturing the surface attribute.

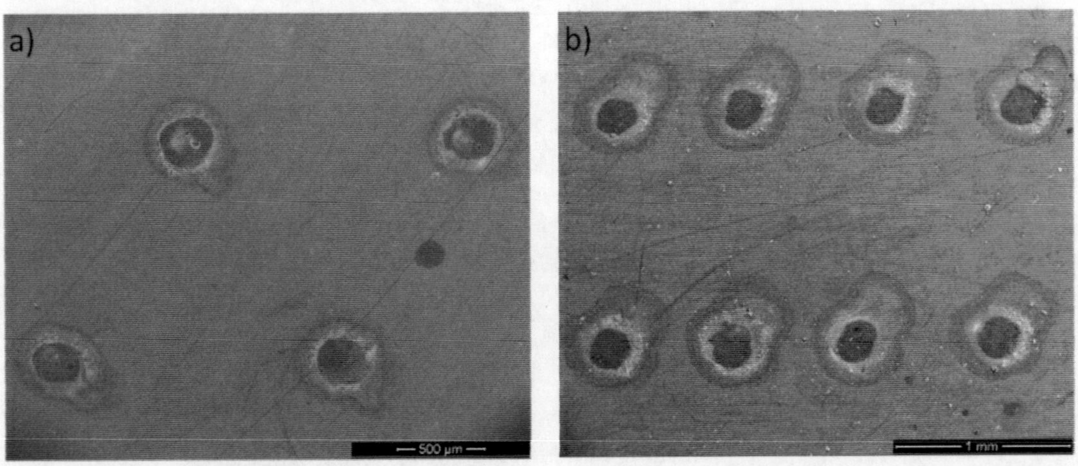

(a) (b)

Figure 3. Patterns on TiO_2 nanotubes morphology (a) pattern 1 and (b) pattern 2.

The hydrophilic/hydrophobic behaviours of the surfaces are shown in the Figure 4. Ti6Al4V smooth surface have a hydrophilic behaviour while the anodized surface had a hydrophobic behaviour. This hydrophobic behaviour has been widely reported [18,19]. The TiO_2 nanotube surfaces texturized by laser engraving (NT TiO_2 P1 and NT TiO_2 P2) show a reduction in a contact angle, making these anodized surfaces less hydrophobic. Previously it reported that the laser texturing process on smooth surfaces of Ti6Al4V generates an increase in hydrophobicity [15], which it is contradictory to what is observes in this study.

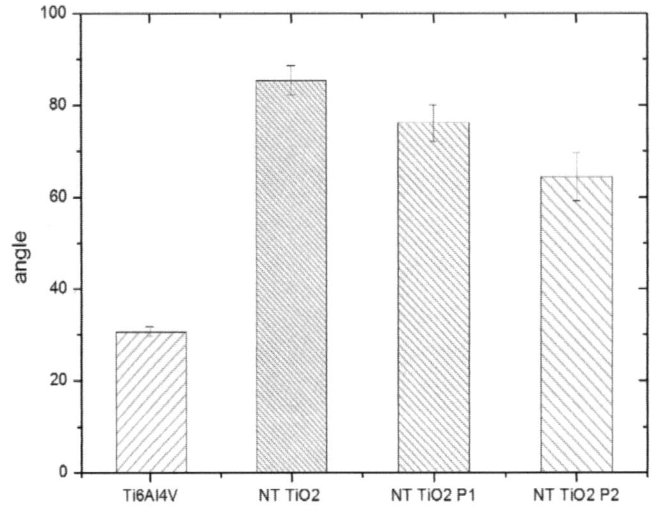

Figure 4. Contact angle.

Ti6Al4V, TiO2 nanotubes and TiO_2 nanotubes texturized by laser engraving result non-toxic to HOS cells, this results consistent with previously reported in the literature [20-22]. Moreover, as shown in Figure 5, the HOS cell adhesion show a high affinity to surfaces with TiO_2 nanotubes. The cell adhesion process is favoured on these surfaces due to the roughness and increased surface area generated by these nanometrics structures [20, 22]. The Figure 6 shows a representative images of the HOS cells during three incubation times, as can be seen the HOS cells were randomly oriented and well dispersed. On the TiO_2 nanotube surface at 5 culture days the HOS cells days is possible appreciate.

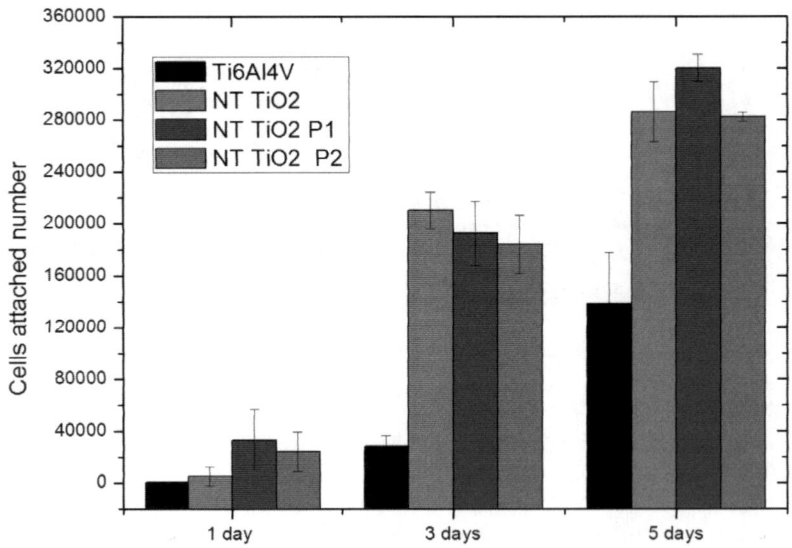

Figure 5. Cell adhesion on the surfaces.

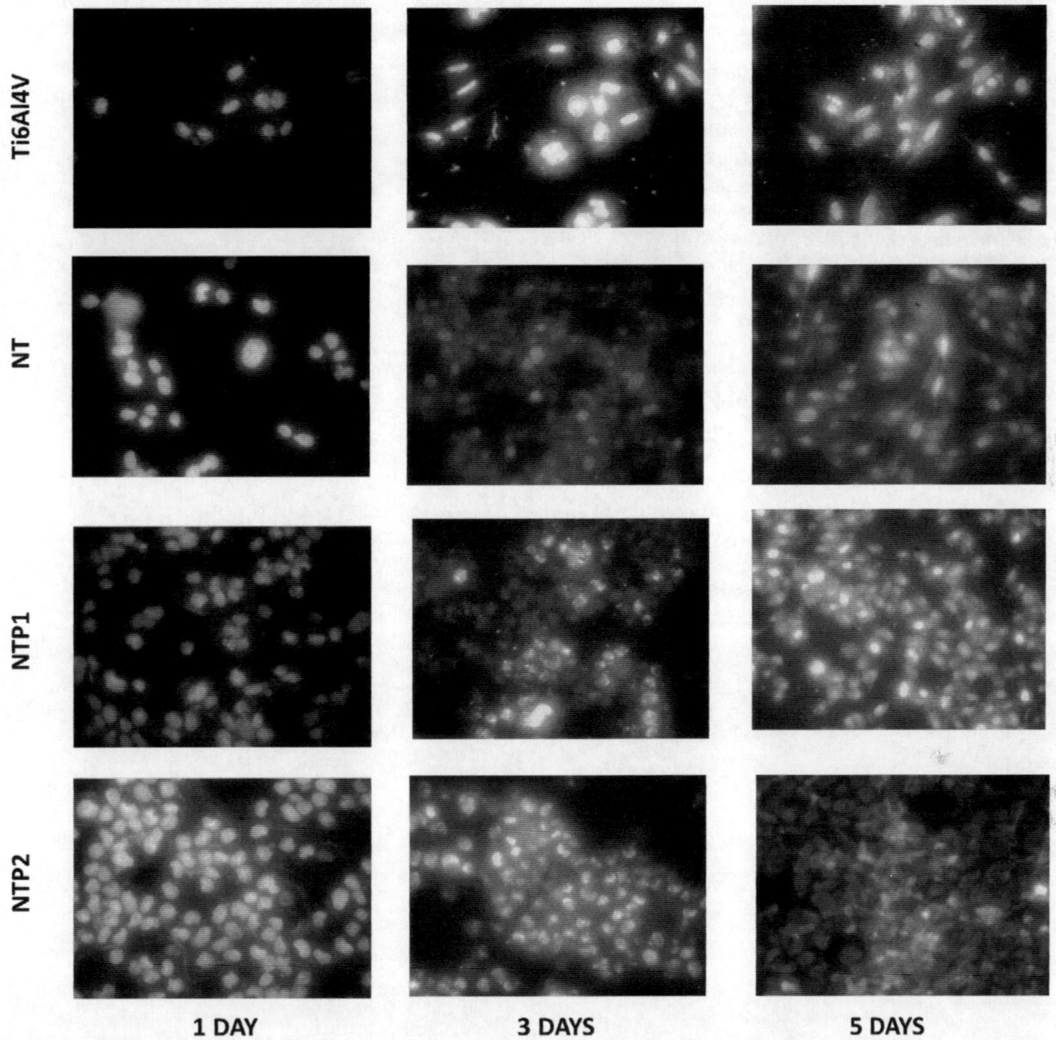

Figure 6. Fluorescence micrographs of HOS cells attached on different surfaces.

4. Conclusions

This study produced nanotube arrays with a uniform tube diameter through the anodization of Ti6Al4V. These surfaces were texturized by laser engraving. In cell culture experiments, the nanotube and nanotube texturized surface exhibited adhesion and spreading that was more rapid than that of untreated surface, despite a higher contact angle. These results provide evidence that nanotube texturized can significantly enhance cell activity. It was shown that the texturized by laser engraving is an effective means of promoting cell functions on nanostructured surface implants, thereby improving the bone-implant interface in vivo.

Acknowledgment

The authors want to give thanks to Universidad Industrial de Santander, Vicerrectoría de Investigación y Extension and Grupo de Investigaciones en Corrosión for funding this project under internal code 1880.

IOP Conf. Series: Journal of Physics: Conf. Series **786** (2017) 012010 doi:10.1088/1742-6596/786/1/012010

References

[1] M R Souza et al 2016 *Mater Res Express* **3** 035401
[2] C H Ku et al 2002 *Biomaterials* **23** 1447
[3] A A John et al 2015 *RSC Adv* **5** 39232
[4] E Beltran Partida et al 2015 *Materials* **8(3)** 867
[5] C Yao et al 2007 *J Biomed Mater res* **85(1)** 157
[6] P Sevilla et al 2010 *J. Phys.: Conf. Ser.* **252** 012009
[7] P schumki et al 2011 *Angew. Chem., Int. Ed* **50** 2904
[8] M Kulkani et al 2015 *Nanotechnology* **26** 062002
[9] Menglin Li et al 2013 *Nanotechnology* **24** 305706
[10] M Bereznai et al 2003 Biomaterials **24** 4197
[11] Milan Trtica et al 2006 Applied surface science **253** 2551
[12] A Y Vorobuev 2007 Applied Surface Science **253** 7272
[13] Khosroshahi M E et al 2007 *Applied Surface Science* **253** 8772
[14] N Mirhosseini et al 2007 Applied surface science **253** 7738
[15] A Sandoval Amador et al 2016 *J. Phys.: Conf. Ser.* **687** 012012
[16] Buyng Gwan L 2009 *Trans Nonferrois Met Soc China* **19** 842
[17] M Rossi de Souza 2014 *Revista Materia* **19** 53
[18] E Balaur et al 2005 Electrochem. Commun **7** 1066
[19] J L Rosa et al 2014 Arch Helath Invest **3(5)** 43
[20] A S Zuruzi et al 2006 *Nanotechnology* **17** 531
[21] S Bauer et al 2008 *Acta biomaterialia* **4** 1576
[22] K S Brammer et al 2009 *Acta Biomaterialia* **5** 3215

Journal of Physics: Conference Series **786** (2017) 012011 doi:10.1088/1742-6596/786/1/012011

Characterization of Tungsten Carbide coatings deposited on AISI 1020 steel

A Santos[1], C Gonzalez[1] and Z Y Ramirez[2]
[1] Universidad Pontificia Bolivariana, Bucaramanga, Colombia
[2] Unidades Tecnológicas de Santander, Bucaramanga, Colombia

E-mail: alfonso.santos@upb.edu.co

Abstract. In order to determine the variation in the mechanical properties of AISI 1020 standardized steel, heat treated by a quenching and tempering process and with a Tungsten Carbide coating, was performed a microstructural and chemical characterization of the coating material through electron microscopy scanning and X-ray energy dispersive spectroscopy. The steel received a heat treatment of quenching performed by heating to 850°C, followed by cooling in water and tempering at a temperature of 450°C with air cooling. Tests of a) microhardness with a Wilson-Wolpert Tukon 2100B micro durometer and b) resistance to adhesive and abrasive wear following the ASTM G99-05 "Standard test method for wear testing with a pin-on-disk machine" and ASTM G65-04 "standard test method for measuring abrasion using dry sand and rubber Wheel" standards respectively. The results show that the microhardness of the steel do not vary with the load used to perform the test; in addition, the heat treatment of quenching and tempering improves by 5.5% the property while the coating increase it by 124.2%. Regarding the abrasive wear resistance, it is observed that the amount of material lost increases linearly with the distance covered. It was determined that the heat treatment decreased on average by 17.5% the volume of released material during the tests while the coating recued it by 66.7%. The amount volume of material lost during the adhesive wear tests increases linearly with the distance covered while the heat treatment decreased on average by 10.5% the volume of released material during the trial and the coating reduced it by 66.5%.

1. Introduction

The protection of metallic substrates against wear can be achieved by using abrasive and abradable coatings (tribological) composed mainly of ceramic materials [1,2]; the coatings with tungsten carbide (WC) are increasing its use in applications where is required high hardness, good tribological properties, excellent corrosion resistance and high thermal resistance [3-5].

One of the most used processes for applying coatings on metal surfaces is thermal spraying due to its versatility and variety [6,7]; these procedures use a spray nozzle which handles an oxyfuel flame, electric arc or plasma arc [8,9], and allows deposition of both metallic and nonmetallic materials. The progresses of thermal spraying are widely accepted in the industry, both in the manufacture of parts and maintenance, where the application field is extended and increasingly extensive due to the development of new alloys and processes [10].

There are several thermal spraying processes among which stand out: thermal spray wire, metal powder thermal spraying, plasma, trigger gun, high speed spraying with oxyfuel and arc wire; the coatings obtained have a layered structure of deposited material and can have porosities between 10 and 20% caused by air and oxide particles trapped by high temperatures that are used [8].

In recent years, the manufacture of ceramic nanostructured coatings using nano-crystalline powders has become more technologically attractive due to its excellent properties compared with conventional equivalents [11-13], since by harnessing the properties associated with nanostructures can improve the performance and durability of conventional plasma sprayed coatings that already have a wide variety of applications in aerospace, biomedical, automotive and chemical industry [14, 15]. This paper presents the results of abrasive wear resistance, microhardness and adhesive wear resistance tests conducted on AISI 1020 commercial steel when it is standardized, is heat-treated through a process of quenching-tempering and when is deposited on it a coating of tungsten carbide in powder form of nanometric size.

2. Methodology

The methodology used for the development of the project consisted of 4 stages.

2.1. Morphological and chemical characterization of the WC

It was performed by the technique of Scanning Electron Microscope (SEM) using a microscope of high resolution FE Schottky brand equipped with Field Emission technology for electron emission which has integrated a microanalysis system by energy dispersive X-ray spectroscopy (EDS) which allows to collect X-rays generated by the sample and perform various semi quantitative and distribution of elements in surface analysis.

2.2. Manufacture and preparation of specimens for microhardness and abrasive and adhesive wear resistance testing

The specimens for testing were manufactured from machining process in lathe and milling machine; the geometric shape and dimensions are adapted to the requirements of ASTM E384-11 "Standard test method for Knoop and Vickers hardness of materials", ASTM G65-04 "Standard test method for measuring abrasion using the dry sand/rubber wheel apparatus" [16] and ASTM G99-05 "Standard test method for wear testing with a Pin-on-disk apparatus" [17] and the equipment used in the different tests.

2.3. Deposition and characterization of WC coatings

The coatings were made through thermal spraying by flame process using an Eutalloy 85 BX gun which is adapted to conventional oxyacetylene combustion equipment. The distance used between the torch and the base material was 30cm with an inclination to the horizontal of 60°. The oxygen and acetylene pressure used was 170kPa and 34kPa respectively, which defines a pressure ratio of 1:5. In order to anchor the coating to the base material a fusible nickel-based alloy is used.

2.4. Testing

The tests were performed on three different specimens: the first were manufactured in AISI 1020 standard steel; the second were of the same steel but thermally treated with a quenching process (heating at 850°C and cooling with water) and tempering (heating at 450°C and cooling with air); and the last were of AISI 1020 steel with WC coating of nanometric size, deposited through thermal spraying by flame process. The microhardness tests are performed following the ASTM E384-11 standard at a Wilson-Wolpert Tukon 2100B micro durometer. The tests of abrasive and adhesive wear resistance are made according to the specifications of ASTM G65-04 and ASTM G99-05 standards, respectively.

Table 1. Testing parameters of abrasive wear.

Specified Procedure	Force Against Specimen (N)	Wheel revolutions (RPM)	Linear Abrasion (m)
A	130	6000	4309
B	130	2000	1436
C	130	100	71.8
E	130	1000	718

Table 2. Testing parameters of adhesive wear.

Specified procedure	Force Against Specimen, N	Wheel Revolutions (RPM)	Contact diameter (m)	Linear distance covered (m)
A	10,2	3980	0,02	250
B	10,2	5971	0,04	750
C	10,2	5307	0,06	1000

The parameters of the tests performance are listed in Table 1 for abrasive wear and in Table 2 for adhesive wear. The measurements of wear are reported as volume loss in cubic millimetres for each one of the specimens used on tests.

3. Results obtained

Based on the images of WC obtained through scanning electron microscopy, is analysed a pre-sample of the dimensions of the side lower than 10 different particles and with them is defined a sample size of 40 data to achieve a reliability of 95% and a maximum error of 5%, the average size of the WC particles was 80.8nm and the distribution is related on Figure 1(a). Chemical analysis of the powders used was performed through the values obtained from microanalysis by energy dispersive X-ray spectroscopy (EDS) which are related in Figure 1(b).

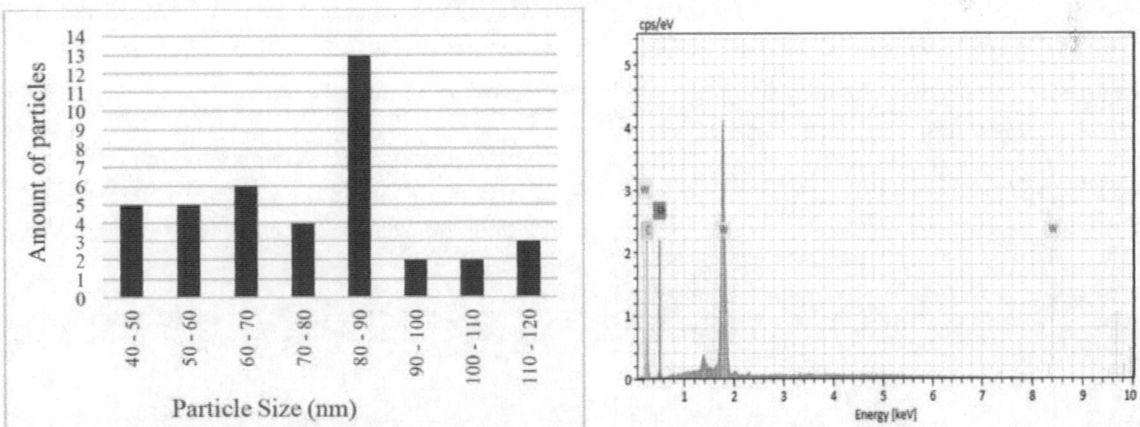

(a) (b)

Figure 1. Chemical characterization of the particles used in the coating a) Distribution of particles size of WC; b) EDS results of WC particles.

The percentage of error that was obtained between the theoretical and experimental atomic weight obtained by EDS analysis was 3.3% which allows to conclude that the material used in the coatings was WC.

The tests of Knoop microhardness related in Figure 2, allow to conclude that thermally treated specimens with a quenching-tempering process improve by 5.5% this property with respect to normalized steel which is due to the low carbon content of AISI 1020 steel. It can also be observed that the coating of WC has a more significant increase in this property, reaching a value of 124.2% compared to the normalized steel, this is due to the great capacity of carbides to resist plastic deformation on its surface. Another appreciation that can be obtained of Figure 2 is that Knoop microhardness, for this material, does not depend on the load with which the test is performed.

Figure 3 obtained from the test of abrasive wear resistance allows to conclude that the specimens thermally treated with a quenching-tempering process decreases an average of 17.5% the material loss during the test, while the coated with tungsten carbide reach an average reduction of 66.7% compared to AISI 1020 standard steel. Figure 3 also shows a linear relation directly proportional between the material loss in the specimen and the linear distance covered which is consistent with the behavior of materials to this wear phenomenon.

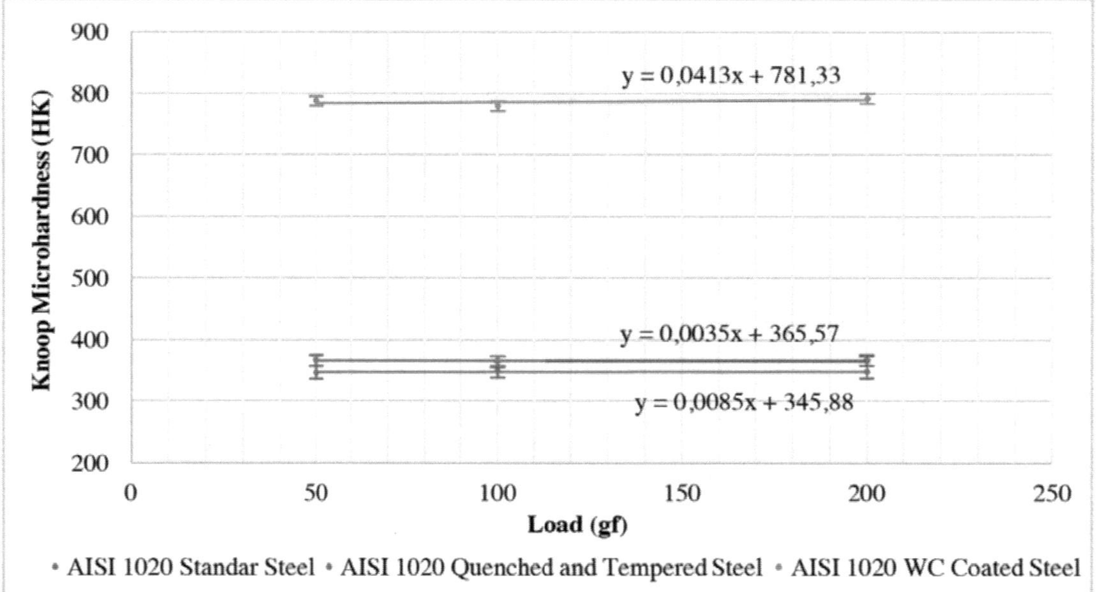

Figure 2. Comparison of Knoop hardness in AISI 1020 steel specimens.

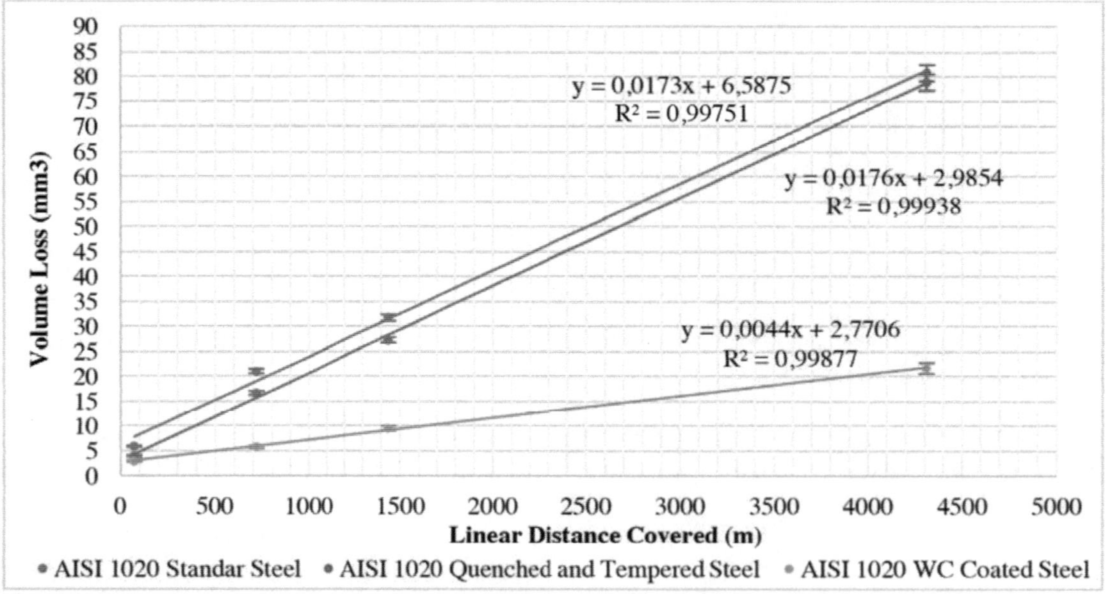

Figure 3. Material loss during abrasive wear resistance test on AISI 1020 steel specimens.

Figure 4 obtained from the adhesive wear resistance test allows to conclude that the specimens thermally treated with a quenching-tempering process decreases in average by 10.5% the material loss during the test while the coated with tungsten carbide reach an average reduction of 66.5% compared with AISI 1020 standard steel. The Figure 4 also shows a linear relation directly proportional between the material loss in the specimen and the linear distance covered.

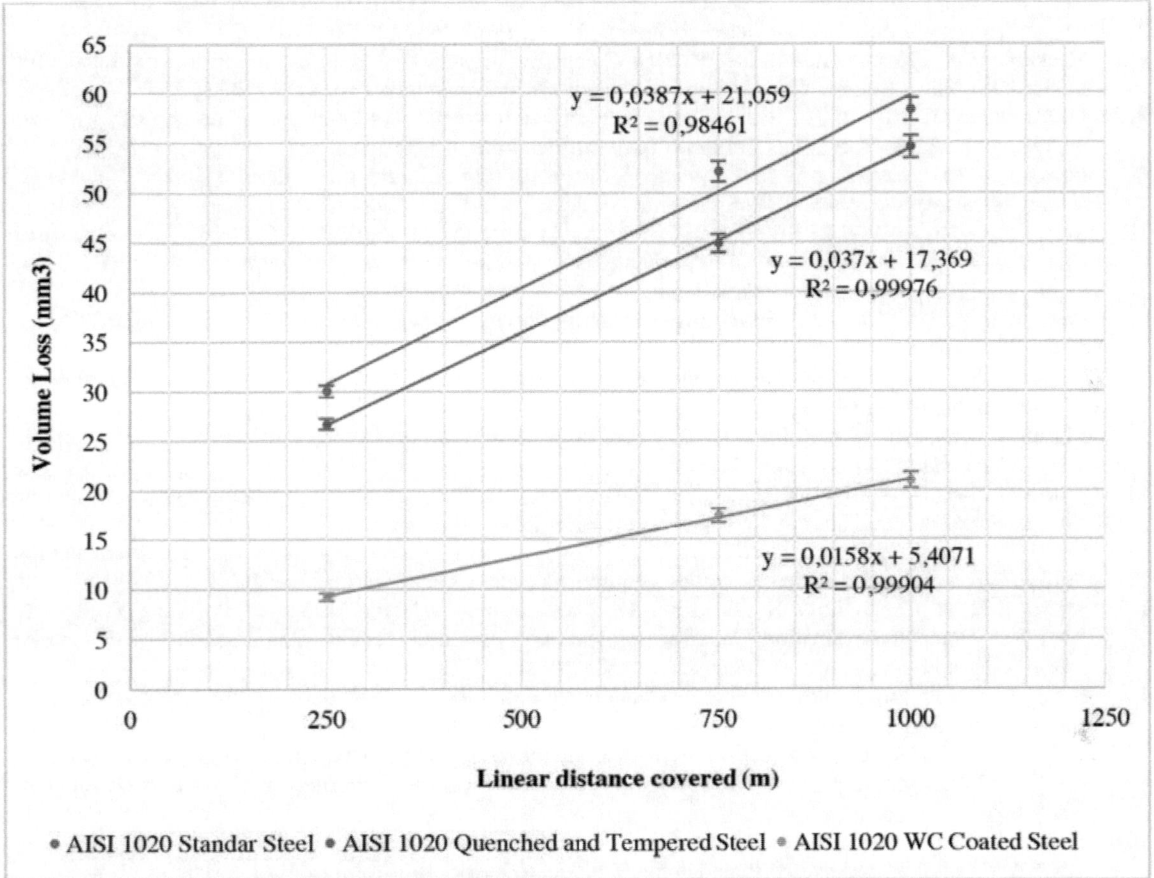

Figure 4. Material loss during adhesive wear resistance test on AISI 1020 specimens.

4. Conclusions

The thermal spraying by flame process allows the deposit of nanometric sized WC on a base material of AISI 1020 steel but is necessary the use of an anchoring material. However the coatings obtained, although retain their nanometric dimensions, have a high porosity and discontinuity which affects the hardness and wear resistance thereof.

The tests of Knoop microhardness allow to conclude that thermally treated specimens with a quenching-tempering process improve by 5.5% this property while the WC coatings has a more significant increase reaching a value of 124.2% compared to standard steel.

The average loss of material per unit of covered length during abrasive wear resistance tests on AISI 1020 standard steel, thermally treated with a quenching-tempering process and coated with nanometric sized WC is of 0.0382mm^3/m, 0.0291 mm^3/m and 0.0151mm^3/m respectively.

The average loss of material per unit of covered length during adhesive wear resistance tests on AISI 1020 standard steel, thermally treated with a quenching-tempering process and coated with nanometric sized WC is of 0.0827mm^3/m, 0.0736 mm^3/m and 0.0271mm^3/m respectively.

References

[1] Rainforth W M 2004 The wear behaviour of oxide ceramics: A review *Journal of Materials Science* **39** 6705-6721

[2] Konyashin I Y 1995 Wear-resistant coatings for cermet cutting tools *Surface & Coatings Technology* **71** 284-291

[3] Ferré F G, Bertarelli E, Chiodoni A, Carnelli D, Gastaldi D, Vena P, Beghi M G and Di Fonzo F 2013 The mechanical properties of a nanocrystalline Al_2O_3/a-Al_2O_3 composite coating measured by nanoindentation and Brillouin spectroscopy *Acta Materialia* **7** 2662–2670

[4] Hackemann S, Flucht F and Braue W 2010 Creep investigations of alumina -based all- oxide ceramic matrix composites *Compos Part A: Applied Science and Manufacturing* **41(12)** 1768–1776

[5] Di Girolamo G, Brentari A, Blasi C, Serra E 2014 Microstructure and mechanical properties of plasma sprayed alumina-based coatings *Ceramics International* **40(8B)** 12861–12867

[6] Davis J R 2001 *Introduction to Thermal Spray Processing Handbook of Thermal Spray Technology*. (Ohio: ASM International and Thermal Spray Society) pp 3–13

[7] Steeper T J, Riggs II W L, Rotolico A J, Nerz J E, Varacalle D J and Wilson G C 1992 Thermal Spray Coatings: Properties, Processes and Applications *Proceedings of the Fourth National Thermal Spray Conference* (Pittsburgh: ASM International) p 13

[8] Pawlowski L 2008 *The Science and Engineering of Thermal Spray Coatings* (Chichester: John Wiley & Sons Ltd) pp 67-89

[9] Grainger S, Blunt J 1998 *Engineering Coatings-Design and Application 2nd edition* (Cambridge: Woodhead Publishing Ltd) pp 1-7

[10] Marulanda J L, Zapata A & Isaza E 2007 Protección contra la corrosión por medio de rociado térmico *Scientia et Technica* **34** 237-242

[11] Wang Y, Jiang S, Wang M, Wang S, Xiao T D, Strutt P R 2000 Abrasive wear characteristics of plasma sprayed nanostructured alumina/titania coatings *Wear* **237** 176–185

[12] Gell M, Jordan E H, Sohn Y H, Goberman D, Shaw L, Xiao T D 2001 Development and implementation of plasma sprayed nanostructured ceramic coatings *Surface & Coatings Technology* **146–147** 48–54

[13] Jordan E H, Gell M, Sohn Y H, Goberman D, Shaw L, Jiang S, Wang M, Xiao T D, Wang Y, Strutt P. 2001 Fabrication and evaluation of plasma sprayed nanostructured alumina–titania coatings with superior properties *Materials Science & Engineering* **A301** 80–89

[14] Singh V P, Sil A, Jayaganthan R 2012 Tribological behaviour of nanostructured Al_2O_3 coatings *Surface Engineering* **28(4)** 277-284

[15] Yang Y, Yan D, Dong Y, *et al.* 2011 In situ nanostructured ceramic matrix composite coating prepared by reactive plasma spraying micro-sized Al–Fe_2O_3 composite powders *Journal of Alloys and Compounds* **509** 90–94

[16] Standard ASTM G65–04 2010 *Standard Test Method for Measuring Abrasion Using the Dry Sand/Rubber Wheel Apparatus* (Philadelphia: American Society for Testing and Materials) pp 1-14

[17] Standard ASTM G99-05 2010 *Standard Test Method for Wear Testing with a Pin-on-Disk Apparatus* (Philadelphia: American Society for Testing and Materials) pp 1-5

Influence on the wear resistance of the particle size used in coatings of Alumina

A Santos[1], R Guzmán[1] and Z Y Ramirez[2]
[1] Universidad Pontificia Bolivariana, Bucaramanga, Colombia
[2] Unidades Tecnológicas de Santander, Bucaramanga, Colombia

E-mail: alfonso.santos@upb.edu.co

Abstract. In the literature, it is common to find that the size of the particles used in coatings through thermal spraying processes influences the hardness and wear resistance thereof; this project aimed to quantify the importance of this parameter in the adhesive and abrasive wear resistance when aluminium oxide is deposited on a substrate of AISI 1020 steel, through a thermal spraying by flame process. The methodology consisted of: a) morphological characterization of the powder used in the coatings by scanning electron microscopy, b) deposition of coatings, c) testing of adhesive and abrasive wear (ASTM G99-05 Standard test method for wear testing with a pin-on-disk apparatus and ASTM G65–04 Standard test method for measuring abrasion using dry sand/rubber wheel *apparatus*), and d) statistical analysis to determine the influence of particle size on wear resistance. The average size of the powder used for coatings was 92, 1690, 8990 and 76790nm. The obtained results allow to identify an inversely proportional behaviour between particle size and wear resistance, in both types of wear (adhesive and abrasive) is shown a logarithmic trend indicating an increase in loss mass during the test as the particle size is also increased and therefore a decrease in wear resistance of the coating.

1. Introduction

One of the most used techniques in recent years, to improve the wear resistance and hardness of materials consists of the deposition of coatings on its surface [1,2], which can be obtained from ceramic, polymeric or metal materials [3,4] available as wire, rod or powder and deposited through different physical, chemical or thermal procedures [5,6].

Among the ceramic coatings, the coatings made of Al_2O_3 are the most commonly used; one of the processes used to apply them is thermal spraying on metallic surfaces through a spray nozzle which handles an oxyfuel flame, electric arc or plasma arc. The thermal spraying is one of the advanced techniques of hard coatings of engineering for the preparation and surface protection [7]. The technology has been widely used as a remedy to combat wear, corrosion, heat corrosion and other problems that occur across the spectrum of manufacturing and engineering industries [8]. The diversity of thermal spraying processes used for hard coatings is due to the variety of applications and required properties, as well as consideration of economic aspects [9]. The coating material may have the form of wire, rod or powder, and the droplets of particles collide with surfaces at speeds of 100 to 1200m/s. The surfaces to spray are cleaned and roughed first, to improve the adhesion strength (which depends on the process and the particular technique used). The coating has a layered structure of ceramic deposited and may have porosity caused by air and oxide particles trapped by the high temperatures that are used [10].

In the conventional processing, the ceramic coatings are produced with powders of micron sized particle and their properties have been studied in depth, showing that the wear resistance is related to hardness and toughness of the coating [11-13]. In recent years, the manufacture of ceramic nanostructured coatings using nano-cryistalline powder has become more technologically attractive because the mechanical properties of the nanometric coatings exhibit significant improvements over their conventional counterparts, because the properties associated with the nanostructures can improve their performance and durability [14-16].

This paper presents the results of the abrasive and adhesive wear resistance tests of a commercial AISI 1020 steel when is deposited over it an Alumina coating through a thermal spraying by flame process, using as projection material Al_2O_3 powder of different particle size (92, 1690, 8990 y 76790nm); additionally, the influence of the particle size on the amount of removed material during the abrasive and adhesive wear tests performed, is determined.

2. Methodology

For the development of the project, it was started with the morphological characterization of the powders sued to make the coatings, for which the technique of Scanning Electron Microscopy (SEM) was used through the equipment FE Schottky of high resolution equipped with Field Emission technology for electron emission which has integrated a microanalysis system by X-Ray Energy Dispersive Spectroscopy (EDS) that allows to collect the rays generated by the simple and performing semi-quantitative analysis and element distribution on surfaces. In order to quantify the size of the particles with a reliability greater than 95%, it was performed a statistical analysis with a random pre-sample of 10 magnitudes, from the images obtained from MEB and a Digital image processing software.

The specimens for testing were made from a traditional machining process, the geometric shape and dimensions are adapted to the requirements of ASTM G65-04 (Standard Test Method for Measuring Abrasion Using the Dry Sand/Rubber Wheel Apparatus) and ASTM G99-05 (Standard Test Method for Wear Testing with a Pin-on-Disk Apparatus), and the equipment used in the different tests. The coatings were made through the thermal spraying by flame process using an Eutalloy 85 BX gun, which is adapted to a conventional equipment of oxyacetylene combustion, the distance used between the torch and the base material was 30cm with an inclination of 60° relative to horizontal, the oxygen pressure was 170kPa and the acetylene pressure was 34kPa which defines a pressure ratio of 1:5; To anchor the coating to the base material is used a fusible nickel-based alloy.

The tests were conducted on specimens of AISI 1020 steel with Alumina coating, obtained from particles of 92, 1690, 8990 and 76790nm in size; The abrasive and adhesive wear resistance tests are made following the ASTM G65 – 04 and ASTM G99-05 standards respectively, which are listed in Table 1 for abrasive wear and in Table 2 for adhesive wear. The wear measurements are reported as the volume loss in cubic millimetres for each of the specimens used in the tests.

Table 1. Testing parameters of abrasive wear.

Specified Procedure	Force Against Specimen (N)	Wheel Revolutions (rpm)	Linear abrasion (m)
A	130	6000	4309
B	130	2000	1436
C	130	100	71.8
E	130	1000	718

Source: ASTM G65 – 04 standard (REAPPROBED IN 2010) [17]

Table 2. Testing parameters of adhesive wear.

Specified Procedure	Force Against Specimen (N)	Wheel Revolutions (rpm)	Contact diameter (m)	Linear distance covered (m)
A	10.2	3980	0.02	250
B	10.2	5971	0.04	750
C	10.2	5307	0.06	1000

Source: ASTM G99-05 standard [18].

3. Results

The tests performed in the scanning electron microscope result in images as the seen in Figure 1(a) which allow determining the morphology and size of the particles used for the realization of coatings.

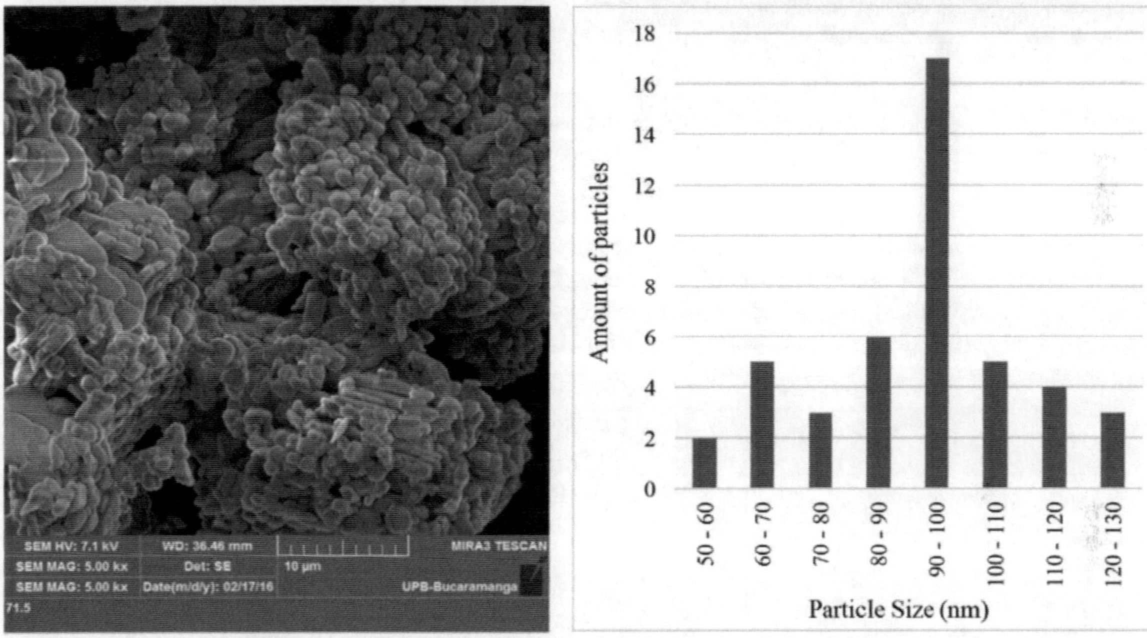

(a) (b)

Figure 1. (a) Image of alumina particles of micrometric size obtained by SEM, (b) Size distribution of the alumina particles of nanometric dimensions.

Based on the images, a pre-sample of dimensions of smaller side of 10 different particles is analysed, and with them a sample size of 45 measurements is defined in order to obtain a reliability of 95% and a maximum error of 5%. The distribution of measurements for the nanometric powder is shown in Figure 1 (b), the average value of the particles size was 92.88nm and the higher frequency are in the range of sizes of 90 to 100nm corresponding to 37.8% of the total, besides the 73.3% of the particles has a size less than 100nm. Other powders used are in the range of micrometric size and showed average values of 1690, 8990 and 76790nm under the same considerations used for nanometric size powder.

The process and the final result of the coating obtained through thermal spraying by flame process is related in Figure 2(a) and 2(b) respectively, it can be observed how the particles are projected onto the base material and the alumina deposits adhered on the anchor material used to attach them to the substrate. It can be concluded that the coating is not uniform on the surface which is characteristic of this deposition process.

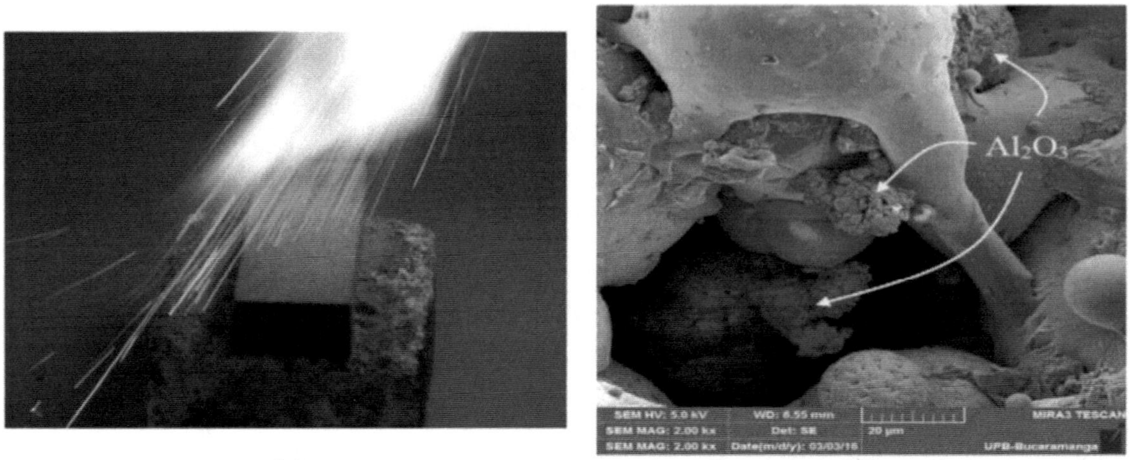

(a) (b)

Figure 2. (a) Image of thermal spraying by flame process used to perform the coatings, (b) Image obtained by SEM of the coating made by thermal spraying by flame process.

The results of the tests of abrasive and adhesive wear made to alumina coatings are related in Figure 3 and 4 respectively; in both a linear behaviour of the volume of lost material during the test with regard to the linear distance covered, is observed. Additionally, it is displayed that in so far as the size of the particles used to make the coatings increases, the volume of material lost during the test also increases, which describes a directly proportional behaviour between these two parameters.

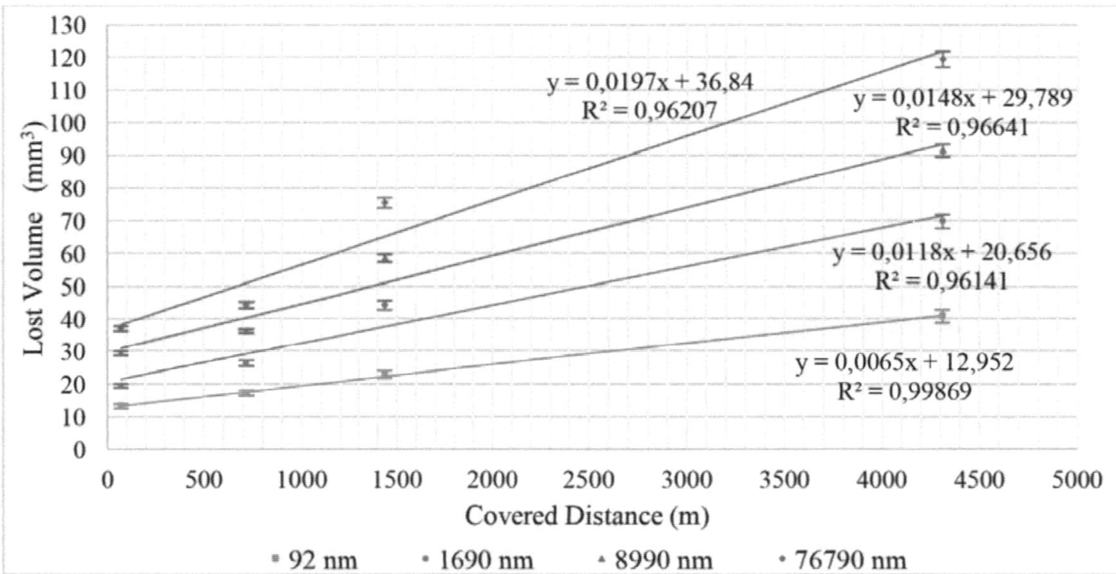

Figure 3. Lost material during abrasive wear tests conducted on alumina coatings obtained from particles sizes of 92, 1690, 8990 and 76790nm.

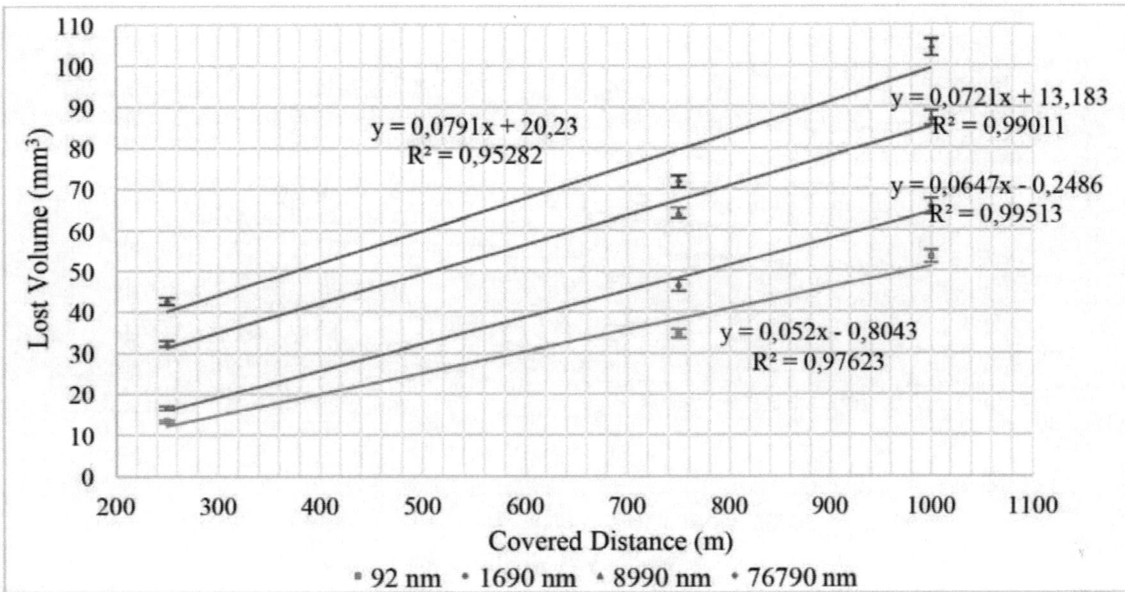

Figure 4. Lost material during adhesive wear tests conducted on alumina coatings from particles sizes of 92, 1690, 8990 y 76790nm.

In Figures 5 and 6 is related the influence of particle size used in conducting the alumina coatings in the volume of material lost during the abrasive and adhesive wear resistance tests. It can be seen in both figures that the greater the size of the particles used, the volume of material lost during the tests is also increased which allows to conclude that there is an inverse proportional ratio between the particles dimension and the wear resistance, that is among smaller the particle size, the abrasive and adhesive wear resistance of the coating are greater.

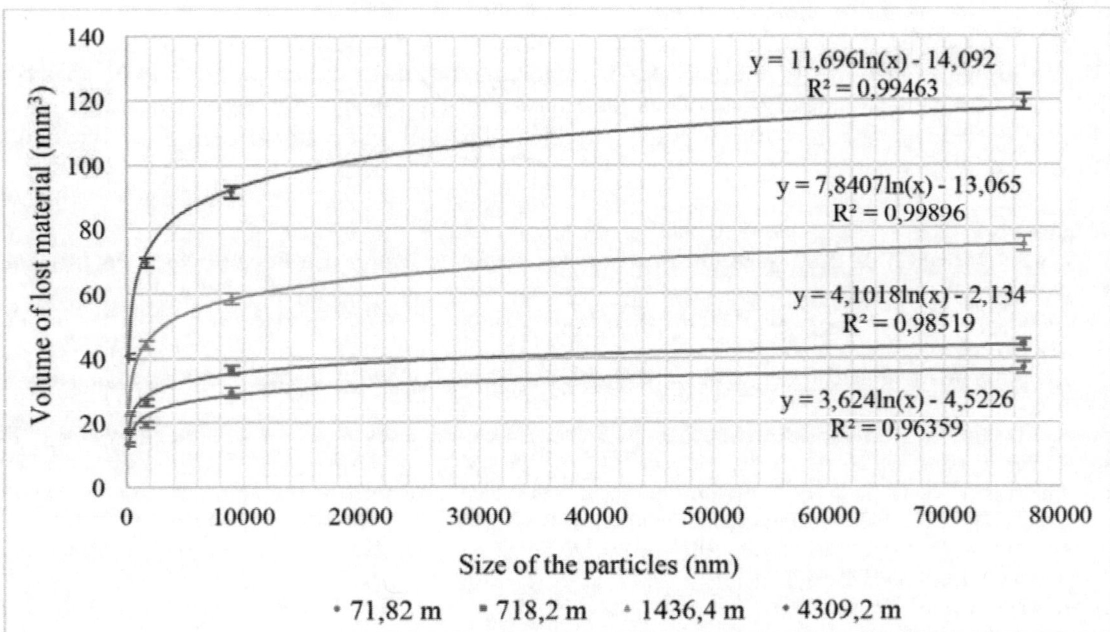

Figure 5. Influence of the particles size used on making the alumina coatings in the volume of material lost during abrasive wear resistance tests.

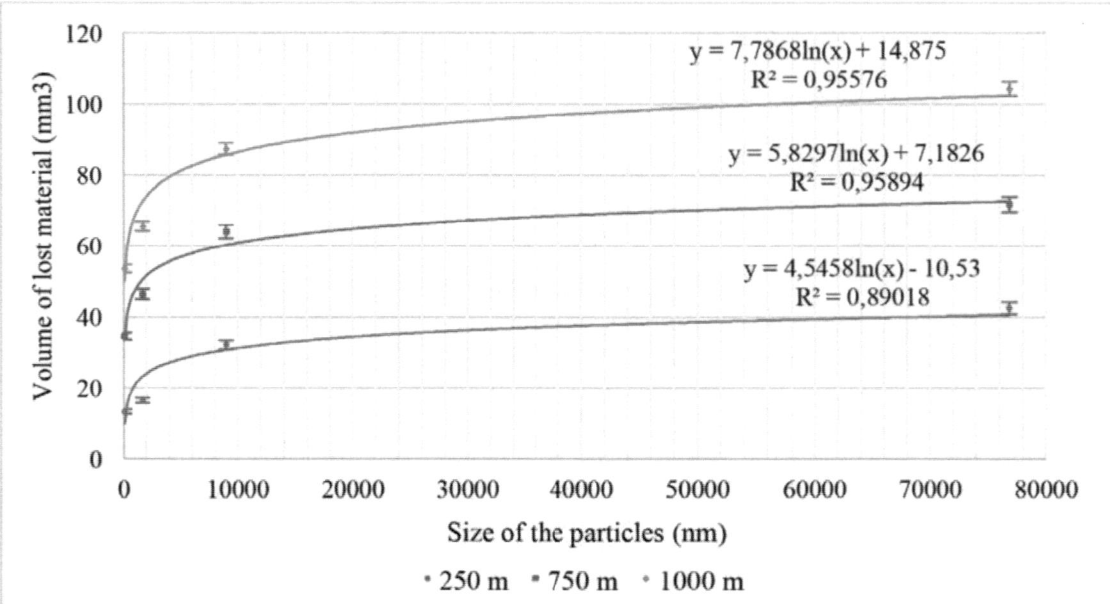

Figure 6. Influence of the particles size used on making the alumina coatings in the volume of material lost during adhesive wear resistance tests.

It can also be seen in Figures 5 and 6 that the influence is much more accentuated when the particle size passes from the nanoscale (<100nm) to the micrometric scale as much so that for particles larger than 10μm practically the volume of lost material during the test is not affected by this parameter.

4. Conclusions

The particle size used for making alumina coatings through thermal spraying by flame process influences inversely proportional in the abrasive and adhesive wear resistance of the coating.

The influence of the size of the particles used for making the alumina coatings is much more accentuated when this parameter passes from the nanoscale (<100 nm) to micrometric scale, as much so that for particles larger than 10 μm practically the volume of lost material during the test is not affected by this parameter.

References

[1] He J, Schoenung J M 2002 Nanostructured coatings *Materials Science and Engineering* **A336(1-2)** 274-319

[2] Tjong S, Chen H 2004 Nanocrystalline materials and coatings *Materials Science and Engineering* **R45(1-2)** 1-88.

[3] Rainforth W M 2004 The wear behaviour of oxide ceramics-A Review *Journal of Materials Science* **39** 6705-6721

[4] Konyashin I Y 1995 Wear-resistant coatings for cermet cutting tools *Surface & Coatings Technology* **71** 284-291

[5] Thankur A, Gangopadhyay S 2016 Influence of tribological properties on the performance of uncoated, CVD and PVD coated tools in machining of incoloy 825 *Tribology International* **102** 198-212

[6] Nilsson M, Olsson M 2011 Tribological testing of some potential PVD and CVD coatings for steel wire drawing dies *Wear* **273** 55-59

[7] Knotek O 2001 *Thermal spraying and detonation gun process Hand Book of Hard Coating* ed. R F Bunshah (New Jersey: Noyes Publications) pp 77-107

[8] Harrisson K A-Metco Ltd. 1996 Thermal spraying: An overview *Surface Engineering Casebook: Solutions to corrosion and wear-related failures* ed. Burnell J S, Datta P K (England: Woodhead Publishing Limited, Cambridge) pp 73-92

[9] Grainger S, Blunt J 1998 *Engineering Coatings: Design and Application* 2nd ed. Grainger S, Blunt J (UK: William Andrew Publishing) pp 1-7

[10] Kalpakjian S, Schmid S R 2002 *Manufactura, ingeniería y tecnología* 4ta (México: Editorial Pearson Educación,) pp 906-911

[11] Shipway P H, McCartney D G, Sudaprasert T 2005 Sliding wear behavior of conventional and nanostructures HVOF sprayed WC-Co coatings *Wear* **259** 820-827

[12] Yang Q, Senda T, Kotani N, Hirose A 2004 Sliding wear behavior and tribofilm formation of ceramics at high temperatures *Surface & Coatings Technology* **184** 270-277

[13] Stewart D A, Shipway P H, Mccartney D G 1999 Abrasive wear behaviour of conventional and nanocomposite HVOF-sprayed WC-Co coatings *Wear* **225-229(4)** 789-798

[14] Gell M, Jordan E H, Sohn Y H, Goberman D, Shaw L and Xiao T D 2001 Development and Implementation of plasma sprayed nanostructured ceramics coatings *Surface & Coatings Technology* **146-147** 48-54

[15] Jordan E H, Gell M, Sohn Y H, Goberman D, Shaw L, Jiang S, Wang M, Xiao T D, Wang Y and Strutt P 2001 Fabrication and evaluation of plasma sprayed nanostructured alumina-titania coatings with superior properties *Materials Science & Engineering A*. **301** 80-89

[16] Taimin G, Pingping Y, Xiaoting Z, Zhongyi Z, Yelong X, Lin Z, Haibin Z, Minwen D, Qi W, Aiwen Z 2016 Influence of WC carbide particle size on the microstructure and abrasive wear behavior of WC-10Co-4Cr coatings for aircraft landing gear *Wear* **362-363** 135-145

[17] Standard ASTM G65–04 2010 *Standard Test Method for Measuring Abrasion Using the Dry Sand/Rubber Wheel Apparatus* (Philadelphia: American Society for Testing and Materials) pp 1-14

[18] Standard ASTM G99-05 2010 *Standard Test Method for Wear Testing with a Pin-on-Disk Apparatus* (Philadelphia: American Society for Testing and Materials) pp 1-5

Induction hardening treatment and simulation for a grey cast iron used in engine cylinder liners

E L Castellanos-Leal[1], D A Miranda[1], A E Coy[1], J G Barrero[1], J A González[1] and O P Vesga Rueda[2]
[1] Universidad Industrial de Santander, Bucaramanga, Colombia
[2] Industrias LAVCO, Floridablanca, Santander, Colombia

E-mail: dalemir@uis.edu.co

Abstract. In this research, a technical study of induction hardening in a grey cast iron used in engine cylinder liners manufactured by LAVCO Ltda., a Colombian foundry company, was carried out. Metallurgical parameters such as austenitization temperature, cooling rate, and quenching severity were determined. These factors are exclusively dependent on chemical composition and initial microstructure of grey cast iron. Simulations of induction heating through finite elements method were performed and, the most appropriate experimental conditions to achieve the critical transformation temperature was evaluated to reach a proper surface hardening on the piece. Preliminary results revealed an excellent approximation between simulation and heating test performed with a full bridge inverter voltage adapted with local technology.

1. Introduction

Now metalworking industry in Colombia is going through a critical situation due to external factors such as the economic crisis that no occurs only in Colombia, if not worldwide, and other internal situations as poor development in the production processes innovation, low flexibility and diversification in the manufacture lines offered to the market [1]. Adding to this, inequality in the cost and excellent quality of foreign products inevitably leads to the purchase of imported raw materials [2]. LAVCO Ltda. requires expanding their business opportunities through the production of engine cylinder liners with improved properties to those already offers its customers. Engine cylinder liners are made from a grey cast iron, with acceptable mechanical properties for correct operation of the pistons in the engine. However, these properties must be improved to increase the half-life of the piece and avoid continuous replacements.

Grey cast irons are alloys of iron, carbon, and silicon, with flake graphite formed during solidification. This graphite is dispersed in a matrix whose microstructure is determined by the chemical composition and solidification process [3,4]. The induction hardening is a promising option in surface thermal treatment to improve mechanical properties such as hardness and wear resistance in auto parts for automotive industry [5,6]. The adoption of induction hardening can let to harden inside parts of the cylinders, representing an attractive option that would increase sales of the company, increasing competitiveness and representation in national and international markets.

Induction heating is a method wherein a magnetic field is generated via an electrical conductor (coil) on the workpiece, alternating current passes through the conductor, creating an alternating magnetic field that produces the Eddy current, this current dissipates energy as heat [7]. In comparison to other conventional techniques, induction heating provides energy savings and other advantages, such as, easy automation and control, reducing the physical requirements for installation, the ability of self-monitoring and short operating times, among others. Temperature pattern generated for this process is especially affected by some variables like geometry and electric-magnetic properties of the workpiece, coupling distance and coil design, frequency selection and operating current and heating time [8,9]. For these reasons, COMSOL Multiphysics was used in this work to create a physical model that predicts the induction heating process in a grey cast iron through simulation and modelling by finite elements method. This modelling allows to select the heat treatment parameters to generate the minimum energy consumption for proper temperature distributions in the grey cast iron surface when a hardening process is performed, and thus achieve improved mechanical properties. Finally, results obtained in the heating process simulation were evaluated and compared with experimental results.

2. Experimental

2.1. Conventional quenching of grey cast iron

Initially, a chemical and microstructural characterization of the casting was realized using the optical emission spectrometry technique on an equipment Bruker brand, reference Q4 Tasman with arc spark under an argon atmosphere and a microscope OLYMPUS GX71 at 500X and 1000X respectively. Subsequently, a quenching heat treatment was applied, considering that its silicon content mainly influences the critical temperature range transformation in cast irons. The lowest temperature of the critical range for casting was estimated by the following expression [10]:

$$\textit{Critical Temperature}, {}^{\circ}C = 730 + 28.0(\%Si) - 25.0(\%Mn) \tag{1}$$

On the other hand, determination of quench severity (H) is another important factor in treatment control. Usually, for grey cast irons, a cooling oil is recommended instead of performed in water due to the good response observed on the hardness and a cracking reduction [11]. Therefore, a range of 50°C above the value obtained by the theoretical calculation was used for austenitizing temperature (870°C) and oil with vigorous agitation as a cooling medium on round bars of 25.4mm in diameter and 20mm in height.

2.2. Modelling and simulation of induction heating

After preliminary testing of conventional quenching, modeling and simulation of the heating process was performed using COMSOL Multiphysics. As conductor (coil) "Copper [solid, annealed (12 micron GS)]" was used and to model the alloy "AAR Grade C steel casting wheels" was employed, the parameters of the alloy were selected to coincide with the LAVCO engine cylinder liner. At this stage, several numerical simulations for heating of cylinder manufactured by the company, and further a Jominy test piece made of the same cast iron and a copper coil of 13 turns with a diameter of 5mm were used for comparison of experimental results performed in the induction heating prototype adapted using technology locally.

2.3. Induction heating of grey cast iron

Induction heating was carried out with an electric system consisting of the following components: a rectifier; responsible for receiving the alternating current from the mains and transform it to DC, which was used in the inverter power circuit. Full bridge inverter; which it provides an alternating signal with a frequency and amplitude determined to the coil. Finally, the series RLC resonant load that allowed the heating of the workpiece that is located within the coil.

3. Results

3.1. Characterization and conventional quench of grey cast iron

The chemical composition of the grey cast iron reveals small amounts of alloying elements (Cr, Ni, Mo, and Mn), which could positively influence on the hardenability [12] (see Table 1). The presence of these elements generates different effects on the hardening process. For example, silicon content between 0.6 and 3.0% causes displacement on the phase transformation curves [13]. Additions of 1 to 2-wt% of nickel or copper are effective in increasing the hardened layer [14]. Moreover, the presence of chromium and molybdenum improve the mechanical strength of the parts in the case of products with simple, such as cylinder liners, piston rings, among other sections, due to the formation of chromium carbides and molybdenum [15].

Table 1. The chemical composition of the grey cast iron in weight percent.

Element	C	Si	Mn	Cu	Cr	S	Ni	P	Mo
Grey cast iron [wt%]	3.87	2.72	0.39	0.281	0.2	0.15	0.097	0.074	0.025

The sample studied has a pearlite matrix with steatite (ternary eutectic cementite, iron phosphide, and ferrite), see Figure 1(a). These morphological characteristics could provide rapid austenitic, which is beneficial in the thermal hardening treatment [16]. Conventional quenching was performed and the micrograph shown in Figure 1(b) was obtained, which reveals a typical microstructure of a typical hypereutectic grey cast iron, mainly composed of martensitic (fine needle-like structure) together with a small amount of retained austenite (white shade) and undissolved flake graphite [17]. This shown that it is possible to make a suitable quench procedure of the material, giving a final hardness of 54 HRC, exceeding the initial hardness of the sample (23 HRC). Finally, it can be inferred that hardening heat treatment is technically feasible to obtain engine cylinder liners with improved mechanical strength.

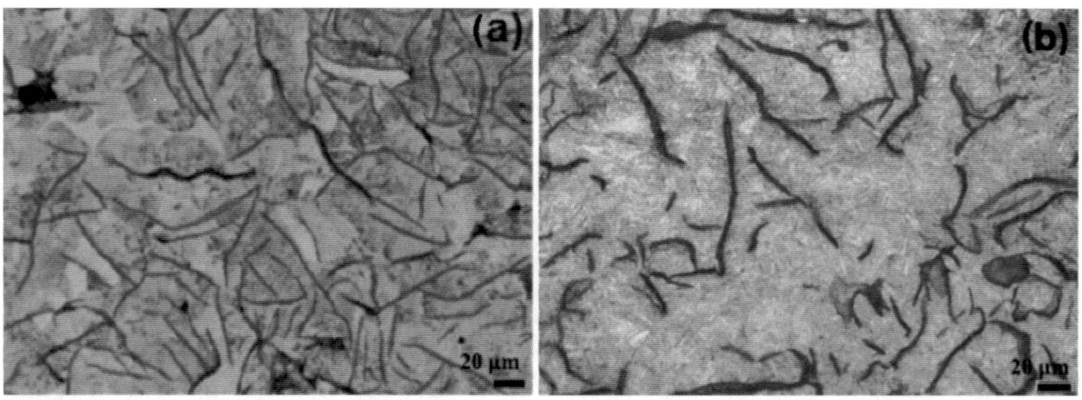

Figure 1. Optical micrographs of the grey cast iron with a chemical etching (2% nital solution) to 500X. Before quenching, (a), and after treatment, (b).

3.2. Modelling and simulation of induction heating for a grey cast iron

Initially, equations that model the magnetic induction process and Eddy current generation by Maxwell's equations and Ohm's law for the electric field were determined [9] to prepare the model in COMSOL Multiphysics. Using sinusoidal harmonic excitations, Ampere's Law and conductive and capacitive effects of the material where currents are induced, the following expression is obtained:

$$j\omega\sigma^{*}\vec{A} + \vec{\nabla}\, x \left(\mu^{-1}\vec{B}\right) = \vec{J_e} \tag{2}$$

Where $\vec{A}$ is the vector potential; $\vec{B} = \vec{\nabla} x \vec{A}$, magnetic field; $\vec{J_e}$, current density produced by external sources; ω, frequency of the harmonic excitation; $\sigma^* = \sigma + j\omega\varepsilon$, complex electrical conductivity of the material and μ, magnetic permeability of the material [18]. Boundary conditions are determinate by hybrid conditions, which Neumann ($\vec{J}_{surface} = \mu^{-1}\hat{n} \, x \, \vec{B}$) and Dirichlet $(\hat{n} \, x \, \vec{A} = 0)$ boundary conditions are combined. Furthermore, heat transfer is determinate by the first law of thermodynamics, which study case can be written as follows [19]:

$$\rho C_p \frac{\partial T}{\partial t} + \rho C_p \vec{u} \cdot \vec{\nabla} T = \vec{\nabla} \cdot (k \vec{\nabla} T) + Q \tag{3}$$

Where T is temperature; $\vec{u}$, velocity vector; Q, heat produced by heat sources different to viscous sources; ρ, density; C_p, heat capacity at constant pressure and k, thermal conductivity of the material. Boundary conditions of heat transfer correspond to the thermal insulation on the boundary $[-\hat{n} \cdot (-k\vec{\nabla}T) = 0]$ and heat conduction conditions between the materials [11]. Electromagnetic and thermodynamic equations are coupled via electromagnetic power equation associated with Eddy currents induced in the material, which corresponds to a heat source as follows:

$$Q = \sigma\omega |\vec{A}|^2 \tag{4}$$

Figure 2(a) shown a cross section of engine cylinder liner and parametric mesh multiform implemented for calculations and Figure 2(b) shows the temperature as a function of radial position for different time (in seconds). In 3s a maximum temperature of 800°C is obtained, which decays with increasing distance from the centre. This result is consistent with the expected temperatures for phase transformation required for induction hardening in LAVCO cylinders.

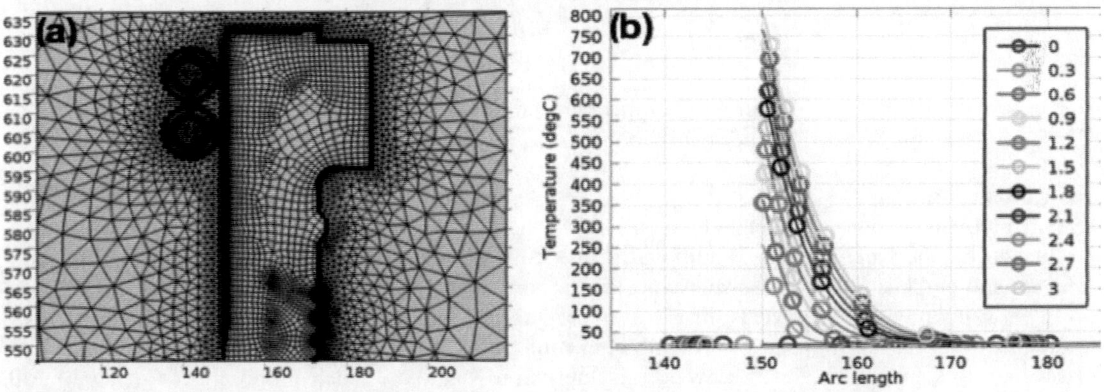

Figure 2. Cross section of LAVCO engine cylinder liner and meshing for the model, (a), and temperature distribution for different radial positions (abscissas, in millimetres), (b).

3.3. Heating of grey cast iron

For that, the full-bridge inverter works keeping load oscillation, it must produce commutations in synchronism with the oscillation, i.e., its switching frequency can be the same to the natural frequency of the load [20]. Preliminary assessments were performed with the local prototype. However, in this work only could make an evaluation of heating using a maximum current of 25A and a working frequency of 30kHz, due to the inconvenience of switching transistors used in the inverter circuit. Temperature profile obtained under these conditions in the simulation is presented in Figure 3. It can be seen that a maximum temperature of 93.8°C is reached on the surface of the specimen. Moreover, when the specimen under the same conditions model was heated, a surface temperature of 83°C was obtained, which corresponds to 10.8°C less than the value expressed in the simulation. The overall efficiency of these systems is about 94% (near the resonance). The losses in the rectification stage can be assumed as 1% of the total efficiency. While remaining 5% may be related to inverter stage [21]. In our case, losses can be associated with an additional 5% to slight gap between the square voltage signal and the sinusoidal signal of the current, which did not allow the RLC load operated in the resonance frequency, consequently, less energy transfer was achieved. Despite this, the experimental results are very close to the values obtained from the physical model. For this reason, it is advisable to improve the prototype system of induction heating and deepen aspects concerning the type of piece, induction coil and operating conditions in the computational tool to have control and quality assurance in induction hardening of engine cylinder liners or any other piece.

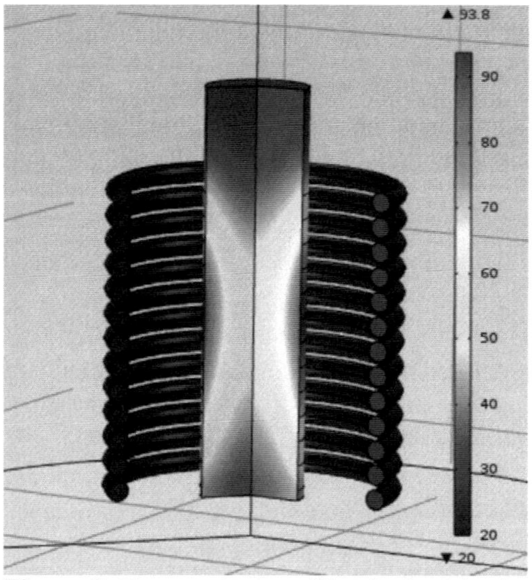

Figure 3. Temperature profile obtained in the cylindrical specimen.

4. Conclusions

Small amounts of alloying elements such as Cr, Ni, Mo and Mn in the cast iron have a positive influence on the hardenability. Additionally, the grey cast iron has a matrix of fine pearlite with steatite and graphite flakes with size, shape, and fair distribution, which allows a rapid austenitic.

The simulation study reveals that by employing moderate currents (90 and 500 A) at high frequencies (300 to 1000 kHz), the critical temperature profiles thinner but less uniforms are obtained. However, to achieve a minimum power consumption, it was found that a minimum frequency of 150 kHz operation is necessary. The experimental results obtained in the specimen of 25.4mm (diameter) at low temperature, showed great similarity with the simulation.

The possibility of surficial hardening on these materials controlling the operating parameters, minimizing energy consumption and ensuring effective hardening depth, allow adding value to the quality of parts that require more precise specifications surface hardness. The deepening of this type of research will allow real solutions to the metalworking companies that are interested in improving the quality of their products.

Acknowledgments

The authors would like to acknowledge the financial support of COLCIENCIAS (Project 467859937197).

References

[1] Ferreira J C 2002 *J. Mat. Process. Tech.* **121** 94
[2] Tejeko I, Adrian H, Skalny K, Pakiet M and Stasko R 2009 *J. Ach. Mat. & Manuf. eng.* **37** 2
[3] Balachandran G, Vadiraj A, Kamaraj M and Kazuya E 2011 *J. Materials & Design* **32** 4042
[4] Yeh C P, Hwang W and Lin C 2009 *J. Materials Transactions* **50** 2584
[5] Barglik J, Smalcerz A, Przylucky R and Dolezel I 2014 *J. Comp. & Appl. Math.* **270** 231
[6] Joseph R D 1996 *ASM Specialty Handbook: Cast Irons* (Ohio: ASM International) p 9
[7] Cunningham J L, Medlin D J and Krauss G 1999 *J. Mat. Eng. & Perform.* **8** 401
[8] Smolijan B, Tomašić N, Iljkić D, Felde I and Reti T 2006 *J. Ach. Mat. & Manuf.* **17** 281
[9] Sadeghipour K, Dopkin J A, and Li K 1996 *J. Computers in Industry* **28** 195
[10] D.M. Stefanescu 2008 *ASM handbook: Castings* vol 15 (Ohio: ASM International) p 914
[11] Cajner F, Smoljan B and Landek D 2004 *J. Materials Processing Technology* **158** 55
[12] Gliner R E and Vybornov V 2014 *J. Metal Science and Heat Treatment* **56** 424
[13] Valencia G A 2009 *Technology of metal heat treatment* vol 1 (Medellín: Colombia) p 87
[14] Grossman M 1942 Hardenability Calculated from Chemical Composition *Transactions AIME* **150** 227
[15] Ferhathullah S, Viquar M, Laxminarayana P and Krishnaia A 2015 *Int. J. Res. Eng. Tech.* **04** 114
[16] Aleksandrov N N and Il'icheva L V 1963 *J. Metal Science and Heat Treatment* **5** 646
[17] Vadiraj A, Balachandran G, Kamaraj M and Kazuya E 2011 *J. Materials and Design* **32** 2438
[18] Zhu S, Wang Z, Qin X, Mao H and Gao K 2016 *J. Materials Processing Technology* **229** 814
[19] Peram S, Ramesh V and Ranganayakulu J 2013 *J. Eng. Res. and Appl.* **3** 66
[20] Kifune H, Hatanaka Y and Nakaoka M 2004 *IEE Proceedings-Electric Power Appl.* **151** 150

IOP Conf. Series: Journal of Physics: Conf. Series **786** (2017) 012014 doi:10.1088/1742-6596/786/1/012014

Impedance spectroscopy in photovoltaic materials of Cu_2ZnSnS_4 (CZTS) and use of the KK transform

M Patarroyo[1], E Vera[1], Y Pineda[1], J Gómez[1], G Soracá[1] and D Sánchez[1]

[1]Universidad Pedagógica y Tecnológica de Colombia (UPTC), Tunja, Colombia

E-mail: miguel.patarroyo@uptc.edu.co

Abstract. The article presents some details about the synthesis and evaluation of photovoltaic materials related with the Cu_2ZnSnS_4 system (abbreviated CZTS), using a hydrothermal route that provide the optimal way to synthesize the proposed materials. The ceramic was obtained starting from corresponding metal nitrates and thiourea as sulphur source in stoichiometric amounts. Corresponding reagents were dossed in a steel Teflon lined vessel and treated at different temperatures to evaluate the effect of external variables in synthesis process. The structure was evaluated by means scanning electron microscopy (SEM) and X-ray diffraction. The electrical characteristics were evaluated by solid state spectroscopy using a statistical analysis coupled with a simple model fitted to the data of electrical conductivity of the material as function of synthesis temperature, for this, the mathematical formulation of the impedance was analyzed, with the use of the Kramers-Kronig transform (KK), (mathematical equations that describe the relationship between the real and imaginary parts of certain complex functions analytic) as well as documentation and research related with the subject of this article. The results show a good behavior of the material, showing that the higher synthesis temperatures promotes a corresponding increase in the electrical conductivity in accordance with previous works [1].

1. Introduction

For the synthesis of photovoltaic cells of low cost and good performance is necessary a constant research and laboratory test, which in many cases are expensive due to consumption of analitical reagents of high purity, which could be improved by the use of mathematical modelling and techniques of formulation that can validated the processes. So, in this work, we explore the correspondece between experimental and modeled data related with the synthesis of Cu_2ZnSnS_4 system, in terms of influence of the time and synthesis temperature, using as control parameter the electrical impedance as dependant variable by means a Kramer-Kronig transform equations.

The electrical impedance spectroscopy is a technique to collect experimental data, which can be given a proper interpretation regarding the material under study, this interpretation depends on how the data are analyzed. This technique consists in apply an electrical disturbance sinusoidal (voltage or current) to the material under study, this makes always is present some strange factor or noise that can influence the measurements [2, 3] is here where the transformed Kramer-Kronig do their work, as these purely mathematical equations help validate the reliability of the data and correct decision making. The Nyquist plot represents the impedance in a complex plane, and The Bode plot represents the impedance in function of the frequency. To consider the impedance as a vector, it allows that you can manipulate as a complex number, so its magnitude will be $|Z| = \frac{V_0}{I_0}$ and phase angle φ.

1.1. Obtaining mathematical impedance

Suppose that a current flows through a circuit formed by a resistance, an inductor and a capacitor in series, the circuit is fed with a sinusoidal voltage, and there is not presence of transient phenomena, wherewith the system is linear and the current the permanent regime is also sinusoidal with the same frequency as that of the original source, well now, if the current amplitude and the phase gap that may have regarding the voltage of fed, is not known, then, supply voltage is $V = V_0cos(\omega t)$, the current will be of the form $I = I_0cos(\omega t + \varphi)$, where φ is an unknown gap. With this, the equation to be solved is: $V_0cos(\omega t) = V_R + V_L + V_C$.

V_R, V_L, V_C are the voltages between the ends of the resistance, inductance and capacitor. Applying Ohm's law, the definition of inductance and condenser is obtained:

$$V_0cos(\omega t) = RI_0cos(\omega t + \varphi) - \omega LI_0sin(\omega t + \varphi) + \frac{1}{\omega C}I_0sin(\omega t + \varphi) \quad (1)$$

Where I_0 y de φ they must be found and the equation must satisfy all values of t.

Suppose now that an identical circuit is fed by another source of sinusoidal voltage which begins with a quarter period delayed, then the voltage will be $V = V_0cos(\omega t - \frac{\pi}{2}) = V_0sin(\omega t)$, then, the solution will also have the same delay and the current will be: $I = I_0cos(\omega t + \varphi - \frac{\pi}{2}) = I_0sin(\omega t + \varphi)$. The equation of the second circuit will be:

$$V_0sin(\omega t) = RI_0sin(\omega t + \varphi) + \omega LI_0cos(\omega t + \varphi) - \frac{1}{\omega C}I_0cos(\omega t + \varphi) \quad (2)$$

Performing some mathematical calculations you get to: (*) $I_0e^{j\varphi} = \frac{V_0}{R+j\omega L+\frac{1}{j\omega C}}$.

The current amplitude and phase gap is observed in (*) and must be equal to the module and argument of complex number on the right side.

2. Methodology

The CZTS material was obtained by the implementation of a hydrothermal route, starting from nitrate solutions in 1.0mol L^{-1} concentration and solid thiourea as S source, all from Merck. The systems in each case were treated under magnetic stirring conditions (100rpm) and subjected to different time and thermal treatments (200, 220, 240, and 260°C), in a teflon lined stainless steel autoclave. The obtained material as a black precipitate was washed several times using absolute ethanol and acetone to achieve a rapid drying of it. As second stage, the solids were calcined at 400°C in a tubular furnace under helium flow (20ml min^{-1}) for two hours, to remove volatile impurities and consolidate the crystalline phase. The obtained solid was characterized by X-ray diffraction (XRD), on a PANalytical X′pert PRO MPD equipment provided with an ultra-fast X′celerator detector in Bragg-Brentano configuration, using the Cu Kα radiation (λ=1.54Å). The diffractograms were taken between 10 and 90° 2θ and the results were analyzed using the X′Pert High Score software.

The morphological evaluation of solid was performed by scanning electron microscopy (SEM), in a Leica-Zeiss LEO 440 electron gun equipment with an accelerating voltage of 30kV and the electrical and conductivity properties of the material was evaluated using the solid state impedance spectroscopy technique (IS), in a GAMRY potentiostat-galvanostat instrument between 1 and 10MHz for which a set of pellet samples, were obtained using a pressure of 5.0Mpa and a constant diameter of 10mm and a thickness of 0.8mm. Each sample was tested five times to avoid mistakes in the numerical information of impedance.

3. Analysis and results

The morphological characteristics of solid, initially was evaluated by means scanning electron microscopy at different magnifications, the images in Figure 1, confirm the obtention of homogeneous aggregates of regular morphology in accordance with works in which is clear that hydrothermal treatment configure specific characteristics of similar compositions based on CZTS materials.

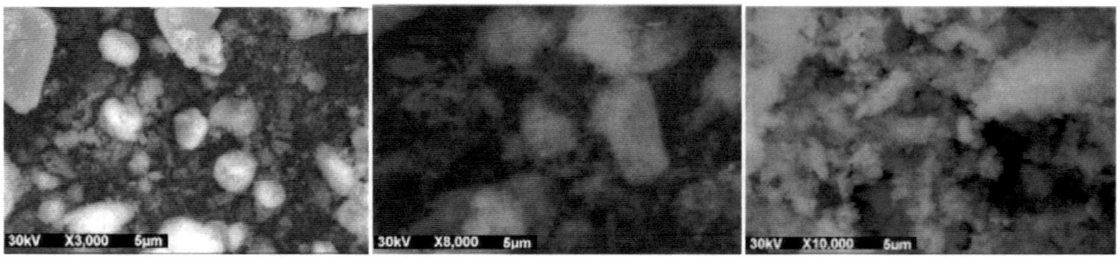

Figure 1. Scanning electron microscopy (SEM), of CZTS materials at different magnifications.

The structural characterization by XRD measurements confirm the obtention of a pure CZTS phase related with a tetragonal calssification in accordance with 00-026-0575 PDF card with crystal group I-42m(121) and cell parameters a=5.427 angstrom and b=10.848 angstrom. As shown in Figure 2, the XRD pattern confirm that in all cases the second stage of calcination provide an optimal way to obtain the crystalline phase.

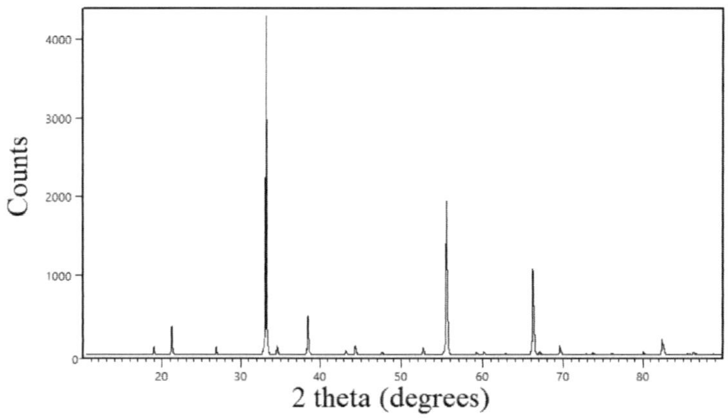

Figure 2. Structural characterization by XRD of CZTS materials.

The Figure 2 XRD is the pattern of representative CZTS system obtained by hydrothermal method in two reaction stages.

The electrical characterization by solid-state impedance spectroscopy shown in Figure 3, confirm a tipical behavior of materials in terms of Nyquist and Bode plots.

The obtained impedance data took into account the different temperatures of synthesis of the material (independent variable), and the maximum values of the impedance modules of each test, Table 1, finding with them the material conductivity (inverse of the maximum impedance) Table 2, leaving as fixed effects the synthesis time and material (Cu_2ZnSnS_4). To these data were made a statistical analysis of variance, obtaining a favorable correlation factor as shown in the Table 3.

As Fo>1 the treatments provided (temperature levels) are actually the factor greatest influence on the dependent (electrical response) variable, also as Fo>F, the independent variable (T) of the experiment is 95% probability of success the most influential on the value of the dependent variable.

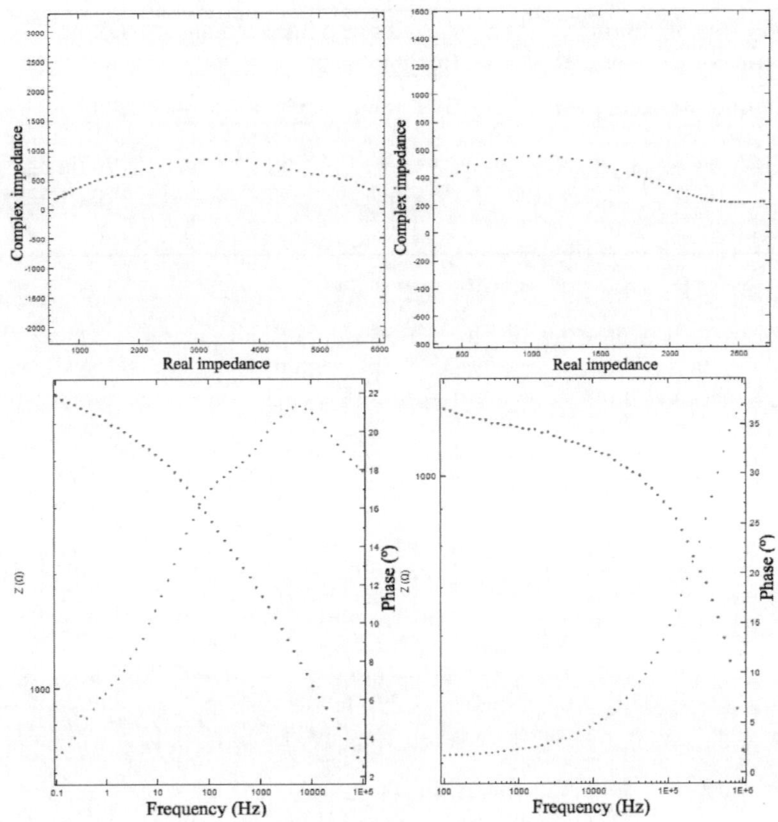

Figure 3. CZTS material impedance spectra for 220 and 260°C temperatures (Nyquis and Bode).

Table 1. Impedance spectroscopy (CZTS) to different temperatures.

T C	Registry 1	Registry 2	Registry 3	Registry 4	Registry 5
220	5977.77	7080.79	7472.18	7325.91	7540.15
240	3610.41	4316.57	4750.21	4811.20	5112.52
260	1243.05	1552.35	2028.25	2296.49	2684.89

Table 2. CZTS conductivity material at different temperatures in ^{o}C.

T C	Registry 1	Registry 2	Registry 3	Registry 4	Registry 5
220	1.67	1.41	1.34	1.36	1.33
240	2.77	2.32	2.11	2.08	1.96
260	8.04	6.44	4.93	4.35	3.72

4. The KK transform

The Kramers-Kronig relations are matematical equations that describe the relationship between the real and imaginary parts of complex analytic functions (Defined functions on an open subset of the complex plane $\mathbb{C}$ with values in $\mathbb{C}$ and differential at each point, that further, they can be

IOP Conf. Series: Journal of Physics: Conf. Series **786** (2017) 012014 doi:10.1088/1742-6596/786/1/012014

Table 3. Statistical analysis of the electrical conductivity of the material CZTS.

Source of variation	Sum of squares	Degrees of freedom	mean squares	F_0	Value P
Temperature levels	4.65×10^{-7}	2	2.32×10^{-7}	22.05	< 0.01
Error	1.26×10^{-7}	12	1.05×10^{-8}	F(0.05)=4.47 F(0.01)=5.95	
Total	5.91×10^{-7}	14			

described by its Taylor series) in the complex upper half plane [4]. These relationships are used to calculate the real and imaginary parts of responses in physical systems in a linear, causal, estable process and finite impedance [5, 6].

$$Z'(\omega) = Z'(\infty) + \frac{2}{\pi}\int_0^\infty \frac{xZ''(x) - \omega Z''(\omega)}{x^2 - \omega^2}dx \tag{3}$$

$$Z'(\omega) = Z'(0) + \frac{2\omega}{\pi}\int_0^\infty \left(\frac{\omega}{x}\right)\frac{Z''(x) - Z''(\omega)}{x^2 - \omega^2}dx \tag{4}$$

$$Z''(\omega) = -\frac{2\omega}{\pi}\int_0^\infty \frac{Z'(x) - Z'(\omega)}{x^2 - \omega^2}dx \tag{5}$$

$$\Phi(\omega) = \frac{2\omega}{\pi}\int_0^\infty \frac{ln|Z(x)|}{x^2 - \omega^2}dx \tag{6}$$

Where $Z'(\omega)$ and $Z''(\omega)$ are the real and imaginary parts of the impedance at a given frequency; ω and x are the angular frequencies; $Z'(x)$ and $Z''(x)$ are continuous functions that provide the real and imaginary parts of the impedance, respectively, in function of the angular frequency (x) with $0 < x < \infty$; $Z'(0)$ and $Z''(\infty)$ everyone, is the component real at zero and infinite frequency, that is to say, when $x \to 0$ and $x \to \infty$; $|Z(x)|$ is the module and $\Phi(\omega)$ It is the phase angle for the angular frequency $x = \omega$ [7]. well now, using Equations (3) to (6) It can transform the real part of the impedance in the imaginary part and viceversa. [8, 9].

4.1. Testing process for data reliability

The appropriate procedure for use the KK and can validate the data, is calculate the imaginary part of the impedance from the values of the real part measured experimentally, in this case the Equation used is (5), the results obtained from this equation are compared with the results of the imaginary part experimentally measured, confirming the validity of the data. It also can find the real part of the impedance from the data of the imaginary part experimentally measured, in this case the Equation would be used (3). The process just mentioned have a difficulty, it is that you must first adjust to one or several polynomials experimental data of Z' o Z'' with ω, later replace in the corresponding equation KK and integrate numerically, however, there is another way to validate the data without having to perform the previously mentioned process, that, further, always satisfies equations KK, this method is known as measurement models, which they are used by general equivalent circuit. (Maxwell circuits or Voigt). One advantage is that it can be used in a general model for a wide variety of systems. The impedance components at a given frequency for parallel RC circuit are given by

$$Z'(\omega_i) = R_1 + \sum_{k=2}^{M} \frac{R_k}{1+(\omega_i \tau_k)^2} \tag{7}$$

$$Z''(\omega_i) = -\sum_{k=1}^{M} \frac{\omega_i R_k \tau_k}{1+(\omega_i \tau_k)^2} \quad \text{con} \quad t_k = [\omega_k]^{-1} \tag{8}$$

The procedure for calculating the components 'transformed", is as follows: for obtain the imaginary components from the real components, the experimental values of the real component Z', the angular frequency and the time constant τ_k for each frequency in the Equation (7), are substituted and a system of linear equations is established for R_k, which is solved to obtain the values of R_k, later, these are substituted in the Equation (8) for obtain at the same time the imaginary components transformed $(Z''(trans))$, these values $(Z''(trans))$ are compared with the experimental data (Z''), through the imaginary residual calculus (Δ'') for each requency.

$$\%(\Delta'')_i = \frac{Z''(trans)_i - Z''_i}{|Z_i|} * 100 \tag{9}$$

5. Mathematical model for data conductivity of the CZTS material

Realized the analysis with the data described in Table 2, a mathematical model of the electrical conductivity of the material was adjusted, (inverse of the average of the maximum impedance) in function of the synthesis temperature, using the least squares method [10]. The model describes the electrical behavior of photovoltaic CZTS material in function of the temperature with good approximation as shown in Figure 4.

$$\sigma(T) = 8.219 \times 10^{-3} - 7.667 \times 10^{-5} T + 1.812 \times 10^{-7} T^2 \tag{10}$$

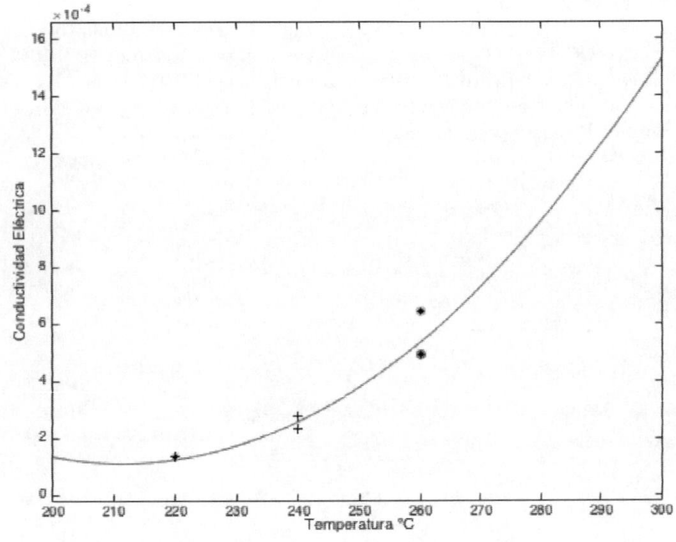

Figure 4. Curve fitted to the conductivity data of CZTS material at 220, 240 and 260°C.

6. Conclusions

The characterization of the material properties, by SEM and XRD techniques, allow to validate the obtention of a pure phase of Cu_2ZnSnS_4 material in terms of a two stage hydrothermal synthesis process. The electrical characterization of solid obtained at different synthesis temperatures confirm a semiconductor behavior in which the impedance response may be modelled by a fitted set of equations in terms of kramers-kronig equations, confirmng that exist a dependence between synthesis temperatures and impedances responses in obtained solids. So, its clear that conductivity of CZTS materials with the characteristics and process described above, show a strong dependence with the synthesis temperature, according with nyquist and bode plots, which permit us to validate the impedance data and confirm the accuracy of the methodology with excellent approximations. Finally is clear that the implementation of Fourier transforms, can improve the model and interpret the dynamics associated with the system.

References

[1] Suryawanshi M P, Shin S W, Agawane G L, Gurav K V, Ghorpade U V, Hong C W, Gaikwad M A 2015 A Promising Modified SILAR Sequence for the Synthesis of Photoelectrochemically Active Cu2ZnSnS4 (CZTS) Thin Films *Israel Journal of Chemistry* **55** 1-6

[2] Boukamp B A 1993 Practical application of the Kramers-Kronig transformation on impedance measurements in solid state electrochemistry Solid State Ionics **62** 131-141

[3] Antaño L Q R 1997 Aplicación de un Algoritmo Basado en un Modelo de Medición para la Detección de errores en las Medidas Experimentales de Impedancia (México: Universidad Autónoma Metropolitana-Iztapalapa)

[4] Urquidi-Macdonald M Real S, Macdonald D D 1986 Application of Kramers-Kronig Transforms in the Analysis of Electrochemical Impedance Data II Transformations in the Complex Plane *Journal of the Electrochemical Society* **133 (10)** 2019-2024

[5] Matthew E, Orazem M, 1991 On the Application of the Kramers-Kronig Relations to Evaluate the Consistency of Electrochemical Impedance Data *The Electrochemical Society Inc* **138 (1)** pp 67-76

[6] Dambrowski J 2013 Validation of Impedance-Data and of Impedance-Based Modeling Approach of Electrochemical Cells by Means of Mathematical System Theory *IECON - The 39Th Annual Conference Of The IEEE Industrial Electronics Society* (Austria)

[7] Boukamp B A 1995 A linear Kronig-Kramers Transform Tes for Immitance Data Validation *The Electrochemical Society Inc.* **142** 6

[8] Bayer M H, Schneider I A 2013 Application of the Kramers Kronig relations to locally resolved impedance data of polymer electrolyte fuel cells *Journal of Electroanalytical Chemistry* **689** 42-45

[9] Len C, Martn J M, Santamara J, Skarp J, Gonzalez D G, Snchez Q F 1996 Use of Kramers-Kronig transforms for the treatment of admittance spectroscopy data of p-n junctions containing traps (United States: American Institute of Physics) pp 7830-7837

[10] Montgomery D C 2004 Diseño y análisis de experimentos (United States: Limusa Wiley)

Design and implementation of a web application to estimate the surface density of implanted ions in solid substrates

B R Pérez Gutiérrez[1], F H Vera-Rivera[1] and E D V Niño[2,3,4]
[1] Universidad Francisco de Paula Santander, San José de Cúcuta, Colombia
[2] Universidad Politécnica de Madrid, Madrid, España
[3] Universidad Industrial de Santander, Bucaramanga, Colombia
[4] Foundation of Researchers in Science and Technology of Materials, Colombia

E-mail: borisperezg@ufps.edu.co

Abstract. In the present research a web application is designed and implemented allowing us to calculate the dose of implanted ions on solid substrates from experimental parameters (repetition rate and pulse duration of current, potential difference, work pressure and treatment time) and current pulses acquired during start-up and shutdown of high voltage electrical discharges at low pressures. By physical and mathematical processing, it is achieved at first estimate the value of the ionic charge, and then the areal density. This study will provide more accurate and suitable developments in relation to applications of ion implantation as an alternative technique to increase the service life of the surfaces of the materials used in various industrial sectors developments.

1. Introduction

The surface modification of a substrate involves modifying the composition and structure of materials, either by creating coatings (few microns wide) or by introducing new elements on its surface (dopant ions) in nanometric depths [1].

The three-Dimensional Ion Implantation technique (3DII) [2] performed in the JUPITER reactor [3] is a process of implanting dopant ions into metallic and nonmetallic solid substrates [1-6]. The parameters that characterize this implementation are: the ion's type, the ion's energy and the implanted dose [1,7,8]. Therefore, modify surface and sub-surface layers (without to change geometric dimensions) of solid substrates are subject of increasing interest in several sectors like energy, transport and manufacturing [9-12].

According to [13-14], the 3DII is generated from hybrid discharges of high voltage and electric arc at low pressures. This implementation makes an impact with sufficient energy (between 5 and 60 kV) ions of non-metallic species on the surface of a solid substrate. These ions penetrate a few layers (nanometric level) by changing the physical and chemical properties of the surface [13-14]. This penetration provokes a lot of collisions with the atoms composing the solid substrate, causing a set of collective processes which influence the final distribution of ions, such as the formation of defects, ions tunneling, cascade collisions, sputtering, etc. [15-19].

According to the above and considering that the final concentration of implanted ions is a very important feature of the surface modification by ion implantation [20-22], it is very important to characterize the calculation of superficial density of ions implanted in some material, moreover, there is a need to understand the complex phenomena that occur during the process. This calculation depends mainly of time of treatment, density of ionic current, repetition frequency and pulse duration, the secondary emission ion-electron coefficient of material and the area of the cathode.

Therefore, the development of a web application and a computer simulation are both really important, especially because we can obtain information that today could not be obtained by other means. That is why it is necessary to develop computational tools that estimate more quickly, accurately and precisely the estimate the value for implanted dose on the surface of solid substrates [23].

In this paper, we introduce EDIIS - Estimation of the Surface Density of Implanted Ions in Solid Substrates, a free web application with a user-friendly interface. EDIIS was developed in a joint effort between the Software Engineering Development Research Group (GIDIS) of the Universidad Francisco de Paula Santander and the Physics and Plasma Technology and Corrosion Research Group (FITEK) of the Universidad Industrial de Santander. The web application is the result of a technological applied research, which can be used at any time, from anywhere, in different electronic devices (computers, tablets, mobile phones, etc.).

2. Methodology

The estimation of superficial implanted dose of a material using the web application is carried out according to the scheme presented in Figure 1.

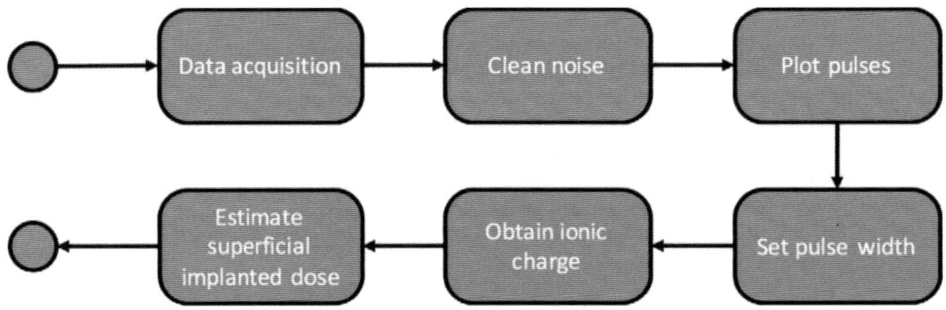

Figure 1. Methodology of estimation superficial implanted dose.

Current pulses versus time during electric high voltage discharge at low pressures were obtained from the experimental parameters set in a process of nitrogen ion implantation into metallic solid substrates. The pulses recorded represent the total electronic charge transported. After that, the pulses (data and image) are introduced into the web application where they are scanned and digitized. Each one of these datasets corresponds to the pulses of current that can be captured on an oscilloscope during turn on of the electric high voltage discharge (in this case we use the oscilloscope Tektronix TDS 2002 B).

Graphics are displayed and then the user proceeds to set the pulse width. It depends on the region of interest. Then, the ionic charge is obtained using the difference between both pulses [24]. Finally, the superficial implanted dose in solid substrates is estimated.

3. Web application

A web application was developed using a two-tier architecture. Web application was developed with Hyper Text Markup Language (HTML), Cascading Style Sheet (CSS), Java Server Pages, BootStrap and Google Charts. The latter help us to plot and digitize the pulse of the implementation process.

The process begins with the acquisition of data and image pulse, after the user upload the information into the web app, he has the possibility of set the pulse width and estimate the ionic charge (Q_i). After that, user provide the input data (experimental parameters of the ion implantation process) which are: frequency (f) of the discharge pulse (Hz), total time (t) of the discharge (s), total area (A) exposed (cm^2) and number of ion of the charge (Ze). The web application calculates the total

IOP Conf. Series: Journal of Physics: Conf. Series **786** (2017) 012015 doi:10.1088/1742-6596/786/1/012015

ionic charge (Qit) and the dose approximate of ions implanted superficially. Equations 1 to 3 show how the process was done.

$$n = f * t \tag{1}$$

$$Qit = Qi * n \tag{2}$$

$$D = \frac{Qit}{A * Ze} \tag{3}$$

Where, n: pulses number. Qit: total ionic charge transported, Q_i: ionic charge.

Figures 2 present in detail how the calculation of the ions implanted superficially through the web application is performed.

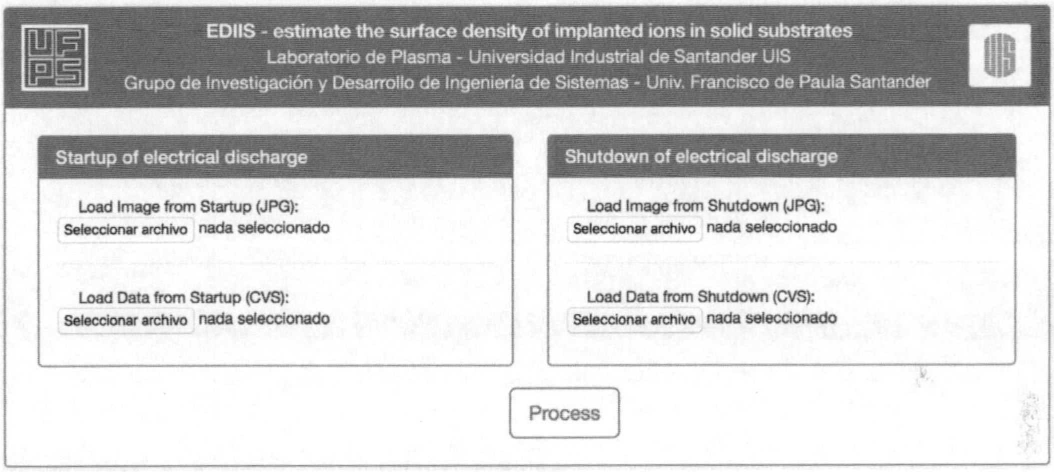

Figure 2. Section to load the image and the pulse data.

After user load the required data and clicked in Process button, the web app automatically performs a process of digitization of datasets and graphics. The screen is splitted in two sides, each one presenting the graphic along with the image of the pulses. The web app allows the user to set the maximum and minimum values to determine the pulse width necessary to calculate the area under the curve and estimate the transported load for each pulse (see Figure 3).

On top of Figure 3 is shown both pulses images obtained from Tektronix TDS 2002 B oscilloscope. Left one corresponds to the total charge transported during the electric high voltage discharge and the right one corresponds to the electric charge carried. At the bottom both pulses are graphed in different charts using our web application (the graph obtained is very similar the acquired with the oscilloscope). User can adjust the pulse width to work just with some data. In this case, user enter the required data corresponding to the X axis (Figure 4).

After that, user clicked the Next button to show both charts mixed in one chart, so the user can quickly identify the area to be calculated, corresponding to the ionic charge. Values entered in pulse width are kept in this chart. The estimated value of the ionic charge is estimated and presented (see Figure 5).

Finally, the web application proceeds to estimate the superficial implanted dose according to Equations 1 to 3 and whose value is reported at the bottom in Figure 5.

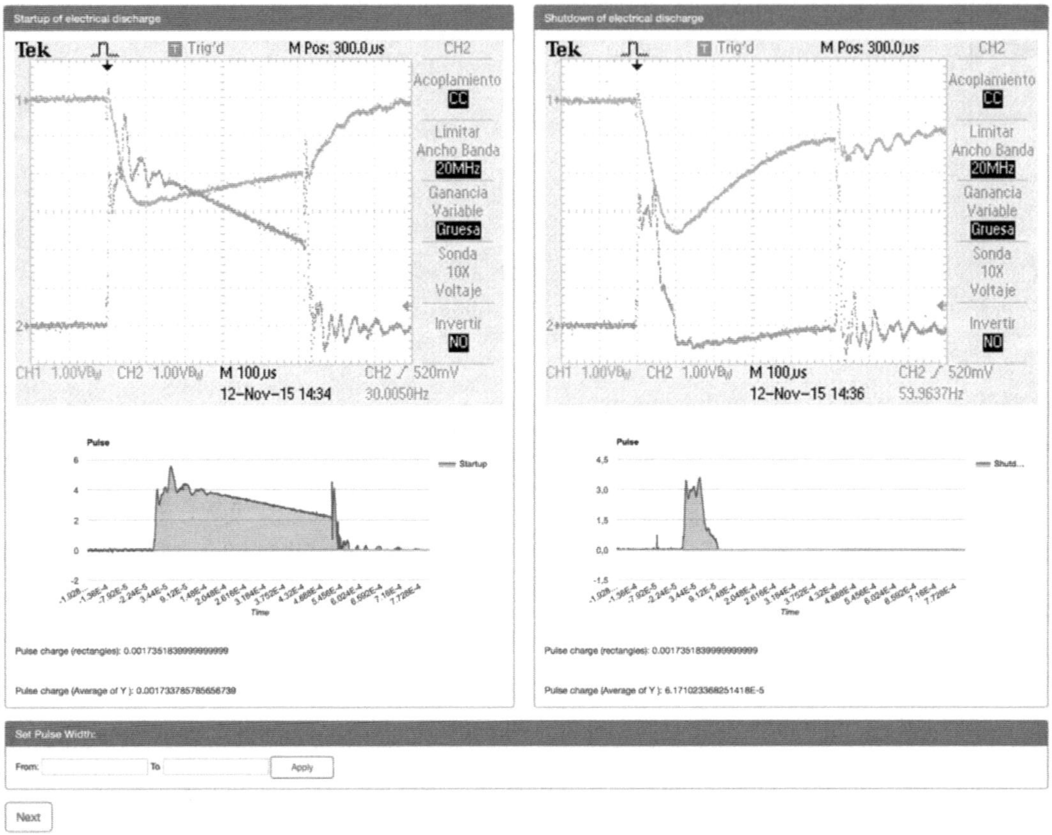

Figure 3. Images and graphics of pulses in the web application.

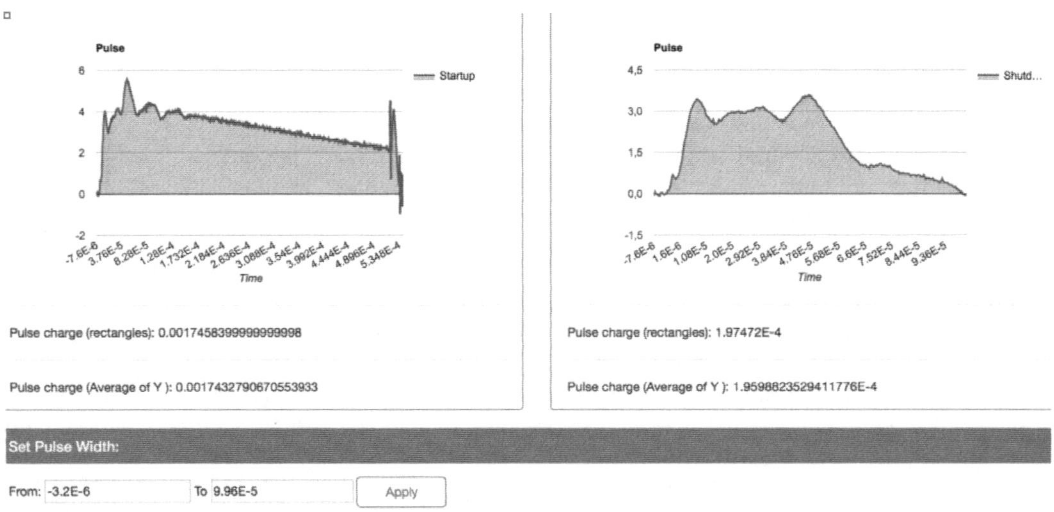

Figure 4. Pulse width adjusted by user.

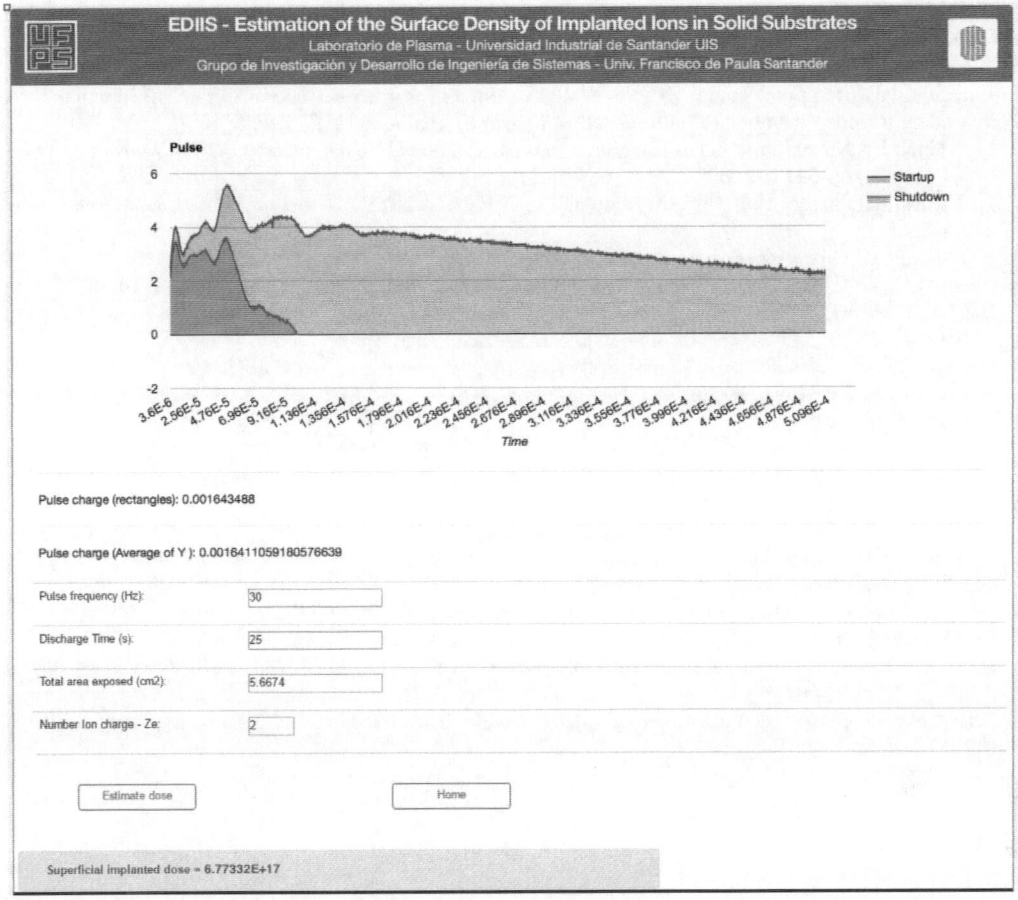

Figure 5. Experimental parameters of the ion implantation process.

4. Conclusions

A computational web tool was built to work as an aid to the estimation of the dose of ion concentration implanted in solids, bringing benefits to experimental researches in the UIS' plasma laboratory. These calculations were faster and more precise than traditional methods. Researchers can focus on the results and not spending time making complex calculations which can lead them to inaccurate and faulty implementations.

The web application developed can be used anywhere any time, only an Internet connection is required and the pulses data to be processed. The format of the data received by the system can be downloaded from the web application.

The results for the estimation of the surface dose of ions obtained from the web application were very similar to the theoretical results, ensuring that the implemented web application works correctly.

References

[1] Dougar-Jabon V D, Dulce Moreno H J and Tsygankov P A 2003 Implantación Iónica *Revista Respuestas* **8** 19-30

[2] Dougar-Jabon V D, Castro B J, Dulce Moreno H J, Tsygankov P A 1999 *Rev Colom de Física* **31(2)** 181

[3] V I Khvesyuk, P A Tsygankov 1997 *Surface & Coatings Technology* **96** 68

[4] Conrad J R 1988 *Method and apparatus for plasma source ion implantation* (USA: Patent No 4764394)

[5] Conrad J R, Radtke J L, et al. 1987 *J Appl Phys* **62** 4591

[6] J R Conrad 1987 *Mater Sci Eng* **116A** 197

[7] H J Dulce-Moreno, P A Tsygankov, V Dugar-Zhabon, E D V-Niño, F F Parada-Becerra 2011 Estudio de los haces electrónicos en la descarga eléctrica de alto voltaje a bajas presiones *Puente Revista Científica* **5**(**2**) 21

[8] Valbuena Niño E D and Dugar-Zhabon V 2006 Comportamiento del acero SAE-4140 implantado con iones de nitrógeno en ambientes hidrogenados *Revista Colombiana de Física* **38**(**1**) 61

[9] Ely Dannier V Niño, Hernán Garnica, Veleriy Dugar-Zhabon and Genis Castillo 2014 *Journal of Physics Conference Series* **511** 012080

[10] J J Castro-Maldonado, H J Dulcé-Moreno and E D V-Niño 2013 *Journal of Physics: Conference Series* **466** 012005

[11] E D V-Niño, V Dugar-Zhabon, H A Garnica 2011 *Revista Iteckne* **8**(**1**) 37

[12] Héctor J Dulcé M, Alejandro Rueda V and Valeri Dougar-Jabon 2005 *Phys Stat Sol* (c) **2**(**10**) 3778

[13] H Jaime Dulce M, Valeri Dougar J and Carlos R Cabrera 2003 *Revista Colombiana de Física* **35-2** 332

[14] V Dougar, J Dulcé and P A Tsygankov 2002 *Rev Sci Instr* **73** 828

[15] Edmanuel Torres, Valeri Dougar-Jabon 2006 *Revista Colombiana de Física* **38**(**1**) 89

[16] K Krantzman, Z Postawa, B Garrison, N Winnograd, S Stuart and J Harrison 2001 *Nucl Instr and Meth in Phys Res B* **180** 159

[17] K Nordlund, J Keinonen, M Ghaly and R S Averback 1999 *Nucl Instr Meth Phys Res B* **148** 74

[18] J F Ziegler, J P Biersack, U Littmark 1985 *The stopping and range of ions in matter* (New York: Pergamon)

[19] Yu V Gott 1978 *Interaction of Particles with substances in plasma research* (Moscow: Atomizdat)

[20] Towsend P D, Chandler P J and Zhang L 1994 *Optical effects of ion implantation* (Cambridge: Cambridge University press)

[21] Wolfgang Eckstein 1991 *Computer simulation of ion-solid interactions* (Berlin: Springer-Verlag)

[22] Burcham W E 1979 *Elements of nuclear physics* (New York: Longman Group Limited)

[23] F H Vera-Rivera, V D Dugar-Zhabon, H J Dulce-Moreno and E D V-Niño 2016 *Journal of Physics: Conference Series* **687** 012058

[24] B R Pérez Gutiérrez, F H Vera-Rivera and E D V Niño 2016 *Journal of Physics: Conference Series* **743** 012017

IOP Conf. Series: Journal of Physics: Conf. Series **786** (2017) 012016 doi:10.1088/1742-6596/786/1/012016

Computational simulation of a diffusion process of carbon nanotubes in aluminium to improve the mechanical properties of a composite material

G Suárez Guerrero[1], M F Valencia-García[2], H V Martínez Tejada[1] and O J Toro Cadavid[1]
[1] Universidad Pontificia Bolivariana, Medellín, Colombia
[2] Universidad EIA, Medellín, Colombia

E-mail: gustavo.suarez@upb.edu.co

Abstract. A computational thermo-mechanical insertion model was implemented considering the incorporation of carbon nanotubes (CNTs) into semisolid aluminium. A shell surface of CNTs within an aluminium matrix was obtained using the Particles Dynamic Method (PDM). Also, energy absorption simulations were performed through computational impact tests in order to characterize the behaviour of the nanocomposite under high strain rates. Theoretical results are useful in the design of nanocomposites and the experimental processing of Al/CNTs nanocomposites for different applications.

1. Introduction

Matter diffusion processes in which diffusion coefficients are mainly dependent on temperature and other transport and substance-specific properties such as viscosity, size and shape of solute, have been reported extensively [1]. Many reports in this topic involve modelling of phenomena related to substances with well-known physicochemical properties [1]. On the other hand, reports of diffusion processes involving nanomaterials are much more unusual.

In this paper, a computational mathematical model that describes the diffusion of carbon nanotubes into aluminium in semisolid state is presented and diffusion coefficients are introduced to allow consideration of the translational and rotational effects, as well as other geometric parameters of nanotubes in mixture [2,3]. In this way, modelling of diffusion and dispersion of CNTs is attainable, allowing for far more realistic physical conditions [2-4].

Computational tests for ballistic impact energy absorption were developed on Al/CNT nanocomposites with the goal of estimating the improvement of the material's mechanical properties by the contribution of the CNTs [5-7].

2. Mathematical modelling and computational simulation

Diffusion of CNTs in Al was mathematically modelled following Fick's law in transient state [2]. Random movement of particles that complied with thermodynamic principles was guaranteed by addition of physical interactions for different relations of CNTs to Al.

Additional parameters such as rotational coefficient, translational coefficient, static flexion persistence length, static bend radius, intrinsic viscosity and mean spatial distance were included in the diffusive coefficient with the in order to have an appropriate representation of the behaviour of thermo-mechanical diffusion processes [2,3].

The results of concentration functions for mixture in transient state were achieved following the implementation of a numerical model in Matlab®. Simulations of the composite material were executed in order to determine mechanical resistance and overall contribution of nanotubes to composite material. Boundary conditions related to impact energy on Al/CNTs composite were also implemented.

2.1. Mathematical descriptions and constitutive laws

Rotational effects of CNTs were determined using equation (1) [3]:

$$D_r = \left(\frac{kT}{n_s}\right)\left(\frac{2}{l_p L^2}\right)\left[0.253\left(\frac{L}{4l_p}\right)^{\frac{1}{2}} + 0.159\ln(2L_P) - 0.387 + 0.160\right] \tag{1}$$

In (1), L is the contour length of CNTs, T is the temperature at which diffusion occurs, k is the number of static flexion points, n_s is the viscosity of the solvent, L_P is the spacing between friction elements along the contour and l_P is the length of persistent dynamic flexion.

Equation (2) was used to denote translational effects of CNT's [3]:

$$D_T = \left(\frac{kT}{3\pi L n_s}\right)\left[1 + \ln(2l_{ap}) - 2.431 + 1.843\left(\frac{N}{2l_{ap}}\right)^{\frac{1}{2}} + 0.138\left(\frac{N}{2l_{ap}}\right)^{-\frac{1}{2}} - 0.305\left(\frac{N}{2l_{ap}}\right)^{-1}\right] \tag{2}$$

In (2), N is the number of friction elements and l_{ap} is the apparent persistence length.

Fick's law mathematical model was applied [1] (see Equation 3) to represent diffusive flux on insertion of Al/CNTs. Random movement of CNTs reached homogeneity when equilibrium between forces related to concentration gradient and friction forces related to medium viscosity was achieved [1].

$$\frac{\partial C}{\partial t} = D\frac{\partial^2 C}{\partial x^2} \; ; \; 0 \le x \le L \wedge t \ge 0 \tag{3}$$

In (3), C is the concentration of CNTs in aluminium; $D=f(D_r, D_T)$ is the diffusion coefficient.

Boundary and initial conditions in the diffusion system were defined by:

$$\frac{\partial C}{\partial x}(0,t) = K(C_s - C) \rightarrow K = g(D, A_T,) \; ; \; x = 0 \wedge t \ge 0 \tag{4}$$

$$\frac{\partial C}{\partial x}(L,t) = 0 \; ; \; x = L \wedge t \ge 0 \tag{5}$$

$$C(x,0) = C_0 = 0 \; ; \; 0 < x \le L \wedge t \ge 0 \tag{6}$$

Schematically, the process can be represented as seen in Figure 1.

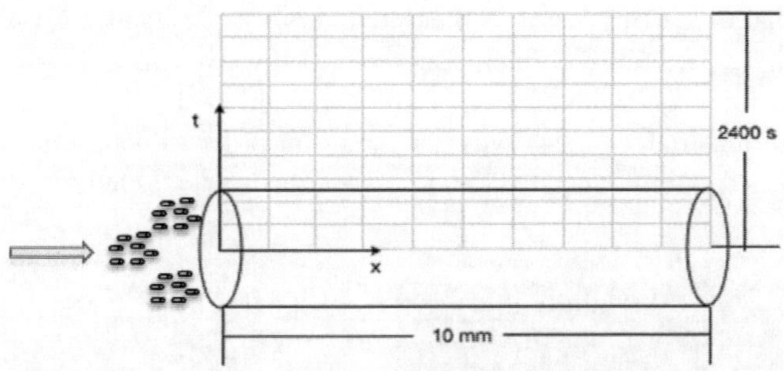

Figure 1. Insertion process of CNTs into Al domain.

For the implementation of computational tests on the material exposed to projectile impact, Al/CNTs was configured by means of generalized Hooke's model for anisotropic materials [7] and Johnson-Cook's model was used to determine material failure and propagation of energy [7,8].

In Hooke's model, composite density was calculated from mass proportions [7]:

$$\rho_c = \frac{1}{\frac{m_{AL}}{\rho_{AL}} + \frac{m_{NTC}}{\rho_{NTC}} + \frac{V_V}{\rho_C}} \tag{7}$$

m_{NTC}, m_{AL} are the mass proportions of the constituent and Vc is the proportion in void volume.

Johnson-Cook's Constitutive Model describes the relationship between stress, strain, strain rate and visco-elastic material temperature. This model is appropriate for situations in which strain rate varies between $10^2 s^{-1}$ and $10^6 s^{-1}$ and temperature varies according to plastic deformation caused by thermal softening. Stress flux model is represented as in (8) [8,9]:

$$\sigma = (A + B\varepsilon^n)(1 + C\ln\dot{\varepsilon}^*)(1 - T^{*m}) \tag{8}$$

In (9) T is the system temperature, $\dot{\varepsilon}^*$ is the velocity of equivalent plastic deformation, ε is the equivalent plastic deformation, A is the initial cadence stress (MPa), B is the hardening module, n is the strain hardening exponent, C is the strain-rate dependent coefficient and m is the thermal softening coefficient.

The materials' properties values used during diffusion process and ballistic computational tests for Al/CNTs composite with a material type Al 1100 and for CNTs were given a stiffness module of 1.8×10^{12} and a Poisson ratio of 0.45 [9-11].

2.2. *Numerical-computational description*

Fick's law was modelled through progressive divided differences approximation for transient behaviour applications. General scheme is described in (10) [12]:

$$\frac{C_i^{(n+1)} - C_i^{(n)}}{\Delta t} = D\frac{C_{i-1}^{(n)} - C_i^{(n)} + C_{i+1}^{(n)}}{(\Delta x)^2} \tag{9}$$

Numeric mesh size in time interval Δt should guarantee:

$$\Delta t \le \frac{0.5\Delta x^2}{D} \tag{10}$$

IOP Conf. Series: Journal of Physics: Conf. Series **786** (2017) 012016 doi:10.1088/1742-6596/786/1/012016

In (10) D is the diffusion coefficient, Δt is the time interval or step and Δx is the spatial interval.

During computational tests, finite elements analysis (FEA) was used to study the behaviour of Al/CNTs composite. A flat plaque with an area of 20x20cm and thickness of 5.2mm was elaborated to represent the composite material. A mesh was elaborated in order to determine the behaviour of the plaque when subjected to dynamic impact force. Uniform quadrilateral elements were used to develop the mesh (Figure 2).

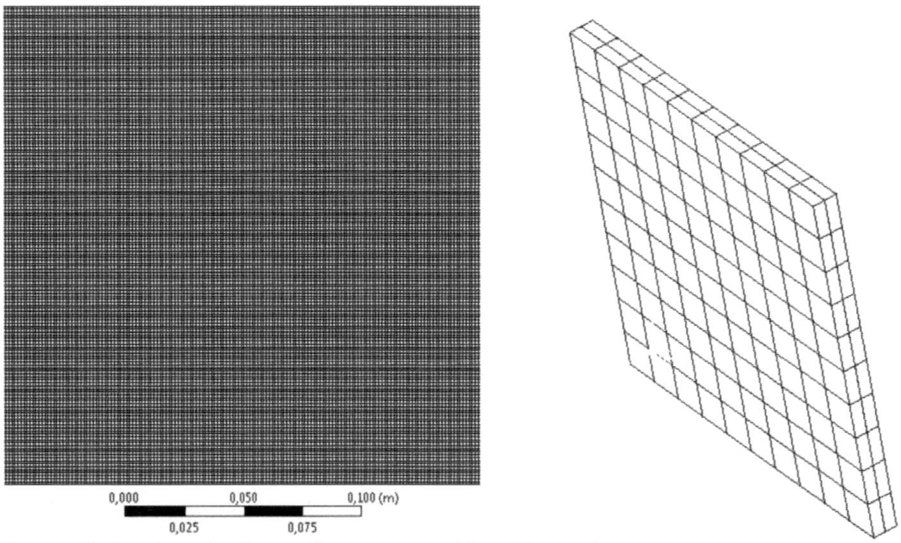

Figure 2. Lattice of a flat-surface plaque with uniform elements.

In the same way, another numerical mesh for the 7.62mm-caliber projectile was elaborated. This projectile impacted the flat plaque, Figure 3.

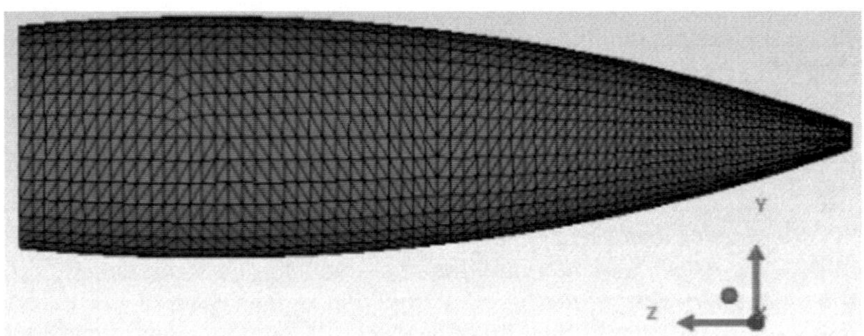

Figure 3. Framework of the 7.62mm-caliber projectile with controlled numerical elements.

Based on international standards for shielding and ballistic testing [13], criteria and reference data for computational tests were determined. Projectile speed equalled 847m/s±9.1m/s, dimensions and geometric characteristics were selected as show the Figure 4.

NOM-166-SCFI-2005 [13] was the main guideline for the development of simulations and the determination of ballistics-specific characteristics such as: protection level, gun type, projectile calibre, work velocity, dimensions of target and others.

IOP Conf. Series: Journal of Physics: Conf. Series **786** (2017) 012016 doi:10.1088/1742-6596/786/1/012016

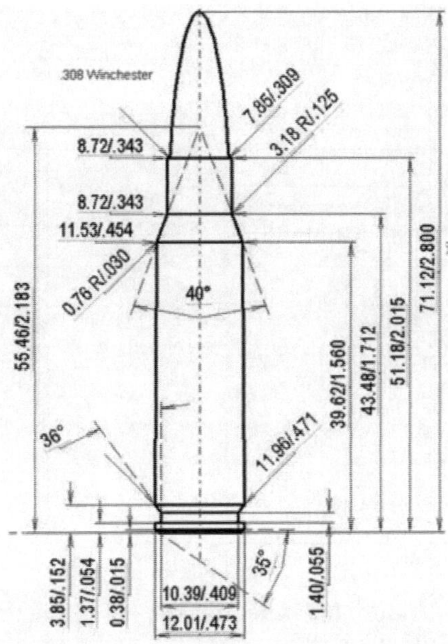

Figure 4. Characteristics of the conical-headed projectile [13].

Using the Particles Dynamic Method (PDM.), a computational algorithm was implemented with the goal of analysing behaviour of dissipation and propagation of energy in a domain comprised of elastic Al and CNTs point particles. Newton's laws and elastic laws were specifically used in this case [14,15].

$$\rho \partial^2 u_t = \sum f \tag{11}$$

P.D.M was designed according to the structure presented in Fig. 5.

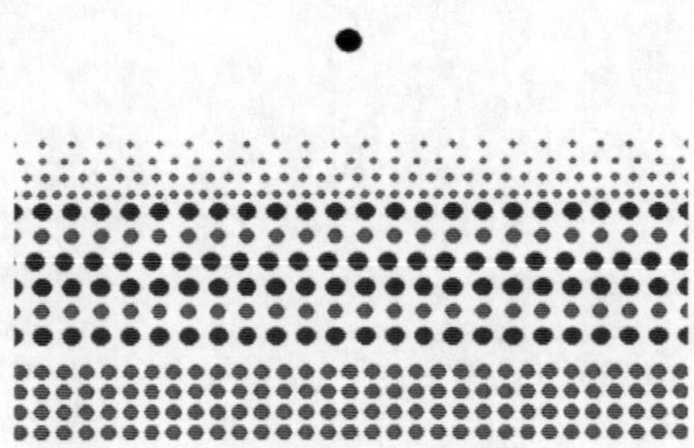
Figure 5. Surface model conformed by Al 1100 and CNTs point particles. (The single dot in the upper half corresponds to the projectile. The dots in the lower half correspond to Al and CNTs point particles).

3. Results

Functions of flux diffusion of CNTs in Al were obtained, with a tendency to homogenize concentration and temperature (Figure 6).

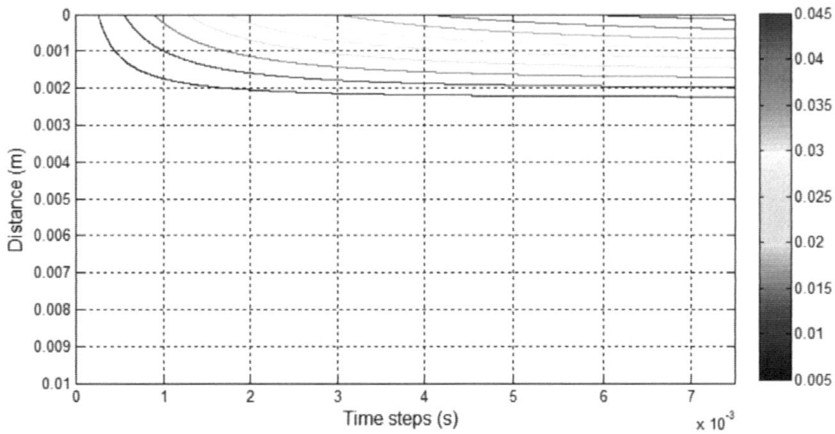

Figure 6. Molecular diffusion behaviour of Al/CNTs (functions indicate level of concentration in a specific time and space).

Mechanical tests performed on the impact-subjected Al/CNTs composites indicated that reinforcement material favour the composite's mechanical properties, achieving energy dissipation-absorption and effectively stopping the projectile trajectory. Impact plaque was 5.2mm thick with 13 layers of Al 1100 and 12 insertions of CNT layers (Figure 7).

Results were validated through PDM simulations. Algorithms implemented with PDM were developed in Matlab's programming language, allowing for modification of different particulate layers, and mechanical properties, projectile velocity, and orientation. In this case, an initial simulation of a particulate surface (plaque) was made using point particles of Al. This surface was impacted by a spherical projectile (Figure 8). Later, an impact subjected plaque, consisting of three different layers of point particles of CNTs and Al 1100 (Figure 9) was simulated. In general, the results showed the composite exhibited kinetic energy dissipation modes and a capacity to diminish impact damage.

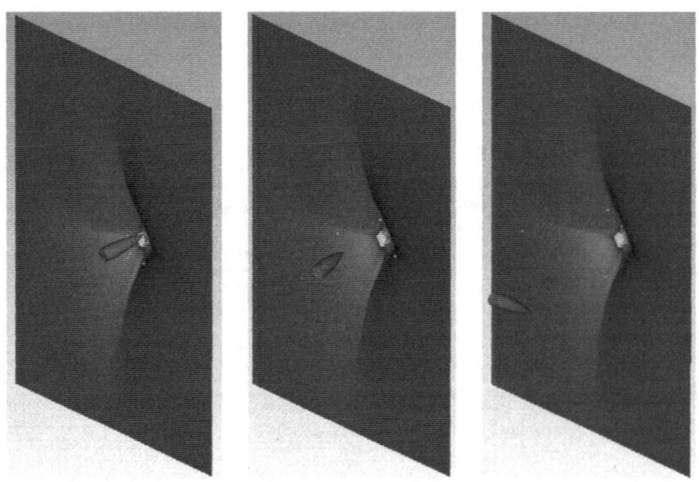

Figure 7. Laminate plaque with 5.2mm thickness made of AL1100 and CNTs.

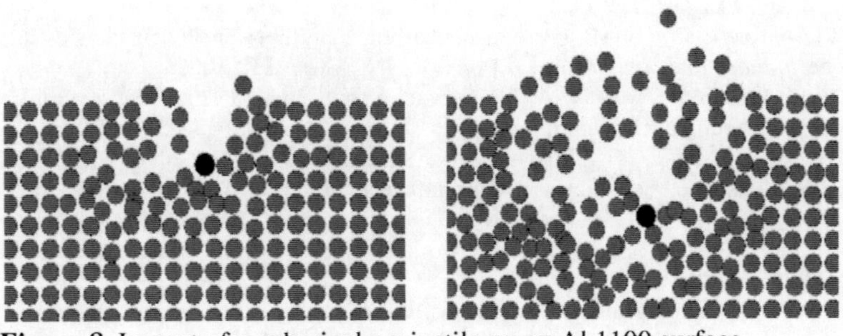

Figure 8. Impact of a spherical projectile on an Al 1100 surface.

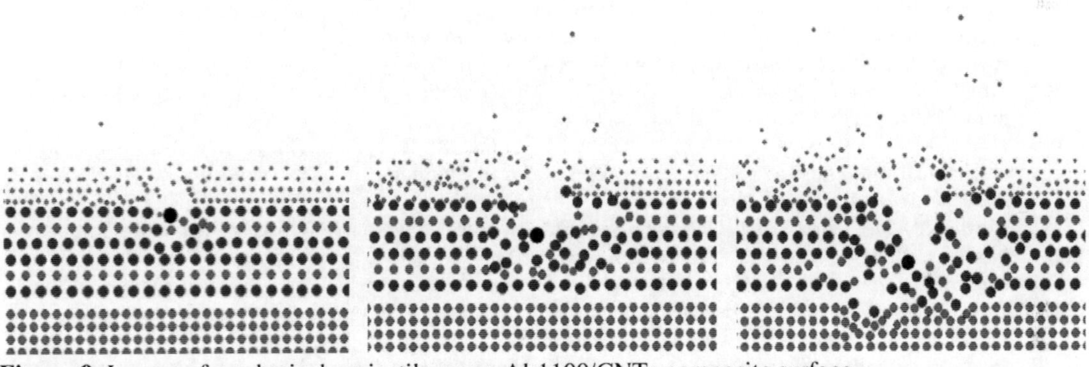

Figure 9. Impact of a spherical projectile on an Al 1100/CNTs composite surface.

4. Conclusions

Initially, a computational algorithm was developed with the purpose of simulating concentration gradient of CNTs on Al matrix. Simulations indicate appropriate distribution of CNTs with respect to the depth achieved during diffusion. Thermal conditions and media concentration can affect the degree of perforation of reinforcement material and the upgrade of the mechanical properties of the matrix.

Several simulations were executed on an Al 1100/CNTs plaque subjected to dynamic impact load until the perforated surface didn't pass the security limit defined by NOM-166-SCFI-2005 guidelines. Simulations considered a solid projectile in order to grant the design a higher safety factor. The most favourable result was achieved using a lamina made of 13 plaques of Al 1100, on which CNTs were deposited. Maximum deformation achieved was 1.5cm deep. Fragmentation of the material was observed in areas close to impact zone.

Modelling of the impact behaviour of a spherical projectile on particulate Al/CNTs surface was also performed. Each particle was characterized using singular mechanical properties of Al and CNTs in order to simulate specific characteristics of materials with continuous constitution. Impact dynamics of elastic particles were considered and fragmentation of the material was observed as the projectile perforated the particulate surface.

This research offers insight on how perforation of composite materials subjected to ballistic impact is generated. Incidence of damage to wearer, as well as possibility of reinforcement improvement and diffusion/dispersion of CNTS in Al is also discussed.

IOP Conf. Series: Journal of Physics: Conf. Series **786** (2017) 012016 doi:10.1088/1742-6596/786/1/012016

References

[1] Itô K 2006 *Diffusion Processes Ency. of Statistical Sciences* (New York: John Wiley & Sons)

[2] Tianlong W Lucien N Brush, and Kannan M Krishnan 2014 A generalized diffusion model for growth of nanoparticles synthesized by colloidal methods *J. of Colloid and Interface Science* (Amsterdam: North-Holland/American Elsevier) **419** 79

[3] Lee H S 2013 Classification of Mass-Produced Carbon Nanotubes and Their Physic-Chemical Properties *Book Nanotechnology and Nanomaterials* (Croatia: IN TECH) **112** 10653

[4] Esawi A M K, Morsi K, Sayed A, Abdel Gawad A and Borah P 2009 Fabrication and properties of dispersed carbon nanotube–aluminum composites *J. Mat. Sc. and Eng. A* **508** 167

[5] Javadi A H, Mirdamadi Sh, Faghihisani M A, Shakhesi S and Soltani R 2012 Fabrication of well-dispersed, multiwalled carbon nano- tubes-reinforced aluminum matrix composites *J. New Carbon Materials* **27(3)** 161

[6] Cleary P W 2010 Elastoplastic deformation during projectile–wall collision *J. Applied Mathematical Modelling* **34(2)** 266–283

[7] Chang F K, Scott R A and Springer G S 1984 Failure strength of nonlinearly elastic composite laminates containing a pin loaded hole, *J. of Composite Materials*, **18(5)** 464

[8] Fernandez D 2008 A new tool based on artificial neural networks for the design of lightweight ceramic–metal armour against high-velocity impact of solids *Int. J. of Solids and Structures* **45** 6369

[9] Raguraman M and Deb A 2006 Accurate prediction of proyectile residual velocity for impact on sigle multilayered steel and aluminum plates *9th International LS-Dyna User Conference, Penetration/Explosive Modeling* (Detroit: DYNAlook) pp 2-37-2-48

[10] James A Pignetti 2016 Metalemen (New York: Metalemen) http://www.metalmensales.com/aluminum.html

[11] Z Y Liu, Xu S J, Xiao B L, Xue P, Wang W G and Ma Z Y 2012 Effect of ball-milling time on mechanical properties of carbon nanotubes reinforced aluminum matrix composites *Compos. Part A Appl. Sci. Manuf.* **43(129)** 2161

[12] Burden R L and Faires D J 2011 *Numerical Analysis. NINTH* (Boston: PWS-Kent Publishing Company)

[13] Norma Oficial Mexicana NOM-166-SCFI 2005 *Seguridad al usuario-Chalecos antibalas-especificaciones y métodos de prueba* (México: Secretaria de Economía Dirección Nacional de Normas)

[14] Duvenbeck A, Hanke S, Weidtmann B and Wucher A 2011 A molecular dynamics investigation of kinetic electron emission from silver surfaces under varying angle of projectile impact *J. Nuclear Instruments and Methods in Physics Research Section B: Beam Interactions with Materials and Atoms* **269(14)** 1661

[15] Johnson B C, Bowling T J and Melosh H J 2014 Jetting during vertical impacts of spherical projectiles *J. Icarus* **238** 13

Preparation of YBCO-BYTO and YBCO-BZO nanostructured superconducting films by chemical method

P Garcés[1], M Coll[2], H Castro[1], T Puig[2] and X Obradors[2]
[1] Universidad Nacional de Colombia, Bogotá D.C., Colombia
[2] Instituto de Ciencia de Materiales, Consejo Superior de Investigaciones Científicas, Bellaterra, Cataluña, España

E-mail: paug76@gmail.com

Abstract. YBCO-BYTO6% and YBCO-BZO10% $YBa_2Cu_3O_{7-d}$-Ba_2YTaO_6 6% (YBCO-BYTO6%) and $YBa_2Cu_3O_{7-}$-$BaZrO_3$ 10% (YBCO-BZO 10%) nanostructured films were grown by the Chemical Solution Deposition method, and compared with YBCO pure films. Films were deposited on YSZ substrates, with $Ce_{0.9}Zr_{0.1}O_2$ and $Ce_{0.6}Zr_{0.4}O_2$ buffer layers. They were characterized by GADDS X-ray diffraction, scanning electron microscopy (SEM) and inductive (SQUID) measurements of the critical temperature (T_c) and critical current density (J_c). It was found that YBCO-BZO10% films presented better superconducting properties (T_c=89.2K and J_c=1.3MA/cm^2), probably due to an enhanced pinning force, originated by BZO nanoparticles. Additionally, it was found that these films have lower reactivity with the buffer layer.

1. Introduction

The development of low-cost superconducting tapes requires the use of multilayer structures where the texture is preserved between the different layers, and reactivity between the superconducting film and substrate is minimized. Growth ex-situ techniques are subject of intense research in order to obtain coated conductors (CC), with YBCO, at low cost and with high critical current density. To make buffer layers with different architectures, some methods, like vacuum deposition (PLD and sputtering), or chemical solution deposition methods (CSD), including sol-gel and MOD, can be used. The production cost is higher in the case of vacuum deposition methods, while in CSD the equipment is simpler; therefore, the cost is very attractive. The main goal of optimizing the production cost to quality ratio is obtained in a CC where all layers are grown by chemical methods, known as full chemical route [1].

YBCO nanocomposite materials have opened new paths to achieve enhancement of vortex pinning, therefore increasing the critical current density, J_c. Fabrication of nanostructured YBCO thin films with incorporation of nanosized oxides as secondary non superconducting phases, has produced strong enhancement of YBCO performance at high magnetic fields, in a wide temperature range. These results boost the potential use of these materials in a broad spectrum of applications [2-5].

Cerium dioxide (CeO_2), has been widely used as buffer layer for growing HTS [6]. Several groups have attempted to use it for YBCO CC because of its excellent chemical compatibility with Ni substrates, as well as a good foundation for the growth of YBCO. Unfortunately, due to the existence of a critical thickness of 50nm, above which cracks and microcracks are formed, CeO_2 rarely has been used successfully. Alternatively, the use of CeO_2 doped with Zr^{4+}, $Ce_xZr_{1-X}O_2$ (CZO), allows the growth of atomically flat and free of cracks epitaxial layers. Such advantage has been attributed to reconstruction of grain boundaries, due to increased atomic mobility, that allows complete eradication of carbon atoms

IOP Conf. Series: Journal of Physics: Conf. Series **786** (2017) 012017 doi:10.1088/1742-6596/786/1/012017

in grain boundaries [7]. CSD is a cost-effective technique for preparation of large area and long length films [6]. Metallorganic decomposition (MOD) of trifluroacetate (TFA) has been widely investigated as a chemical solution route leading to high quality YBCO nanocomposite films, with high J_c [8].

In this work a reference pure YBCO film was made, and nanostructured YBCO films, using secondary phases of BYTO6% (YBYTO) and BZO10% (YBZO), by the TFA route. They were deposited on ytrium stabilized zirconia (YSZ) substrates, with $Ce_{0.9}Zr_{0.1}O_2$ (CZO-0.1) and $Ce_{0.6}Zr_{0.4}O_2$ (CZO-0.4) buffer layers. Films were characterized by GADDS X-ray diffraction, SEM and inductive (SQUID) measurements for the determination of T_c and J_c.

2. Experimental details

YBCO films were prepared by the trifluoroacetates route by using anhydrous precursor solution with 1.5M concentration [9], as described in detail elsewhere [4]. The metallorganic precursor solution was spin coated on crystalline substrates of YSZ/CZO of 5x5mm^2. Then a thermal treatment for pyrolysis was made, under oxygen flow, at 300°C, for 30min [2]. The crystallization process was performed following the thermal cycle shown in Figure 1(a), in a wet N_2 atmosphere, with a water vapor partial pressure of 22 mbar, and O_2 partial pressure of 0.2 mbar. Finally, the superconducting phase was obtained by annealing at 450°C, in oxygen atmosphere. YBYTO and YBZO films were made by adding stoichiometric amounts of tantalum ethoxide and Zirconium acetylacetonate in a precursor TFA-YBCO solution, as described elsewhere [3,10,11]. The metallorganic precursor solution was spin coated on crystalline substrates of YSZ/CZO, 5x5mm^2. The pyrolysis process was to the same performed on samples of pure YBCO. The crystallization process was made following the thermal cycle shown in Figure 1(b). The superconducting phase was obtained by annealing at 450°C, in oxygen atmosphere.

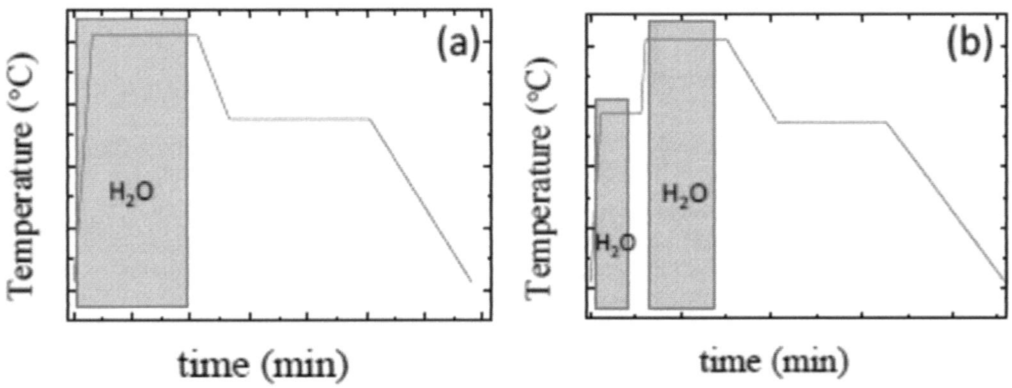

Figure 1. Scheme of the thermal process used to prepare the films: (a) YBCO, (b) YBYTO and YBZO.

The microstructure and phase analysis of the films were studied by two-dimensional X-ray diffraction with a Bruker AXS GADDS diffractometer, and SEM microscopy with a FEI Quanta 200 FEG microscope. The critical current density (J_c) and critical temperature (T_c) were obtained inductively with a Quantum Design MPMS-XL SQUID magnetometer.

3. Results and discussion

In the SEM micrographs in Figure 2 can be observed the morphology of the samples grown on YSZ/CZO-0.1. In all of them there are some precipitate grains of $BaCuO_2$, as confirmed by EDX. There is lower porosity and a smoother surface in the sample YBZO, indicating a better quality film.

Figure 3 presents GADDS images of the samples. For pure YBCO films, deposited on the two kinds of substrates (Figures 3(a) and 3(b)), we do not observe a big difference between the two images. In the YBZO films, for the substrate CZO-0.4 (Figure 3(c)), reflections appear as spots and rings, indicating coexistence of oriented and random YBCO. In contrast, for the substrate CZO-0.1 (Figure 3(d)) reflections are more spot like, indicating a much better crystalline orientation. Something similar happens with the YBYTO films, as shown in Figures 3(e) and 3(f). Thus it can be concluded that the best stoichiometry of the buffer layer for growing nanostructured films is CZO-0.1. From these figures we also conclude that among the two nanostructured films the best epitaxial growth is obtained in YBZO on CZO-0.1. This result agrees with the SEM images of Figure 2, and will be later supported by the superconducting properties. Furthermore, in Figure 3, we observe that reflections of nanoparticles: (220) of BYTO, and (110) of BZO, appear as rings, indicating they are mostly randomly oriented, which is a desirable characteristic [3,5].

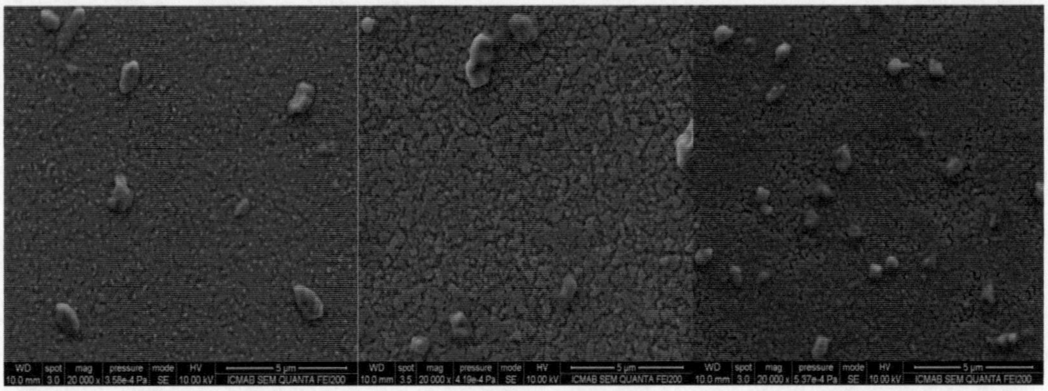

(a) (b) (c)

Figure 2. SEM micrographs of YBCO and nanostructured YBCO films grown on YSZ/CZO-0.1 substrates: (a) YBCO, (b) YBYTO and (c) YBZO. The best quality film, with lower porosity and smoother surface, is YBZO.

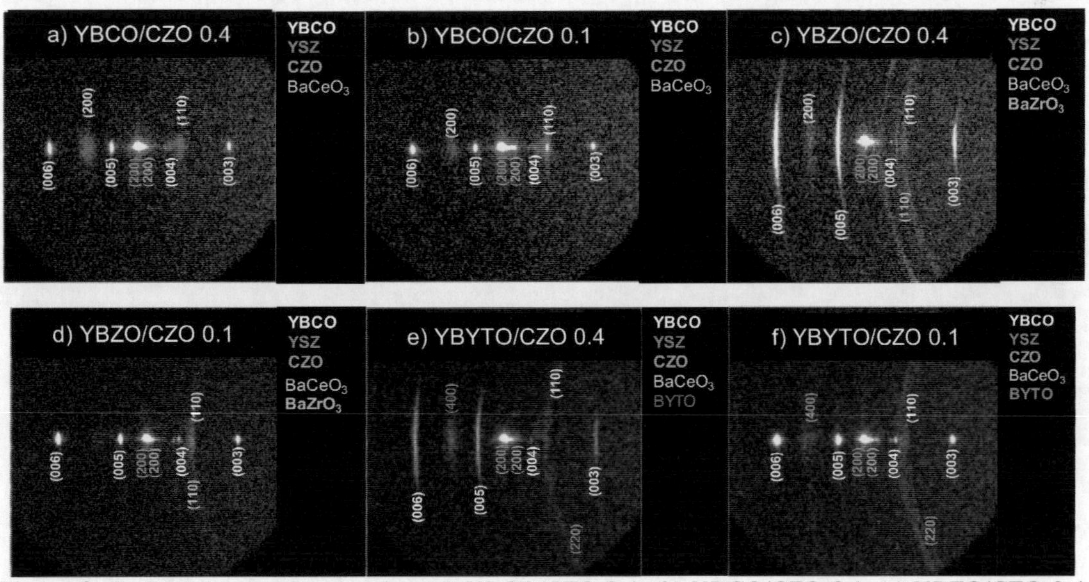

Figure 3. GADDS images for samples: (a) YBCO/CZO-0.4, (b) YBCO/CZO-0.1, (c) YBZO/CZO-0.4 d) YBZO/CZO-0.1, (d) YBYTO/CZO-0.4, (f) YBYTO/CZO-0.1. X-ray reflections are more spot like for films deposited don the CZO-0.1 substrates (right side images).

IOP Conf. Series: Journal of Physics: Conf. Series **786** (2017) 012017 doi:10.1088/1742-6596/786/1/012017

The conventional diffractogram presented in Figure 4 is obtained by integrating the above GADDS data, for the films deposited on CZO-0.1. This allows a better assessment of the intensity relation of the present phases. The diffractogram contains peaks of c-axis oriented YBCO, peaks of the buffer layer CZO, and peaks of the substrate YSZ, as indicated in the figure. In the YBYTO film there are peaks (220) and (400) of BYTO. In all samples there are peaks (110) and (220) of $BaCeO_3$, resulting from chemical reaction with the buffer layer, although in BYTO these peaks have lower intensity.

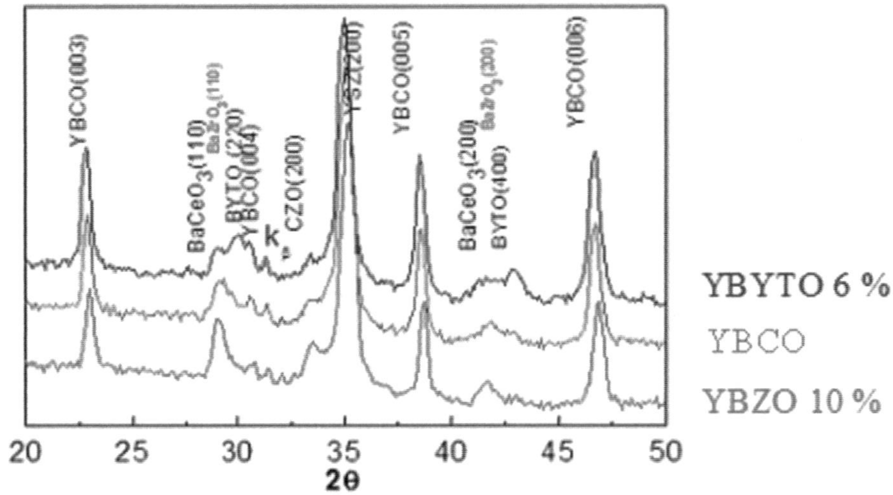

Figure 4. Diffractogram for YBCO, YBZO and YBYTO films grown on CZO-0.1. This graph is obtained by integration of the 2D GADDS information in Figure 3.

Figure 5 presents the inductive measurements of T_c and J_c, for the samples grown on CZO-0.1. Figure 5(a) shows the magnetization as function of temperature, from which we obtain T_c, which is practically the same ~89.5K, for all samples. The transition width, ΔT, in the films YBZO (8.2K) and YBYTO (8.0K), have lower values than pure YBCO (9.0K), indicating higher homogeneity in the nanostructured films. These results are summarized in Table 1. Figure 5(b) shows J_c, as function of temperature. The best behaviour is observed in sample YBZO, followed by YBYTO, with a clear increase of J_c over the pure YBCO sample. J_c was enhanced in YBZO by a factor of ~4, with respect to pure YBCO, over the whole temperature range. Table 1 presents values of J_c at 77K, for comparison. These results are compelling evidence of improvement of the superconducting properties in nanostructured films, compared with pure YBCO.

Table 1. Superconducting properties of the films grown on CZO-0.1/YSZ.

Sample	T_c(K)	ΔT (K)	J_c(77K)MA/cm^2
YBCO	90.0	9.0	0.3
YBYTO	89.5	8.0	1.0
YBZO	89.2	8.2	1.3

The higher J_c values obtained in nanostructured films is due to the pinning centres originated by the nanoparticles. In addition, in these films there is less reactivity with the buffer layer, according to XRD (Figure 4) and SEM (Figure 2) results.

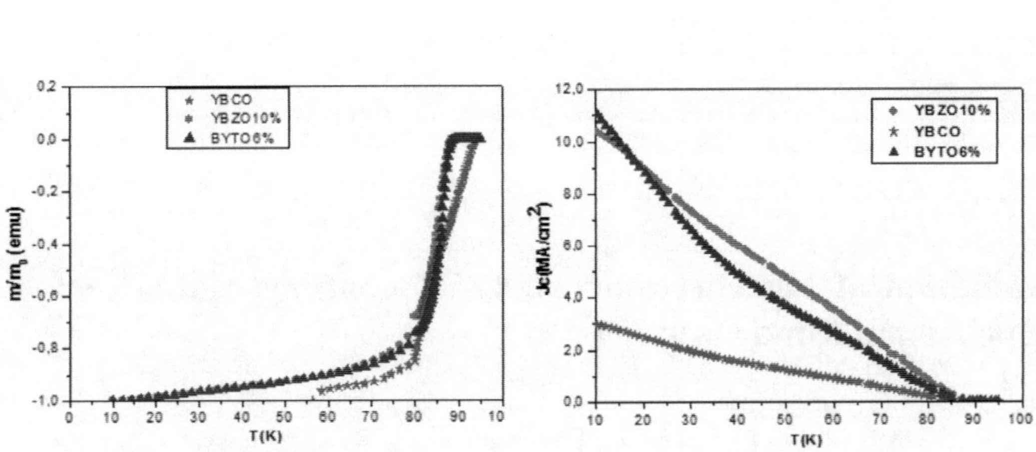

(a) (b)

Figure 5. Inductive measurements of T_c and J_c, for YBCO, YBZO and YBYTO films, grown on CZO-0.1. (a) Normalized magnetization, where T_c is obtained, and (b) J_c. The best sample, according to these superconducting properties, is YBZO.

4. Conclusions

Pure YBCO and nanostructured YBZO10% and YBYTO 6% films were grown on YSZ substrates with buffer layers of CZO-0.1 and CZO-0.4. The buffer layer CZO-0.1 produced better epitaxial growth of the deposited films, than CZO-0.4. Nanostructured films presented better morphology, epitaxial growth and superconducting properties than pure YBCO. The best sample in all respects was YBZO10%, where J_c was enhanced by a factor of ~4 over the corresponding value in pure YBCO, in the temperature range from 10K to 77K. This sample has T_c= 89.2K and J_c = 1.3MA/cm^2 at 10K. These results are quite promising for a sample made by a chemical method, encouraging further research in this direction. A next step in this research is to optimize the thermal process in order to reduce chemical reaction between the nanostructured film and the buffer layer.

Acknowledgements

P. Garces and H. Castro thank the "Departamento Administrativo de Ciencia y Tecnología e Innovación COLCIENCIAS – Colombia, for partial finatial support.

References

[1] M Coll 2006 *Chemical Solution Deposition of Oxide Buffer and Superconducting Layers for YBaCuO Coated* (Bellaterra: Universidad Autónoma de Barcelona)

[2] M Coll, S Ye, V Rouco, A Palau, R Guzman, J Gázquez, J Arbiol, H Suo, T Puig and X Obradors 2013 *Supercond. Sci. Technol.* **26** 8

[3] M Coll, R Guzman, P Garcés, J Gazquez, V Rouco, A Palau, S Ye, C Magen, H Suo, H Castro, T Puig and X Obradors 2014 *Superconduc. Sci. Technol.* **27(4)** 9

[4] X Obradors, T Puig, A Pomar, F Sandiummenge, N Mestres, M Coll, A Cavallaro, N Roma, J Gázquez, J C González, O Castaño, J Gutiérrez, A Palau, K Zalamova, S Morlens, A Hassini, M Gibert, S Ricart, J M Moretó, S Piñoll and D Isford 2006 *Superconduc. Sci. Technol.* **19** S13

[5] P Garcés, M Coll, H Castro, T Puig and X Obradors 2014 *Journal of Physics: Conference Series* **568** 5

[6] J Xiong, B W Tao, W F Qin, J L Tang, X Han and Y R Li 2008 *Superconduc. Sci. and Technol.* **21** 2

[7] M Coll, J Gázquez, F Sandiumenge, T Puig and X Obradors 2008 *Nanothecnology* **19** 7

[8] X Obradors, T Puig, S Ricart, M Coll, J Gazquez, A Palau y X Granados 2012 *Superconduc. Sci. and Technol.* **25** 123001

[9] N Roma, S Morlens, S Ricart, K Zalamova, J M Moreto, A Pomar, T Puig and X Obradors 2006 *Superconduc. Sci. Technol.* **19** 521

[10] Y X Zhou, X Zhang, H Fang, R T Putman and K Salama 2005 *IEEE Trans. Appl. Supercond.* **15** 2711

[11] X Obradors, T Puig, A Palau, A Pomar, F Sandiummenge, P Mele and K Mat sumoto 2011 *Comprehensive Nanoscience and Technology* (Oxford: Academic Press) p 303

Evaluation of hot corrosion of 8YSZ coatings elaborated by suspension plasma spraying

A G González-Hernández[1], H Ageorges[2] and M E López-Gómez[3]
[1] Universidad Industrial de Santander, Bucaramanga, Colombia
[2] Université de Limoges, Limoges, France
[3] Universidad de Antioquia, Medellín, Colombia

E-mail: aggonzal@uis.edu.co

Abstract. In this paper, the evaluation microscopic of hot corrosion of 8mol% Yttria Stabilized Zirconia (8YSZ) coatings was studied in the presence of V_2O_5 and Na_2SO_4 as corrosive molten salt, for 40h at 1050°C. First, the substrates of Inconel 718 super-alloy were sprayed with a NiCrCoAl-Y_2O_3 bond coat by atmospheric plasma spraying (APS). Then this bond coat was polished for elaborated the 8YSZ layer by suspension plasma spraying (SPS). The microstructure of the cross-section and surface of the coating was evaluated by scanning electron microscopy (SEM). After the hot corrosion test, the delamination of 8YSZ coatings was occurred in the ceramic layer due to the creation of stress resulting from the chemical reaction between the molten salts and the yttria (Y_2O_3) of 8YSZ coating at high temperature. According to EDS- SEM analysis, the evaluation of fractured sections of 8YSZ coating showed mainly the formation of crystals composed by Y, V, O and the surface was mainly composed by Zr and O. Those crystals can be related with the tetragonal phase of YVO_4, which they were commonly found by other researchers in studies of hot corrosion of YSZ-based TBCs when its surface reacts with the corrosive salts.

1. Introduction

Thermal barrier coatings (TBCs) systems are widely used to provide thermal and corrosion protections for the metallic hot-section components of gas turbines to achieve extended your durability [1–5]. A typical TBC system is composed by MCrAlY (M=Ni and/or Co) metallic bond-coat as an oxidation resistant layer and the yttria stabilized zirconia (YSZ) topcoat as a thermal insulation layer [1,4,6]. One of the most crucial bond coat properties is to improve the adherence between the metallic component (Ni-based superalloys) and ceramic coating, as well as the resistance to high-temperature oxidation and corrosion of TBC system [1,3]. Yttria stabilized zirconia (YSZ) material has been generally chosen for the ceramic topcoat, due to its low thermal conductivity and high thermal expansion coefficient, with the latter being approximately equal to that of the substrate [3]. TBCs are usually applied either by an atmospheric plasma spray (APS) or electron beam physical vapour deposition (EB-PVD). However, plasma spraying technology has enjoyed widespread acceptance thanks to the high cost-effective deposition conditions and deposition efficiency [1,2]. The other hand, the typical gas turbine operating conditions use of low quality fuels containing impurities such as Na, V and S, which it can lead to premature degradation of TBCs by hot corrosion [7].

Suspension plasma spraying (SPS), as a newly emerged technique of thermal spraying processes, has been intensity investigated to elaborated coatings (TBCs) with enhanced characteristics through submicron or nanometer particles [8–10]. The suspensions contain fine particles that are dispersed in a liquid phase by mechanical or chemical methods. Then the suspension with particles is injected into the plasma through a liquid feedstock system as a liquid stream. Some particles with sizes less than 1μm are easily accelerated and then decelerated in the plasma jet, which requires a shorter standoff distance between the plasma torch exit and the substrate (≈3-7cm) than the conventional processes (≈9-20cm) [8]. Therefore, the substrate experiences a greater heat flux due to the shorter standoff distance. To achieve a desirable coating microstructure, injector properties and spray parameters must be accurately optimized in this SPS process [8]. In addition, previous works on TBCs elaborated by SPS have focused mainly on coating formation, thermal properties, thermal cycling, and isothermal oxidation [7]. 3, that SPS coating offers higher cycling durability that coatings elaborated by APS [7]. However, the hot corrosion behaviour of YSZ coatings elaborated by SPS has not yet been sufficiently studied.

The objective of this work is to evaluate the hot corrosion behaviour of 8mol% YSZ coating elaborated by SPS. The NiCrCoAl-Y_2O_3 bond coats were first deposited on a Inconel 718 superalloy substrate using APS process. Then the YSZ top coat was applied on the polished bond coat surface using the SPS technique. Finally, hot corrosion test was carried out in presence of V_2O_5 and Na_2SO_4 for 40h at 1050°C.

2. Materials and method

Ni-based supper-alloy (Inconel 718) substrates in shape of disk with the dimension of 25mm in diameter and thickness of 5mm were grit blasted with alumina particles to create a desired surface roughness R_a profile of 4-5μm.

Atmospheric plasma spraying (APS) was used to elaborated NiCrAlCo–Y_2O_3 composition (Metco 461NS) bond coat of 250μm thickness using F4-MB torch (Sulzer-Metco, Wolhen, Suiza) with Multicoat system. Thereafter, the YSZ ceramic layer was applied over the polished bond coat through the SPS technique. SPS coating was elaborated using the same torch and Multicoat system with which was equipped with a liquid delivery unit developed by University of Limoges (France). 8mol% YSZ powder with d_{50} of 360nm (d_{10}=230nm and d_{90}=560nm) supplied by UCM Advanced Ceramics was used to achieve the aqueous suspensions. The solid loading of the suspension was fixed at 20wt.% according to published works by researchers at the Université de Limoges (France) [11–15] in which this value was found to be a good compromise between deposition efficiency and required plasma output. The Beycostat C213 with an amount of 2.1wt.% of YSZ powder was used for dispersing the particles in ethanol. Spray parameters for the APS and SPS coatings are shown in Table 1.

Table 1. Plasma spraying parameters.

Parameters	NiCrCoAl-Y_2O_3	8mol% YSZ
Spray distance (mm)	100	30
Intensity current (A)	500	500
Internal diameter of the F4 torch (mm)	7	7
Plasma gas for APS, Ar-H2 flow rate (l/min)	45-15	–
Plasma gas for SPS, Ar-He flow rate (l/min)	–	40-20
APS powder feed rate (g/min)	25-30	–
SPS suspension feed rate (ml/min)	–	24
Inner diameter of the suspension injector (μm)	–	150

Morphological characterization of YSZ powder was carried out by field emission scanning electron microscopy (FE-SEM) microscope (JEOL 7400F equipment). The cross-section of 8YSZ coating was polished according the standard ASTM E1920-03 and was analysed by scanning electron microscopy (SEM) microscope (JEOL JSM-6490LV equipment). In order to evaluate the corrosion behaviour of the coating, 55wt.% V_2O_2 (Alfa-Aesar, 99.6% min) and 45wt.% Na_2SO_4 (Carlo Erba, Ref. 4330007) powders was selected as a corrosive salt. The corrosive salt was strewed over the surface of the coatings in a 25-30mg/cm^2 concentration, leaving approximately 3mm from the edge uncoated to avoid the edge effect. The samples were pushed into the muffle furnace with air atmosphere at 1050°C for 40 hours and then cooled down inside the furnace. An accelerated hot corrosion test was carried out in this study as compared with those in the gas turbine applications; this study is somewhat comparable with similar investigations [10,16,17]. The surface of coating exposed to hot corrosion was analysed by EDS-SEM technique.

3. Results

The morphology of 8mol% YSZ powder is shown in Figure 1. This material presents particles with an irregular and angular shape that are highly correlated with its manufacturing method, which was by high energy milling process. The morphology of the cross-section of 8mol% YSZ coating was observed by SEM, as shown in Figure 2. This morphology is composed by a bimodal structure made of un-molten nanosized particles imbedded in a matrix of molten material, which is a typical characteristic of this kind of coatings [18]. This bimodal structure develops because after the liquid is evaporated in the plasma and the resulting particles or agglomerates may thus be partly melted or melted, yielding the end coating [19]. The thickness of SPS coating was 90±4μm.

Figure 1. FE-SEM image of 8YSZ powder used to prepare the suspension.

Figure 2. SEM images of polished cross- section of the 8YSZ coating.

The macroscopic image of the surface coating sample after 40h exposure to hot corrosion is shown in Figure 3(a). As observed, YSZ coating was delaminated, which occurred when the sample was cooled down inside the furnace until room temperature after 40h of the corrosion test. In other words, the coating was mainly detached by effect of molten salts, thermal stress and differential expansion between the ceramic coating and the bond coat. This depletation is similar to that found in the other work with similar conditions [7]. Figure 3(b) shows the SEM micrograph of a fractured section of YSZ coating after the hot corrosion test. As clearly seen, the rod-type crystals have formed on the coating surface. The EDS analysis of these points show that the coating surface (EDS 1) contains mainly Zr, O and with small amounts of Ni, Cr, V (Figure 3(c)) as well as the rod-type crystals (EDS 2) contain Y, V and O (Figure 3(d)). This rod-type crystal was formed as a result of the reaction of yttria of YSZ coating with the molten salt at elevated temperature.

Furthermore, the porosity and the reactive surface of coating allows the possibility of the penetration of the corrosion salt into the structure of the YSZ layer. Thus, salt can also react with the stabilizer of YSZ layer and cause the destruction and delamination of zirconia coating via thermal stress arising from the formation of rod-crystals and transformation of tetragonal to monoclinic zirconia. This behaviour is similar to that found by other researchers, especially in APS coatings [1,3,7]. In order to understand hot corrosion degradation mechanism of YSZ coatings under molten salts, it's essential to analyse the possible reactions that occur at high temperature. The appearance of rod-crystals of Y-V-0 elements found in EDS 2 (Figure 3) can be explained by the reactions presented in the Equation 1 and 2. This reactions, were also proposed by other researchers [1,3].

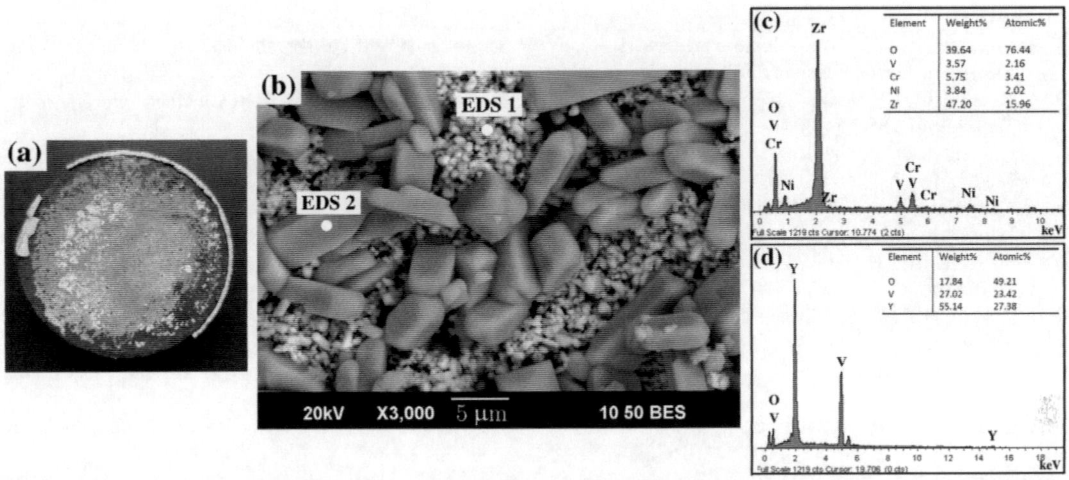

Figure 3. (a) Photograph of 8YSZ coating surface after hot corrosion test, (b) SEM image of fractured section of 8YSZ coating, EDS analysis of (c) point 1 and (d) point 2 in SEM image.

$$V_2O_{5(1)} + Na_2SO_{4(1)} \rightarrow 2NaVO_{3(1)} + SO_{3(g)} \quad (1)$$

$$YSZ_{(s)} + 2NaVO_{3(1)} \rightarrow ZrO_{2(s)(monoclinic)} + YVO_{4(s)} + Na_2O_{(1)} \quad (2)$$

At first, V_2O_5 could be first melting during heating because of its lower melting point (690°C). At≈610°C, the liquid $NaVO_3$ could be formed by the reaction of molten salt (Na_2SO_4+V_2O_5) at the testing temperature of 1050°C, according to Equation 1. Finally, $NaVO_3$ reacted with the stabilizer Y_2O_3 of YSZ to form monoclinic ZrO_2, rod-crystals of YVO_4 and Na_2O (see Equation 2). In this study, Na was not detected in EDS 1. It is possible that Na_2O was sublimated at high temperatures during hot corrosion test.

4. Conclusion

In this work, the evaluation microscopic of the hot corrosion behaviour of 8mol% YSZ coating elaborated by suspension plasma spraying technique was investigated. The hot corrosion test was carried out using a mixture of V_2O_5 and Na_2SO_4, for 40h at 1050°C. The YSZ coating showed a bimodal distribution structure, which was composed of a dense area with semimolten nanometric particles and a lamellar zone with a higher degree melting. This bimodal structure is typical in coating elaborated by SPS.

The mechanism of hot corrosion of YSZ coating elaborated by SPS is similar to the coating deposited by APS. This mechanism consist of the infiltration of molten salt into the YSZ coating elaborated by SPS. Then, the molten salt cause formation of monoclinic-ZrO_2 and YVO_4 phases by interaction between YSZ layer and molten salt. The formation of monoclinic-ZrO_2 induces cracks in the zirconia phase of the top coat leading to the failure and delamination of YSZ layer in this study, especially when the coating was cooling after hot corrosion test.

Acknowledgments

The authors thanks to the "CODI" of the Universidad de Antioquia (Colombia) for financing the project grant number MDC10-1-04, as well as, the economic contributions of the sustainability program of the research group GIPIMME - Universidad de Antioquia. The author Gonzalez thanks to

Colfuturo (Colombia) and Colciencias (Colombia) for its economic support during the development of their doctoral program between Universidad de Antioquia and Université de Limoges (France).

References

[1] Loghman-Estarki M R, Nejati M, Edris H, Razavi R S, Jamali H and Pakseresht A H 2015 *Journal of the European Ceramic Society* **35** 693–702
[2] Pakseresht A, Javadi A, Ghasali E, Shahbazkhan A and Shakhesi S 2016 *Surface and Coatings Technology* **288** 36–45
[3] Hajizadeh-Oghaz M, Razavi R S, Ghasemi A and Valefi Z 2016 *Ceramics International* **42** 5433–5446
[4] Lozovan A A, Betsofen S Y, Ashmarin A A, Ryabenko B V and Ivanova S V 2016 *Journal of Physics: Conference Series* **729** 012017
[5] Baig M N and Khalid F A 2014 *Conference Series: Materials Science and Engineering* **60** 012053
[6] Ito K, Kuriki H, Araki H, Kuroda S and Enoki M 2014 *Science and Technology of Advanced Materials* **15** 035007
[7] Ajay A, Raja V, Sivakumar G and Joshi S 2015 *Corrosion Science* **98** 271–279
[8] Fan W and Bai Y 2016 *Ceramics International* **42** 14299–14312
[9] Fauchais P, Montavon G, Lima R S and Marple B R 2011 *Journal of Physics D: Applied Physics* **44** 093001
[10] Chen X, Ohnuki T, Kuroda S, Gizynski M, Araki H, Murakami H, Watanabe M and Sakka Y 2016 *Ceramics International* **42** 16822–16832
[11] Rampon R, Toma F L, Bertrand G and Coddet C 2006 *Journal of Thermal Spray Technology* **15** 682–688
[12] Rampon R, Filiatre C and Bertrand G 2008 *Journal of Thermal Spray Technology* **17** 105–114
[13] Bacciochini A, Ilavsky J, Montavon G, Denoirjean A, Ben-ettouil F, Valette S, Fauchais P and Wittmann-teneze K 2010 *Materials Science and Engineering: A* **528** 91–102
[14] Bacciochini A, Ben-Ettouil F, Brousse E, Ilavsky J, Montavon G, Denoirjean A, Valette S and Fauchais P 2010 *Surface and Coatings Technology* **205** 683–689
[15] Bacciochini A 2010 Quantification de larchitecture poreuse de dépôts finement structures (sub-micromètre-nanomètre) de zircone yttriée réalisés par projection plasma de suspension (France: Université de Limoges) (Preprint https://www.theses.fr/152605630)
[16] Afrasiabi A, Saremi M and Kobayashi A 2008 *Materials Science and Engineering: A* **478** 264–269
[17] Nejati M, Rahimipour M and Mobasherpour I 2014 *Ceramics International* **40** 4579–4590
[18] Pawlowski L 2009 *Surface and Coatings Technology* **203** 2807–2829
[19] Vicent M, Bannier E, Carpio P, Rayn E, Benavente R, Salvador M and Snchez E 2015 *Surface and Coatings Technology* **268** 209–215

Development of highly open polyhedral networks from vitreous carbon for orthopaedic applications

V Güiza-Argüello[1], M Bayona-Becerra[1], S Cruz-Orellana[1] and E Córdoba-Tuta[1]
[1] Universidad Industrial de Santander, Bucaramanga, Colombia

E-mail: vivragui@uis.edu.co

Abstract. Highly open polyhedral networks were fabricated using an economical and environmentally friendly template route. Recycled cellulose foams were impregnated with a sucrose resin and then pyrolyzed in order to produce reticulated vitreous carbon foams with morphological features that closely resemble trabecular bone. Also, cell sizes ~1mm were achieved, a trait that will allow the mechanical reinforcement of such scaffolds using a biomaterial coating without compromising the pore size that favors osteoblast cell infiltration and growth (200–500μm). Moreover, initial studies showed that carbonization conditions have an effect on the mechanical properties of the synthesized foams and, therefore, such process parameters could be further evaluated towards the enhancement of the mechanical resistance of the scaffolds. The materials developed here are visualized as the porous component of a synthetic bone graft with features that could help overcome the current limitations associated with the medical treatments used for bone defect repair.

1. Introduction

Due to their remarkable surface and light-weight properties, carbon based porous materials have been studied for multiple industrial applications, especially for the fabrication of electrodes [1-4]. Conventionally, carbon foams are fabricated from phenolic resins because of their relatively high carbon yield [5-7]. Recently, sucrose-based resins have been used as an alternative that offers a low-cost and more environmentally friendly process for the preparation of the carbon precursor [8-10]. Lately, a few studies have explored the potential of reticulated vitreous carbon foams (RVC) for bone tissue regeneration applications [11-12]. For instance, Czarnecki and coworkers found that heat-treated RVC foams supported human osteoblast adhesion, mineralization and growth in vitro [12]. Moreover, the in vivo studies conducted by Pec et al. [11] indicated that, even though vitreous carbon is not biodegradable, it can positively integrate with the surrounding tissue upon implantation. This represents an advantage of the use of RVC foams as scaffolds, relative to traditional biodegradable ceramics, since RVC foams could provide a permanent material platform to guide bone remodeling at the defect site.

Nonetheless, one of the major limitations associated with the use of RVC foams for bone repair lies in their low mechanical properties. One way to circumvent this issue is through the reinforcement of the carbon foam with a phase that would allow it to withstand mechanical loading in vivo, while still preserving the morphology and porosity levels that favor osteogenic cell infiltration and proliferation. According to the literature, the optimal pore size for bone tissue growth ranges between 200 500μm [13]. Therefore, the main goal of the present work was to develop vitreous carbon polyhedral networks that: i) show morphological features that resemble trabecular bone, and ii) have cell sizes ~1mm, in order to eventually allow the mechanical reinforcement of such scaffolds using a biomaterial coating without compromising the pore size that favors osteoblast cell infiltration and growth. Toward this goal, vitreous carbon networks were fabricated via template route using a sucrose resin, and cellulose foams that were recycled from common fish tank filters were used as the sacrificial polymeric template. The results showed that it was possible to synthesize highly open polyhedral networks with similar structural properties to expensive, commercial RVC foams. Although the mechanical properties of the prepared foams still remain to be improved, our studies indicate that the effect of process parameters such as carbonization conditions could be further evaluated towards the enhancement of such properties.

2. Experimental

2.1. Fabrication of highly open polyhedral networks

Vitreous carbon polyhedral foams were fabricated from a sucrose resin, which was prepared from an aqueous solution (0.4g/mL sucrose) in a Teflon® beaker. Concentrated HNO_3 was then added at a 1:1 molar ratio relative to sucrose, and the resulting solution was transferred to an oven at 120°C. After 7h, the resin was removed from the oven and immediately used for the impregnation process. Recycled cellulose foams were thoroughly rinsed with water and then cut into 1×1×1cm samples. After this, samples were immersed in the resin for a few seconds, followed by gentle extrusion to remove the excess resin. This procedure was repeated twice in order to ensure proper coating of the template. Samples were then dried at 250°C for 1h prior to carbonization under nitrogen atmosphere at 900°C for 1h. Two different heating ramps were used for the pyrolysis of the impregnated foams: 1 and 3.7°C/min.

2.2. Characterization of the vitreous carbon structures

Structural characterization of the synthesized foams and their commercial counterparts (ERG Aerospace Corp., CA, USA) was carried out using X-ray diffraction (XRD) and Fourier transform infrared spectroscopy (FT-IR). Moreover, scanning electron microscopy (SEM) was employed to characterize the morphology of both types of RVC foams. Finally, a compression test was performed in order to measure the mechanical resistance of the fabricated networks (n=4-6 per sample).

3. Results

3.1. Characterization of the synthesized foams and their commercial counterparts

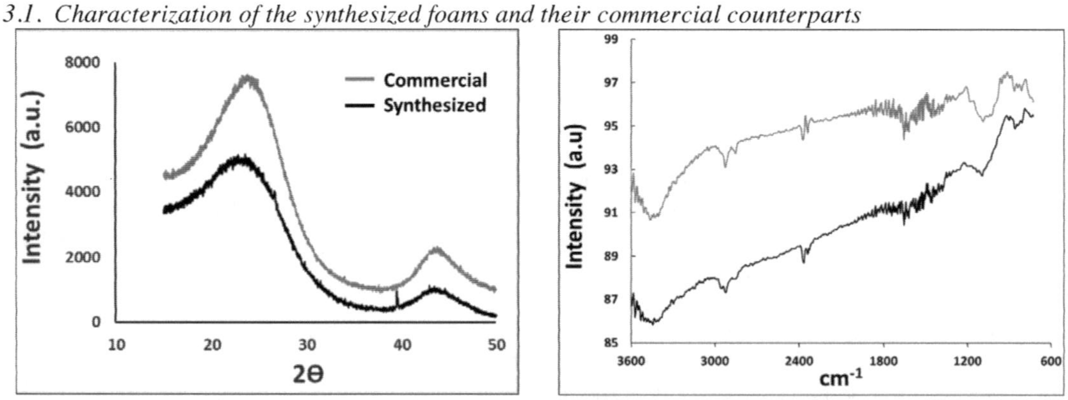

(a) (b)

Figure 1. XRD (a) and FT-IR (b) spectra for the commercial and synthesized RVC foams.

Figure 1 represents the XRD and FT-IR spectra for the synthesized and commercial RVC foams. The XRD spectra show two broad peaks in the 22-25° and 42-46° ranges, which are associated with the semicrystalline reflections from the (00 2) and (1 0 1) planes of graphite [2,8,10]. The weak diffraction peaks confirm the presence of a disordered carbon structure [14], characteristic of vitreous carbon. Moreover, the FT-IR spectra show a broad band at ~3420cm^{-1}, which can be attributed to O-H stretching vibrations from intercalated water [1,15]. The band at ~1625cm^{-1} can be attributed to C-C stretching

IOP Conf. Series: Journal of Physics: Conf. Series **786** (2017) 012019 doi:10.1088/1742-6596/786/1/012019

[16,17], whereas the band at ~1100cm^{-1} can be ascribed to C-O-C stretching vibrations [16,17], a result that corresponds with the model proposed by Kakinoki, which defines vitreous carbon as a combination of tetrahedral and graphitic portions connected by oxygen bridges [18].

Furthermore, Figure 2 shows SEM characterization results for the synthesized and commercial RVC foams. As it can be observed, the morphology of the synthesized foams closely resembles that of trabecular bone. Also, it was possible to achieve cell sizes ~1 mm, which to the best of our knowledge, has not been previously reported in the literature. Such cell size will eventually allow significant mechanical reinforcement of the RVC foams using a biomaterial coating without compromising the pore size that favors osteoblast cell infiltration and growth.

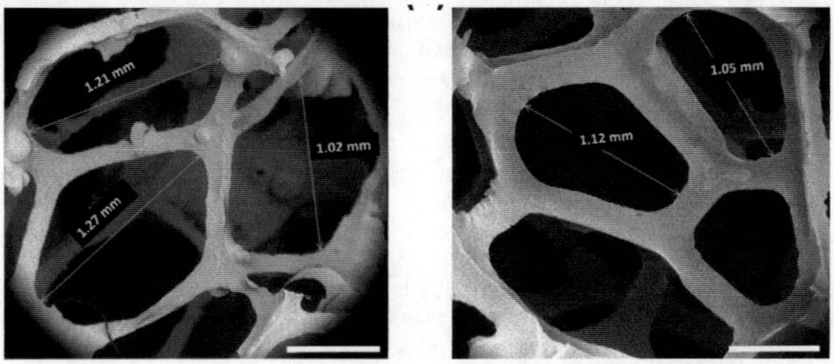

(a) (b)

Figure 2. SEM images for the synthesized (a) and commercial (b) RVC foams. Scale bar=500μm.

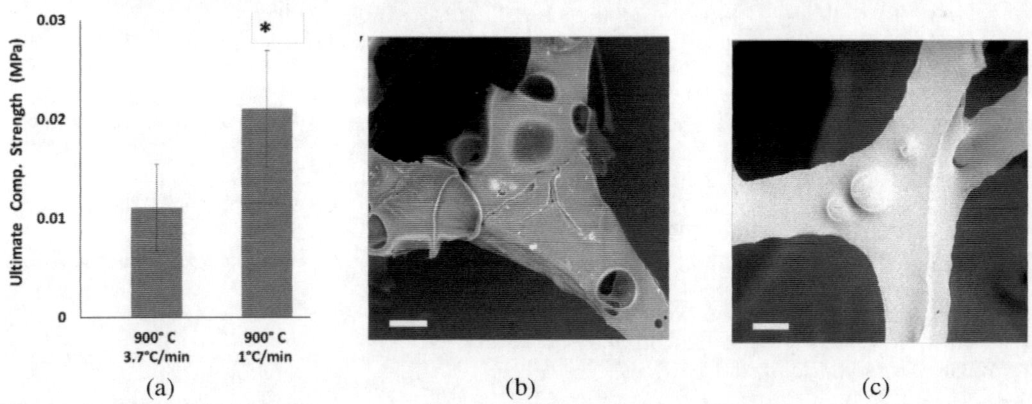

(a) (b) (c)

Figure 3. Effect of the heating ramp during carbonization on the compression strength of the synthesized RVC networks (a). Representative SEM images of foams synthesized using a 3.7°C/min heating ramp (b) versus a 1°C/min heating ramp (c). Scale bar=100 μm*, significant difference (p=0.0205).

3.2. Mechanical properties of the prepared foams: effect of the heat treatment

Figure 3(a) shows compressive strength measurements for the RVC networks synthesized by carbonization at 900°C using two different heating ramps (3.7 and 1°C/min). As it can be noticed, a two-fold increase in the mechanical resistance of the foams was achieved by reducing the heating rate during pyrolysis. In order to gain insight into the possible causes of the observed differences, SEM imaging for both types of foams was performed. Figures 3(b)-(c) reveal that the RVC networks produced using a heating ramp of 1°C/min resulted in a more compact structure with significantly fewer defects relative to the 3.7°C/min sample. This could be due to the fact that a lower heating rate allowed a more gradual release of the gases produced during carbonization (mainly hydrogen, carbon monoxide and carbon dioxide), which in turn lead to a reduced generation of surface vacancies on the synthesized foams. These results are relevant as they provide insight into the mechanisms underlying the conversion of the sucrose resin into vitreous carbon and, thus, will aid the development of fabrication methods to produce RVC foams with improved structural stability using low-cost and more environmentally friendly approaches, relative to the traditional use of highly toxic phenolic resins.

4. Conclusions

In this work, highly open polyhedral networks were fabricated using an economical and environmentally friendly template route. Carbonization of recycled cellulose foams that were impregnated with a sucrose resin resulted in RVC foams with morphological features similar to trabecular bone. The pore size achieved will allow the mechanical reinforcement of such scaffolds with a biomaterial coating without compromising the pore size levels required for bone regeneration. Moreover, our results showed that carbonization conditions have an effect on the mechanical properties of the foams and, therefore, such variables warrant further investigation towards the enhancement of the mechanical resistance of RVC scaffolds that could be used as the porous component of a synthetic graft for bone defect repair.

References

[1] Ntsendwana B, Mamba B, Sampath S and Arotiba O A 2012 *Int. J. Electrochem. Sci.* **7** 3501-12
[2] Kumar R, More V, Mohanty S P, Nemala S S, Mallick S and Bhargava P 2015 *J. Colloid Interf. Sci.* **459** 146-150
[3] Rao A.K and Creager S E 2008 *Anal. Chim. Acta* **622** 1-10
[4] Sarmento E, Cerqueira M, Ferreira M and Gomes N 2006 *Mater. Res.* **9** 147-152
[5] Seo J, Park H, Shin K, Baeck S H, Rhym Y and Shim S E 2014 *Carbon* **76** 357–367
[6] Lv Y, Gan L, Liu M, Xiong W, Xu Z, Zhu D and Wright D S 2012 *J. Power Sources* **209** 152–157
[7] Nam G, Choi S, Byun H, Rhym Y and Shim S E 2013 *Macromol. Res.* **21** 958-964
[8] Ji H, Huang Z, Wu X, Huang J, Chen K, Fang M and Liu Y 2014 *J. Mater. Res.* **29** 1018-25
[9] Jana P, Fierro V and Celzard A 2016 *Ind. Crop. Prod.* **89** 498-50
[10] Prabhakaran K, Singh P K and Gokhale N M 2007 *J. Mater. Sci.* **42** 3894–3900
[11] Pec M K, Reyes R, Sánchez E, Carballar D, Delgado A, Santamaría J, Arruebo M and Evora C 2010 *Eur. Cells Mater.* **20** 282-294
[12] Czarnecki J S, Blackmore M, Jolivet S, Lafdi K and Tsonis P A 2014 *Carbon* **79** 135-148
[13] Okii N, Nishimura S, Kurisu K, Takeshima Y and Uozumi T 2001 *Neurol. Med. Chir.* **41** 100-4
[14] Pradhan B K and Sandle N K 1999 *Carbon* **37** 1323–1332
[15] Yuan Y, Ding Y, Wang C, Xu F, Lin Z, Qin Y, Li Y, Yang M, He X, Peng Q and Li Y 2016 *ACS Appl. Mater. Interfaces* **8** 16852–16861
[16] Wu X, Wen T, Guo H, Yang S, Wang X and Xu A 2013 *ACS Nano* **7** 3589–3597
[17] Yang Y and Lin Z.G 1995 *J. Appl. Electrochem.* **25** 259-266
[18] Kakinoki J 1965 *Acta Cryst.* **18** 578

Silver nanoparticles: Influence of the temperature synthesis on the particles' morphology

S Piñero[1], S Camero[1] and S Blanco[2]
[1] Universidad Central de Venezuela, Caracas, Venezuela
[2] Universidad Industrial de Santander, Bucaramanga, Colombia

E-mail: siblanco@uis.edu.co

Abstract. Silver nanoparticles have a wide range of applications in the medical field, textile and food industries. These and other applications can be found due to the relation between its size and morphology. In this study the influence of bath temperature on the morphology and size of silver nanoparticles are evaluated, which are obtained by chemical reduction of $AgNO_3$ using three reducing agents: sodium borohydride, ascorbic acid and sodium citrate. The evaluation carried out by the traditional UV-vis Spectrophotometric analysis and with High Resolution Transmission Electron Microscopy. The UV-vis spectrum of the silver colloids obtained by chemical reduction using three different reducing agents shows the effect of the temperature change on the growing and aggregative process. The final effect on the morphology, size and aggregation of the particles was confirmed by TEM. The result suggests a change in the growing mechanism, conducted by aggregation of atoms at 5 and 20°C degrees and aggregation of clusters at higher temperatures. Moreover in this work the main synthesis methods of nanomaterials are described.

1. Introduction

Due to the silver nanoparticles properties, there are technological advances in order to bring the development of a wide range of applications. Silver nanoparticles can be used in the medical field, specifically in coated surgical instrument, textile and food industry; and as sensors because of its electrochemical properties. On the other hand, it can be mentioned photoelectrochemical applications as biological markers and actually forming nanocomposites for pipelines coatings [1-3]. These and other applications of silver nanoparticles can be found due to the relation between its size and morphology. According with several studies size and morphology are also related with the reducing agent used in the chemical synthesis [4-6]. Characteristics and considerations can attribute a specific application to the obtained nanoparticle [7]. This study evaluates the influence of bath temperature on the morphology and size of silver nanoparticles obtained by chemical reduction of $AgNO_3$ using three reducing agents: sodium borohydride, ascorbic acid and sodium citrate. The evaluation carried out by the traditional UV-vis Spectrophotometric analysis and with High Resolution Transmission Electron Microscopy (HRTEM).

2. Experimental

Silver nanoparticles have been synthesized with chemical reduction of silver nitrate and three reducing agents: sodium borohydride ($NaBH_4$), ascorbic acid ($C_6H_8O_6$) and sodium citrate ($Na_3C_6H_5O_7$). For the study of the temperature effect a sequential analysis took place, where each temperature condition was selected individually, increasing it on each synthesis and keeping constant the rest of the experimental parameters as shown in Table 1.

Table 1. Experimental parameters used in the synthesis of silver nanoparticles by chemical reduction.

$AgNO_3$	$NaBH_4$	$C_6H_8O_6$	$Na_3C_6H_5O_7$	Temperatures, °C
1mM	2mM			5, 20, 30, 40
10mM		0.5mM		5, 20, 30, 50
1mM			34mM	20, 40, 85

In each case, the synthesis of the nanoparticles occurs by reaction with the reductive agents, according to the following chemical reactions: [4-6].

$$AgNO_3 + NaBH_4 \rightarrow Ag + \frac{1}{2} H_2 + \frac{1}{2} B_2H_6 + NaNO_3 \quad (1)$$

$$2Ag^+ + C_6H_8O_6 \rightarrow 2Ag^\circ + C_6H_6O_6 + 2H^+ \quad (2)$$

$$4Ag^+ + Na_3C_6H_5O_7 \rightarrow 4Ag^\circ + C_6H_5O_7H_3 + 3Na^+ + H^+ + O_2 \quad (3)$$

UV-vis and TEM evaluation are described elsewhere [8].

3. Results and discussion

Figure 1(a-c) shows the UV-vis spectrum obtained from the evaluation of the silver colloids obtained by chemical reduction. Figure 1a shows the characteristic curve obtained using sodium borohydride as reducing agent, the response have a maximum of absorbance near to 400nm.

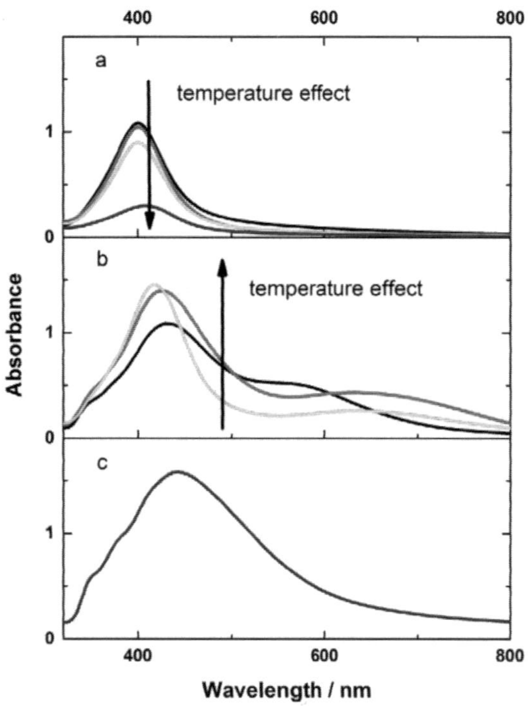

Figure 1. UV-vis absorption spectrum of silver colloids, obtained by chemical reduction using sodium borohydride (a), ascorbic acid (b) and sodium citrate (c) as reducing agent. Black: 5°C, red: 20°C, green: 30°C, blue: 40°C, wine: 85°C.

The evaluation was made changing the bath temperature between 5 and 40°C. The results from the three mentioned temperatures show a similar response, with a small variation in the maximum; the increment of the bath temperature between 30 and 40°C, produce a more relevant change in the response. The shift to higher wavelength values and the reduction of the peak intensity are possible related to a reduction on the number of particles or an increment on size. The increment on the temperature could change the reaction kinetic, the movement of atoms on the solution (Brownian movement) and the aggregative mechanisms of the nanoparticles. In this specific case, where the spectrum shows one maximum with a small shift on the wavelength (less than 8 nm between 5 and 40°C degrees), the effect of the temperature could be related to an increment in the reduction kinetic and the aggregation of atoms to the particle according to the classical LaMer mechanism [9]. Additionally, with the increment of the temperature to 40°C, the sodium borohydride could react to produce hydrogen and boranes and decrease the reduction of silver ions [10]. Figure 1(b) shows the spectrums obtained using ascorbic acid as reducing agent, in this case the responses are different, the spectrums shows two secondary maximums and this characteristic behaviour can be related to the production of non-spherical particles and the possible aggregation of the particles induced by a "glue" effect of the stabilizer [11]. The particles could be produced by the aggregation of clusters, each ones with a specific structure and growing modes. The temperature variation could affect the kinetic of the reaction and the aggregation mechanism (oriented attachment, intraparticle ripening or overgrowth). Figure 1c shows the spectrum of the colloids obtained using sodium citrate as reducing agent at 85°C, the existence of various peaks and the bandwidth are related to a high dispersion of size and shape. This dispersion could be related to a progressive nucleation mechanism, and an aggregative formation of colloids, facilitated by the temperature increase. The UV-vis spectrums of the colloids obtained using sodium borohydride as reducing agent are shown in Figure 1(a). The intensity of the peaks is inversely proportional with the reaction temperature.

Figure 2 shows the images obtained by HRTEM of the nanoparticles synthesized at characteristic temperatures. This result confirms what was suggested from UV-vis analysis. The particles obtained using sodium borohydride at 5 and 20°C, shows and quasi-spherical shape that could be explained by the growing model of LaMer with an atomic aggregation. Instead, the particles obtained at 30 and 85°C degrees show an irregular shape and aggregation of the particles.

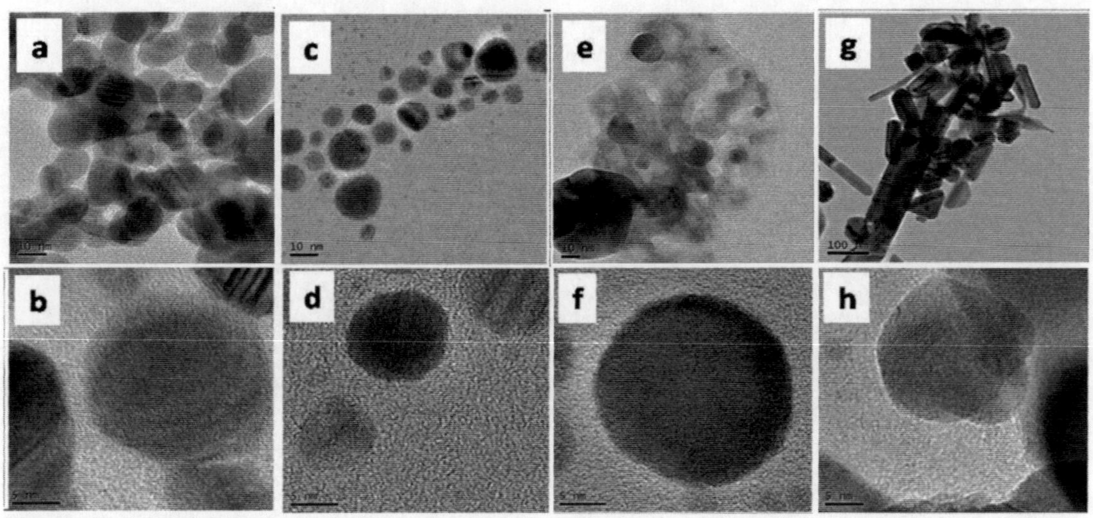

Figure 2. HRTEM images of silver nanoparticles, obtained using sodium borohydride at 5°C (a) 150000X and (b) 500000X; sodium borohidride at 20°C (c) 150000X and (d) 500000X; ascorbic acid at 30°C (e) 100000X and (f) 500000X; and sodium citrate at 85°C (g) 20000X and (h) 400000X.

4. Conclusions

The UV-vis spectra of the silver colloids obtained by chemical reduction using three different reducing agents show the effect of the temperature change on the growing and aggregative process. The final effect on the morphology, size and aggregation of the particles was confirmed by HRTEM. The result suggests a change in the growing mechanism, conducted by aggregation of atoms at 5 and 20°C degrees and aggregation of clusters at higher temperatures. To confirm this proposal an in-situ evaluation of the growing mechanism is needed.

References

[1] Le Ouay L and Stellacci F 2015 Antibacterial activity of silver nanoparticles: a surface science insight *Nanotoday* **441** 1-16
[2] Atta A, Allohedan H, El-Mahdy G and Ezzat A 2013 Application of stabilized silver nanoparticles as thin films as corrosion inhibitors for carbon steel alloy in 1M hydrochloric acid *Journal of Nanomaterials* **V** 1-8
[3] Chikwendu C 2014 The benefits and applications of nanocomposites *International Journal of Advanced Engineering Technology* **V(IV)** 12-18
[4] Solomon S and Bahadory M 2007 Synthesis and Study of silver nanoparticles *Journal of Chemical Education* **84(2)** 322-5
[5] Sondi I, Goia D and Matijevic E 2003 Preparation of highly concentrated stable dispersions of uniform silver nanoparticles *Journal of Colloid and Interface Science* **260** 75-81
[6] Silekaite A, Prosycevas I, Puiso J, Juraitis A and Guobiene A 2006 Analysis of silver nanoparticles produced by chemical reduction of silver salt solution. *Materials Science* **12(4)** 1320-92
[7] Raimondi F, Scherer GG, Kotz R and Wokaun A 2005 Nanoparticles in energy technology: examples from electochemistry and catalysis Angew *Chem., Int. Ed.* **44** 190–209
[8] Piñero S, Blanco S, Camero S and Cañizales E 2015 Síntesis química y caracterización por MET de coloides con nanopartículas de plata *Acta Microscópica* **24(A)** 1-2
[9] LaMer V and Dinegarm R 1950 Theory, production and mechanism of formation of monodispersed hydrosols *Journal of the American Chemical Society* **72** 4847-54
[10] Pacioni N, Borsarelli C, Rey V and Veglia A 2015 *Synthetic Routes for the Preparation of Silver Nanoparticles A Mechanistic Perspective* ed Alarcon, Emilio, Griffith, May, Udekwu, Klas I (Switzerland: Springer International Publishing) pp 13-46
[11] You H and Fang J 2016 Particle-mediated nucleation and growth of solution-synthesized metal nanocrystals: a new story beyond the LaMer curve. *Nano Today*, 11, 145-67

Synthesis and characterization of nanometric zinc oxide for a stationary phase in liquid chromatography

F Gordillo-Delgado[1], C C Soto-Barrera[1] and J Plazas-Saldaña[1]
[1] Universidad del Quindío, Armenia, Quindío, Colombia

E-mail: fgordillo@uniquindio.edu.co

Abstract. The increasing demand for equipment to remove organic compounds in industry and research activity has led to evaluate nanometric zinc oxide (ZnO). In this work, we present the ZnO nanoparticles synthesis for reusing of discarded columns, as a low-cost alternative. The compound was obtained by sol-gel technique using zinc chloride and sodium hydroxide as precursors and a drying temperature of 169°C. An X-ray diffractometer was used to estimate the average particle size at 20.3±0.2nm; the adsorption capacity was 0.0144L/g and the chemical resistance was tested with HCl and NaOH. The ZnO nanopowder was packed with 100psi pressure in an empty C-18 column cavity. The column packing resolution was evaluated using a high performance liquid chromatographer (HPLC-Thermo Scientific Dionex UltiMate 3000); using a caffeine standard, the following parameters were established: solvent flow: 1.2mL/min, average column temperature: 40°C, running time: 10 minutes, mobile phase acetonitrile-water composition (9:1). These results validate the potential of ZnO nanopowder as a column packing material in HPLC technique.

1. Introduction

The stationary phases of silica are the most common packing material of columns used in high performance liquid chromatography (HPLC). However, its usage is limited to pH ranges from 2.5 to 7.5 due to the structural chemical instability of this material. For this reason, some ceramic materials have been studied for replacing the silica [1]. The development of new instruments and stationary phases for the HPLC technique is a subject of interest, but they are being produced with complex manufacturing and are expensive [2].

Several inorganic oxides have a high potential in the development of new columns as alternative to the silica such as titanium dioxide (TiO_2), zirconium oxide (ZrO_2) and zinc oxide (ZnO) [3]. ZnO has similar properties to both TiO_2 and silica, but unlike TiO_2, it is cheaper and more resistant to pH changes than silica [4,5]. In this work, ZnO nanopowder for application as a stationary phase in HPLC was synthesized by sol-gel technique, which is a chemical route widely used because of its simplicity, reproducibility, commercial viability, high yields, product homogeneity, industrial adaptation, and nanometric size control [6,7].

2. Experimental section

2.1. *Zinc oxide preparation*

The reagents were zinc oxide, sodium hydroxide, zinc chloride, hydrochloric acid, analytical-grade nitric and sulfuric acid distributed by Merck KGaA. Acetonitrile, methanol, caffeine and water, with HPLC quality, were acquired from Merck KGaA.

Sol-gel process was used as described by Gel [6]. However, in this work some modifications were taken: firstly, the reactor was placed during one hour under ultrasonic agitation; the reaction temperature was 90°C; methanolic solution concentration was 90% methanol/10% water; 0.1M NaOH was heated at 90°C and 0.25M $ZnCl_2$ (dripped during 30 minutes) was added into the methanolic solution. Finally, the obtained white precipitate was washed with water nine times, under ultrasonic agitation and dried at 200°C.

2.2. *Characterization*

An X-Ray diffractometer (Bruker D8 Advanced) with Cu Ka radiation (λ=1.5418Å) was used to determine the crystallographic phase and crystal size of the synthetized ZnO powder and a scanning electron microscope (SEM-Model JSM-5610 LV, Electron Optics Corporation, Japan) was utilized to observe crystal shapes and size distribution. The initial solid crystallization temperature of the samples was measured using a differential scanning calorimeter (NETZSCH DSC 214 POLYMA) and the decomposition temperature was determined with a thermogravimetric analyzer (NETZSCH TG 209 F1). Finally, an HPLC (Thermo Scientific Dionex UltiMate 3000) was used to identify caffeine, using the nanocrystalline ZnO as column packing material.

2.3. *Column packing*

The packing of the stationary phase of ZnO nanopowder into a stainless steel cylinder (HPLC column, 110mm high with a radius of 2mm), was done using the ZnO nanoparticles in a water/ethanol mixture 30-70%, which was propelled by an air compressor at constant pressure of 100psi.

2.4. *Resistance to alkaline and acidic eluents test*

The sample of zinc oxide (100mg) was deposited in test tubes; 10mL of hydrochloric acid with 0.1M and 1M concentration and sodium hydroxide solution with 50mM and 0.5M concentration were added, respectively. Then, the test tubes were agitated and heated for 48 hours at 40°C; later the supernatant was separated by centrifugation. The residue was washed with distilled water and then centrifuged. Finally, the oxide was dried during 2 hours at 105°C [1].

2.5. *Adsorption isotherm measuring*

The determination of the adsorption isotherms was performed as follows: 20mg of the ZnO sample (adsorbent) were placed into individual test tubes and heated for 2 hours at 105°C. Then, into each tube, 2mL of methylparaben solution were added; this solution was prepared in different concentrations using hexane-2-propanol (9:1) as diluent. Finally, the tubes were placed into an ultrasonic bath during 15 minutes and after 1 hour in a water bath at 40°C [1].

3. Results and analysis

3.1. *Characterization and morphology*

Figure 1 shows the thermogravimetric curve of the sample; it indicates a 24.23% of mass loss at 78.16°C, which corresponds to losing methanol in the synthesis process; the other inflection point in this curve, around 168°C, indicates a mass loss of 16% corresponding to water evaporation. According to DSC curve in Figure 2, the evaporation of solvents that were used in the synthesis is confirmed; the ZnO sample is initially crystalline and it does not need a synthesization process because there is not any exothermic peak in the curve.

In Figure 3, the X-Ray diffractogram (XRD) of the ZnO nanopowder is shown; the representative peaks of ZnO suggest high purity and crystallization degree. The average particle radius of 20.3±0.2nm was calculated indirectly using the Scherrer Equation (1) [8,9]:

$$r = \frac{0.9\lambda}{2\beta \mathrm{Cos}(\theta)} \tag{1}$$

Where 0.9 is a shape factor, λ is the wavelength of X-rays, θ is the Bragg diffraction angle and β is the full width at half-maximum (FWHM) of the diffraction peak corresponding to the plane (1 0 1).

The spherical morphology of the nanoparticles was observed from the SEM micrograph in Figure 4. From this picture, it is possible to see that the nanoparticle size is around 25nm, which is similar to the order of estimated value using the Equation (1).

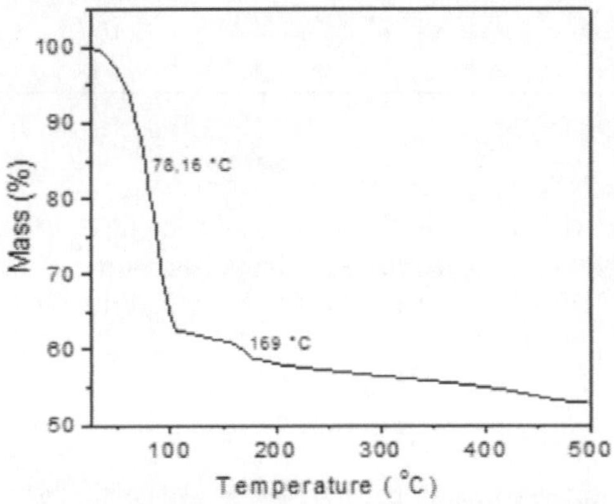

Figure 1. Thermogravimetric curve: mass loss versus temperature for synthesized ZnO powder.

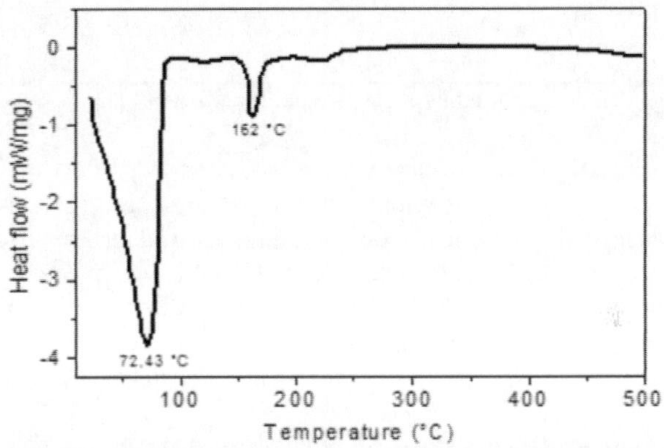

Figure 2. DSC curve: heat flow versus temperature for synthesized ZnO powder.

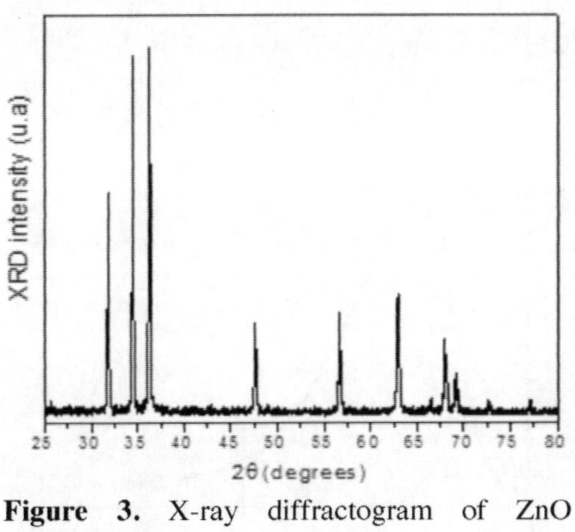

Figure 3. X-ray diffractogram of ZnO nanopowder.

Figure 4. Micrograph of ZnO nanoparticles, magnification X 20000.

3.2. Test of resistance to alkaline and acidic eluents

The endurance of zinc oxide powder to moderately basic and acid pH was registered in Table 1. It was proved that ZnO did not dissolve neither in 0.5 nor in 0.05M NaOH, the ZnO mass recovered completely, whereas it dissolved in 1M HCl solution. The comparison with the percentage recovery values of zinc oxide and silica that was reported by Kawara et al. [1] shows that the ZnO nanoparticles have better stability to contact with higher pH solutions than silica gel.

Table 1. Percentage recovery of the zinc oxide nanopowder after contact with a medium (HCl and NaOH) at different concentration and pH during 48 h at 40 °C.

Material	HCl concentration (M)	pH	Recovery (%)	NaOH concentration (M)	pH	Recovery (%)
ZnO	1	0	0	0.5	14	100
	0.1	1	96.3	0.05	13	100
Silica gel*	1	0	95.5	0.5	14	0
	0.1	1	98.5	0.05	13	73.0

*Values reported by Kawara et al. for silica gel [1].

3.3. Adsorption isotherm measurement

The adsorption isotherm of zinc oxide nanoparticles was measured by shake-flask method; the adsorbed methylparaben per unit of adsorbent weight (Cs, nmol/mg) versus methylparaben concentration in the supernatant (CM, M) was plotted as can be seen in Figure 5. These experimental points were fixed to a straight line using Henry Equation (2) [1,10]:

$$C_s = K_d * C_M \tag{2}$$

Where K_d is the distribution constant. From the fixed straight-line slope, 0.0144Lg^{-1} was estimated for this parameter, which is similar to 0.0188Lg^{-1} reported for titania [1], that have much larger adsorption capacity than silica gel.

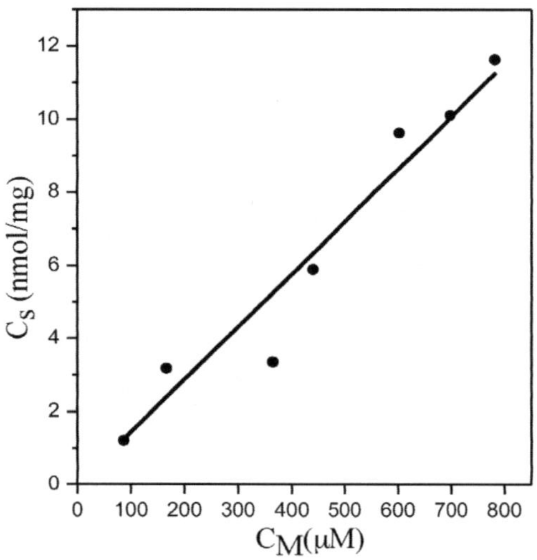

Figure 5. Adsorption isotherm of the zinc oxide nanopowder. The solid line corresponds to the best fit to equation (2).

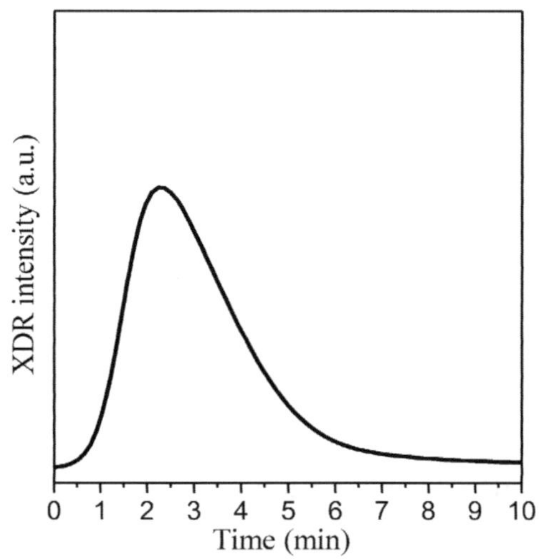

Figure 6. Caffeine chromatogram measured with the column filled with the zinc oxide nanopowder.

3.4. HPLC analysis

The chromatogram obtained from the caffeine in Figure 6 shows a retention time of 2.23 minutes. The analysis conditions for separation profiles of caffeine were column temperature: 40°C, flow-rate: 1.2mL min^{-1}, mobile phase composition (acetonitrile-water): 9:1 and absorption wavelength: 253nm.

4. Conclusion

Zinc oxide spherical nanocrystals with an average radio of 20.3±02nm were synthesized using sol-gel technique without sintering process. The found data suggests that this nanopowder has better properties than silica gel for application in HPLC separation; for instance, it has high resistance to basic pH eluents (12.69-13.69) and 23% more of adsorption capacity that silica gel. Chromatographic analysis with this nanopowder as a stationary phase was acceptable for caffeine separation.

Acknowledgements

The authors thank to the University or Quindío for financial support through the 805th project and the research group's incentive program. Special thanks to Luis Salazar from Chemistry department for the technical assistance with TGA, DSC and HPLC measurements and to Dr. Liliana Tirado and Angélica Marcela Castillo from Interdisciplinary Institute of Sciences for the XRD measurements.

References

[1] Kawara M, Nakamura H and Nakajima 1990 *J. Chrom.* **515** 149
[2] Skook D A, Holler F J, Nieman T A 2001 *Principles of Instrumental Analysis* vol 5 (Madrid: McGraw Hill) p 735
[3] Tena R 2004 *The oxide of zinc: crystalline growth by means of transport in gaseous phase and characterization of physical estates* (Valencia: Universidad de Valencia) p 23
[4] Gel J, Zhao L, Chen L and Yan-Ping 2010 *J. Chrom.* **48** 29
[5] Huang H, Zhou Y, Zhou J and Feng 2014 *J. Chrom* **5** 4
[6] Li Wen, ER Shi, and Tian Ming 2003 *J. Mater. Res.* **4** 1532-1537
[7] Guo M, Zhang Y, Zhang M, Wang X and Du X 2015 *J. Chrom.* **1384** 28-36
[8] Chandrasekaran P, Viruthagiri G and Srinivasan N 2012 *J. Alloys* **540** 89
[9] Talam S, Karumuri S, and Gunnam N 2012 *ISRN. Nanotechnology.* **2012** ID 372505 6
[10] Kawahara M, Nakamura H and Nakajima 1988 *Analytical sciences* **4** 671-673

Systematic study of inorganic functionalization of ZnO nanorods by Sol-Gel method

J K Gamarra[1], C Solano[2,3], I Piñeres[1], H Gómez[2], J Mass[2] and D N Montenegro[1]
[1] Universidad del Atlántico, Barranquilla, Colombia
[2] Universidad del Norte, Barranquilla, Colombia
[3] Universidad de la Costa, Barranquilla, Colombia

E-mail: dianamontenegro@uniatlantico.edu.co

Abstract. A systematic study of the inorganic surface functionalization of ZnO nanostructures by sol-gel method is shown. We have emphasized on the evolution of morphology properties of samples as a function of functionalization parameters. In addition, the effects on thermal stability and some optical properties of samples are discussed.

1. Introduction

During many decades gas sensors have been required for many practical applications, such as explosives and drugs detection, control of air quality and food safety. In addition, more recently these devices have attracted a lot of attention in the medical field. This is because gas sensors can be used to make medical diagnosis and monitoring of some clinical treatments by detecting the different organic volatile compounds (OVCs) content in human breath [1,2]. In the last decades a wide diversity of gas sensing technologies has been developed among them, it is worth noting the gas sensors based on the variation of electrical properties of metal oxide semiconductors (MOS). This is due to MOS based gas sensors offer several technological advantages, e.g., direct electronic transducing, high sensitivity, low cost and suitability to portable devices fabrication [3,4].

In the last years, different investigations have been established in order to optimize some essential properties of MOS based gas sensors such as sensitivity, selectivity and work temperature [3,5]. In this regard, several studies have demonstrated that the sensitivity of these devices can be significantly enhanced by using nanostructures instead thin films [6,7]. However the effective gas response of MOS nanostructures depends on their shape and size. Among a variety of OMS nanostructures, quasi-1D nanostructures (nanowires, nanorods and nanobelts) have shown the best gas response because these nanostructures exhibit higher surface-volume ratio which favours absorption and desorption process of gas molecules on the material surface [6,7]. Nevertheless, in spite of these substantial advances, the enhancing of sensitivity and selectivity parameters continues to have key importance in research and development of gas sensors [5,6]. In this context, surface functionalization of sensing materials has proved to be a promising path to improve these gas sensors characteristics. In the case of MOS nanostructures, functionalization process has been usually done by coating the nanostructures surface with noble metals (nanoparticles or thin films) [8,9]. Noble metals are used to coat OMS surfaces because they are high-effective oxidation catalyst and this characteristic can enhance the reactions on gas sensor surfaces. Although some sensor properties can be improved using this kind of surface modification, e.g., sensitivity [8,9], new functionalization routes are still required in order to enhance additional gas sensor properties as selectivity. Indeed, much effort is doing nowadays with the aim of achieving effective functionalization process with accurate control, reliability, thermal stability and low cost.

In view of the above, in this work we present a systematic study of inorganic surface functionalization of ZnO nanorods. Samples were TiO_2 coated by low-cost method. We have emphasized on the thermal stability and procedure of surface functionalization in order to allow a suitable control on the morphological and physical properties of functionalized nanostructures. ZnO and TiO_2 were selected due to their interesting properties related to gas sensing (thermodynamic stability, surface chemistry, humidity sensitive, suitable electronic configuration and special catalytic activity) which can be enhanced in combination [10]. Moreover ZnO exhibits additional interesting properties such as piezoelectricity, biocompatibility and non-toxicity, making the ZnO nanostructures attractive candidates for biomedical and biosensor applications [11].

2. Experimental

Vertically aligned ZnO nanorods were grown by metal organic chemical vapour deposition system without catalyst on c-sapphire substrates according to the procedure described in our previous work [12]. TiO_2 functionalization was done by wet chemical method using titanium, acetic acid and propylene glycol precursors. In this experiment, a 0.1M aqueous ethanol solution of titanium was kept under constant stirring using a magnetic stirrer for 2 hours at 100°C. Simultaneously aqueous ethanol solution of acetic acid, polyvinyl alcohol and distilled water was also prepared in the same way by stirring during 10 minutes. Both solutions were carefully mixed using magnetic stirrer for 5 minutes at room temperature and subsequently deposited on ZnO nanorods by spin coating. The thickness of TiO_2 layers was obtained by making several coatings on nanorods (e.g., 2, 4 and 6). After each coating, samples were dried at 200°C during 10 minutes. Coated samples were finally annealed between 500-600°C during 2 hours to reach the required TiO_2 structural phase.

In accordance with the samples characterization, morphologies of samples were examined by scanning electron microscopy (SEM-Jeol JSM 5600). Thermogravimetric analysis was performed using a TA Instruments 2050. For these measurements, the samples were heated at rate of 10°C/min and pure Ni gas (99.999%) with a 60mL/min flux was used for all experiments. Crystalline structure was analysed by powder X-ray diffraction (Bruker, D8-Advance) while optical properties were evaluated by UV-visible spectrophotometer Shimadzu UV-2600 equipped with an integrating sphere and 300-800nm wavelength range.

3. Results and discussion

The influence of spin-coating parameters on morphology and optical properties of functionalized ZnO nanorods have been evaluated. This was done by means of systematic studies by varying coating parameters such as spin frequency, number of coatings (i.e., film thickness) and annealing temperature of the deposited TiO_2 layers (see Table 1).

Typical 2θ X-ray diffraction (XRD) scans belong to pristine ZnO nanorods and TiO_2 layers are shown in Figures 1(a) and 1(b), respectively. For pristine ZnO nanorods samples, the diffraction peaks at 34.5 and 71 were indexed to (0002) and (0004) crystal planes of the wurtzite hexagonal structure of ZnO (lattice constants of a=0.324982nm and c=0.520661nm). This diffractogram evidences clearly the preferential c-orientation of nanorods as it is usually reported in literature. Otherwise, exhibited peaks in XRD of TiO_2 layers were ascribed to anatase structure, which is known as the more suitable phase of TiO_2 for gas sensor applications [13]. Figure 2 exhibits the thermogravimetric (TG) measurements of samples before and after functionalization process. For pristine and functionalized samples TG analysis reveals insignificant mass-loss (~0.1%) between room temperature and 400°C. These results show that our samples present only minor desorption of the surface-adsorbed water and the chemical residual species, which can remained after the TiO_2 layers growth. In addition, because chemical reactions between ZnO and TiO_2 are no evidenced in TG scans, TiO_2 functionalization can be seen as thermodynamically stable into the studied temperature range, which is included in the work temperature range of the most gas sensors based on OMS materials [6,7].

Table1. Deposition parameters used for functionalized samples.

Sample	Spin Frequency (rpm)	Annealing Temperature (°C)	Number of Coatings
FZnO1-2C	2000	600	2
FZnO2-4C	2000	600	4
FZnO3-6C	2000	600	6
FZnO4-2C	1000	500	2
ZnO5A	----	600	0

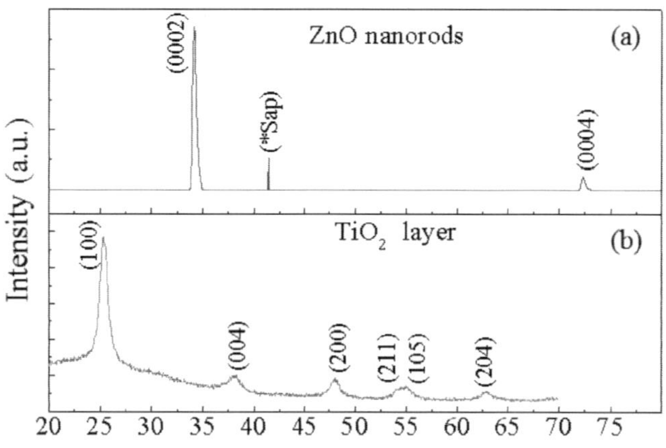

Figure 1. Typical 2θ X-ray diffraction scans belong to (a) ZnO nanorods and (b) TiO_2 layers. The diffraction peak labelled with a star comes to c-sapphire substrate of ZnO nanorods samples.

Morphology properties of samples were examined by scanning electron microscopy (SEM). Figure 3 illustrates the typical SEM images of pristine ZnO nanorods (left column) and functionalized ZnO nanorods (right column). For pristine samples, SEM images show vertically aligned ZnO nanorods with hexagonal facets. Nanorods exhibit diameters between 150nm and 850nm and lengths between 7μm and 15μm. In the case of functionalized samples, samples have been analyzed taking in account the thickness of deposited TiO_2 layers, namely, sample with 2 coatings (FZnO1-C2, Figures 3(g)-(h)); sample with 4 coatings (FZnO2-C4, Figures 3(i)-(j)) and sample with 6 coatings (FZnO3-C6, Figures 3(k)-(l)). TiO_2 layers with smoothed but cracked surface can be observed in the top-view SEM images of all functionalized samples. However, vertically aligned ZnO nanorods are not present in samples with major number of coatings (4 and 6 times), see cross-view SEM images. Indeed, the analysis of results has allowed to determine that ZnO nanorods tend to be detached from the sample surface because spin-coating process and hence the fallen nanorods are agglomerated with TiO_2 layers (insets Figure 4(i) and Figure

4(j)). Moreover, the detaching of ZnO nanorods from surface sample, when TiO_2 films are spin-coated, can be the principal origin of layer cracking observed in our functionalized samples.

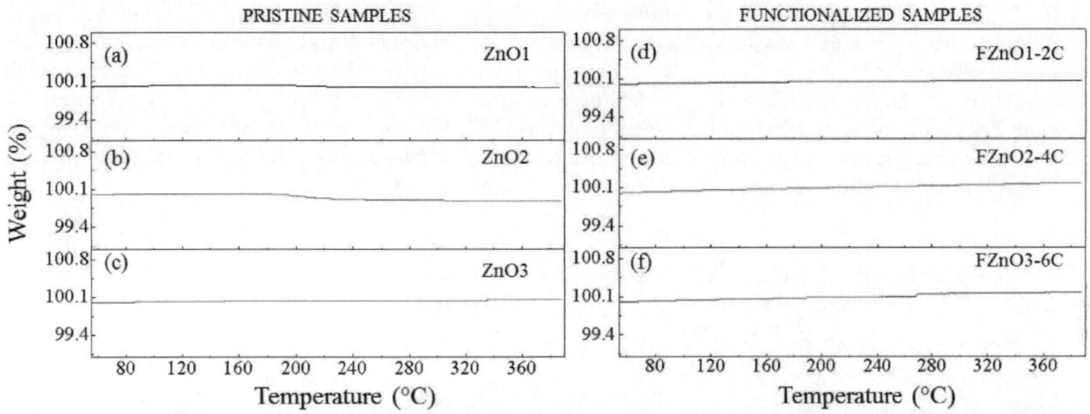

Figure 2. Thermogravimetry measurements of (a)-(c) pristine ZnO nanorods and TiO_2-coated ZnO nanorods (d) 2; (g) 4 and (f) 6 times.

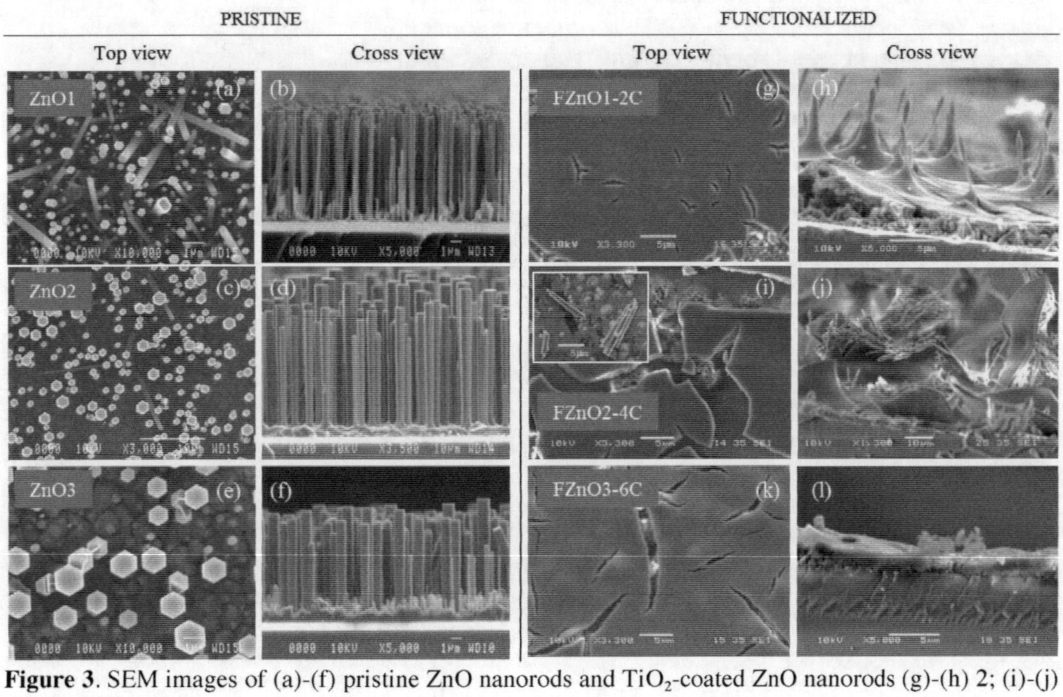

Figure 3. SEM images of (a)-(f) pristine ZnO nanorods and TiO_2-coated ZnO nanorods (g)-(h) 2; (i)-(j) 4 and (k)-(l) 6 times.

In this regard, our systematic study also has enabled to establish that this latter issue can be lessened by varying the spin process during layer deposition. SEM images of surface functionalized samples with minor number of coatings (2 times), i.e., sample with minor full rotation time, evidence that vertically aligned ZnO nanorods can remain on sample after Ti_2O coating, Figure 3(h). Similarly, Figures 4(a) and 4(b) shows that TiO_2 layer can be deposited on the top of nanorods arrays, without losing the vertical alignment of nanostructures, by halving the spin frequency parameter. The annealing process done to TiO_2 layers for its part has revealed to induce undesirable coalescence effects in ZnO nanorods. The analysis of pristine ZnO nanorods annealed between 500°C and 600°C, has shown this thermal procedure tends to favour the lateral growth of nanorods giving rise to a loss of self-assembly ZnO nanorods arrays and the characteristic high surface-volume ratio of nanorods, Figures 4(b) and 4(d).

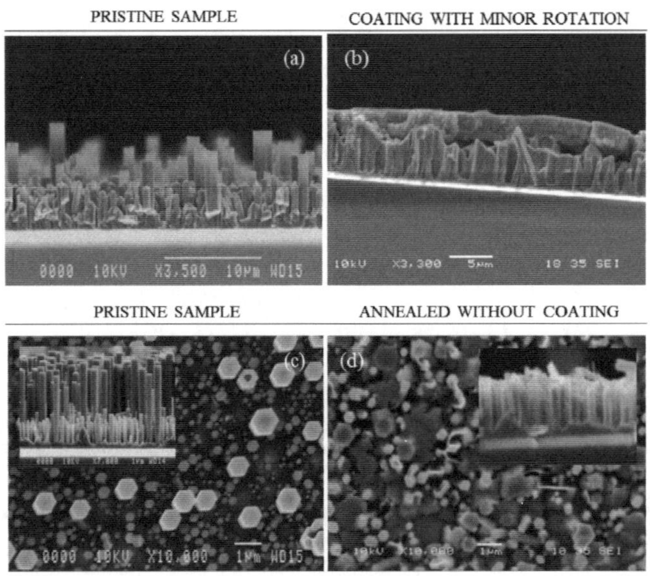

Figure 4. SEM images of pristine ZnO nanorods (a) and (c); TiO_2-coated ZnO nanorods with spin frequency at 1000 rpm (b) and ZnO nanorods annealed at 600°C.

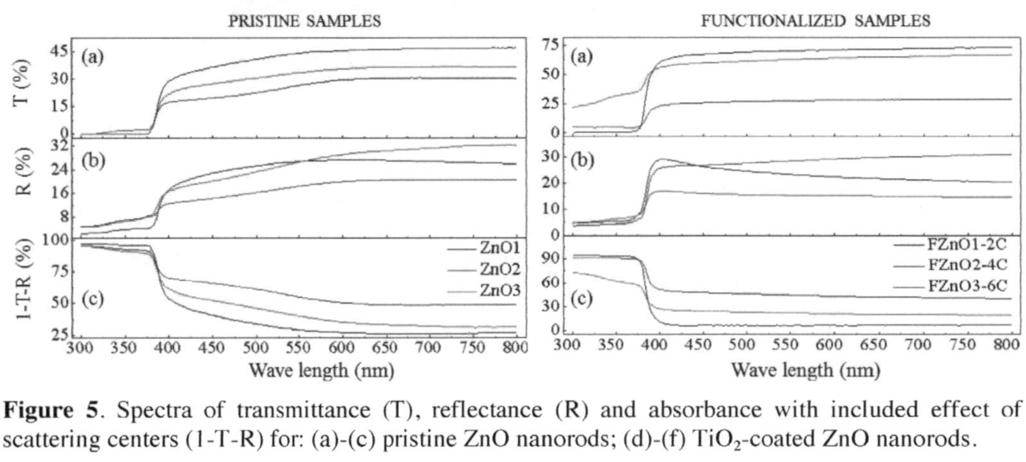

Figure 5. Spectra of transmittance (T), reflectance (R) and absorbance with included effect of scattering centers (1-T-R) for: (a)-(c) pristine ZnO nanorods; (d)-(f) TiO_2-coated ZnO nanorods.

Optical properties of pristine and functionalized ZnO nanorods have been also investigated. Transmittance and reflectance spectra of samples were acquired at room temperature using UV-vis spectrophotometer. Absorbance spectra were calculated from the transmittance data taken in consideration scattering centres due to ZnO nanorod arrays observed in our samples [14]. The transmittance, reflectance spectra in Figures 5(a)-(b) and calculated absorbance in Figure 5(c) for the different pristine samples, reveal a similar behaviour when the energy of excitation photons exceeds the ZnO band gap ($\lambda<380$nm). The major variations of light transmittance, reflectance and absorbance from these pristine samples occur within visible light region ($400<\lambda<800$nm). This can be primarily attributed to the morphology differences between samples, that is, diameter, length, density (#nanorods/area) and vertical alignment of ZnO nanorods (Figures 3 (a)-(f)). In the other hand the transmittance spectra, for functionalized ZnO nanorods, displays major variations and a decrease in visible range. This behaviour is observed to be enhanced for the sample coated with the thickest TiO_2 layer (FZnO1-C6), Figure 5(d). We are certain of this transmittance decrease is caused because reduction of scattering centres in our functionalized samples according with the undergone detaching of ZnO nanorods. That is why the optical spectra of these samples seem to be principally governed by TiO_2 material.

4. Summary and conclusions

Vertically aligned ZnO nanorods were surface functionalized with TiO_2 layers by using optimized spin-coating parameters. Thermogravimetry measurements have allowed showing that TiO_2 functionalization is thermodynamically stable in a wide temperature range, 25-400 °C. The analysis of the functionalization process has enabled to conclude that coating conditions such as spin and annealing process have key importance in development of MOS nanostructures surface-functionalization by spin-coating method. Both parameters have shown to affect significantly some morphology properties of functionalized ZnO nanorods samples (vertical alignment and the aspect ratio (diameter/volume) associated directly with absorption and desorption process of gas molecules on MOS nanostructures surface.

References

[1] Di Natale C, Paolesse R, Martinelli E, Capuano R 2014 *Anal. Chim. Acta* **824** 1
[2] Righettoni M, Amann A, Pratsinis S 2015 *Materials Today* **18** 163
[3] Chengxiang W, Longwei Y, Luyuan Z, Dong X, Rui G 2010 *Sensors* **10** 2088
[4] Korotcenkov G 2007 *Mater. Sci. Eng. B* **139** 1
[5] Liu X, Cheng S, Liu H, Hu S, Zhang D, Ning H 2012 *Sensors* **12** 9635
[6] Barth S, Hernandez-Ramirez F, Holmes J D, Romano-Rodriguez A 2010 *Prog. Mater Sci.* **55** 563
[7] Arafat M M, Dinan B, Akbar S A, Haseeb A S M A 2012 *Sensors* **12** 7207
[8] Wang D, Ma Z, Dai S, Liu J, Nie Z, Engelhard M H, Huo Q, Wang C, Kou R 2008 *J. Phys. Chem. C* **112** 13499
[9] Ruiz A M, Cornet A, Shimanoe K, Morante J R, Yamazoe N 2005 *Sens. Actuat. B* **108** 34
[10] Kalyamwar V S, Raghuwanshi F C 2013 *Adv. Mat. Lett.* **4** 895
[11] Mahbubur Rahman Md, ASaleh Ahammad A J, Jin J-H, Jung Ahn S, Lee J-J 2010 *Sensors* **10** 4855
[12] Montenegro D N, Souissi A, Martínez-Tomás C, Muñoz-Sanjosé V, Sallet V 2012 *J. Cryst. Growth* **359** 122
[13] Tang H, Prasad K, Sanjines R, Levy F 1995 *Sens. Actuat. B* **26** 71
[14] Fekete M, Ludwig W, Gledhill S, Chen J, Patti A, Spiccia L 2014 *Eur. J. Inorg. Chem.* **1** 75

Synthesis, characterization and magnetic evaluation of praseodymium modified cerium oxide

A F Cruz Pacheco[1], J A Gómez Cuaspud[1] and C A Parra Vargas[1]
[1] Universidad Pedagógica y Tecnológica de Colombia, Tunja, Colombia

E-mail: andresfelipe.cruz@uptc.edu.co

Abstract. This paper report the synthesis, characterization and magnetic evaluation of $Ce_{0.6}Pr_{0.4}O_2$ system using a chemical route that involves the combustion of citrate species. The precursors were characterized by Fourier transform infrared spectroscopy (FTIR), to identify, the main vibrational bands associated with the organic material after combustion process. The structural analysis by X-ray diffraction (XRD) revealed the obtaining of an Fm-3m (225) cubic phase with a particle size around 10nm. Finally, the measurement of the magnetization as a function of applied magnetic field and of susceptibility showed a perfect paramagnetic behaviour.

1. Introduction

With the progress of technology, there are new challenges for the development of advanced ceramic materials, in which the rare earth elements have shown unusual and interesting properties for a wide application field. In this context, the cerium is particularly one of the most important and abundant elements in nature and just as praseodymium are the only elements of this group that are stable in two oxidation states Ce^{4+}, Ce^{3+}, Pr^{4+} and Pr^{3+}. The CeO_2 and Pr_6O_{11}, are the most common oxides of these elements and their importance focuses on the possibility to participate in redox reactions and the storage capacity and controlled release of oxygen through the conversion between the main oxidation states respectively. The oxygen vacancies that arise from this conversion can be chemically manipulated by the combination of the two compounds originating materials with electronic, structural, mechanical, optical, catalytic and magnetic properties for technological applications. Some of the most important properties of these oxides have focused on the field of catalysis and photocatalysis [1,2], in development of oxygen sensors and biomaterials [3,4], and design of components for solid oxide fuel cells SOFC [5,6]. Commonly these materials are prepared by solid state reaction, however the low homogeneity of products and the restricted kinetic associated with the synthesis process and the loss of textural, surface and morphological properties, make the current trend is to use alternative routes leading to a better control of composition, structure and purity of phase at nanometric scale. In this sense, the combustion method is one of the most viable for its simplicity, speed and economy, reason why, present work pretend to show the advantage of combustion process in the obtaining of nanostructured oxides based on $Ce_{0.6}Pr_{0.4}O_2$ system [7–9].

2. Experimental

Stoichiometric amounts of $Ce(NO_3)_3.6H_2O$ (99.9%), $Pr(NO_3)_3.6H_2O$ (99.9%) and citric acid monohydrated (99.9%) were used in the preparation of the $Ce_{0.6}Pr_{0.4}O_2$ system, by the combustion method. The oxide was synthesized starting from solutions of nitrates of Ce^{3+} and Pr^{3+}, (Previously validated [10]), using a glass reactor equipped with magnetic stirring (150rpm), temperature control and reflux, as described elsewhere [11]. Once adjusted the corresponding dissolutions, the citric acid was added in a molar ratio 1:0.5, with respect to the total concentration of cations. The system was kept under reflux at 140°C for 12 hours until formation of a viscous liquid, which was then heated at 200°C under air flux in an oven until complete solvent evaporation and treated at 310ºC to initiate the combustion process. The precursor formed was treated at 500°C to remove the remaining carbon under oxygen flow for 2 hours.

The metal-organic precursor was analysed by infrared spectroscopy FTIR, using a Shimadzu FTIR prestige-21 equipment in ATR configuration (attenuated total reflectance), in order to assess the formation of citrate complexes and allow the verification of the effectiveness of the synthesis process and the chemical composition. The crystalline structure of the solid after calcination of precursor, was evaluated by X-ray diffraction (XRD), on a PANalytical X'Pert PRO MPD diffractometer, equipped with Ultra-Fast X´celerator detector and Bragg-Brentano configuration, using Cu Kα radiation (λ=1.54186Å) between 20° and 90°. The measurements were developed with a voltage of 40kV and a current of 20 mA in all cases. The crystallite size was calculated using the highest diffraction signals by the Debye-Scherrer equation, taking the value of half peak width set by a Lorentzian function and using a constant of 0.89 as reference [12]. The magnetization hysteresis (M-H) loop and zero field cooling (ZFC) studies, of the $Ce_{0.6}Pr_{0.4}O_2$ system, was carried out by using a vibrating sample magnetometer VersaLab (VSM). The magnetization measurements were performed at several magnetic fields (500Oe–5000Oe), and the magnetic hysteresis measurements were done at different temperatures (100 and 300K) [13].

3. Results and discussion

FTIR spectrum of the $Ce_{0.6}Pr_{0.4}O_2$ precursor before and after combustion process is shown in Figure 1.

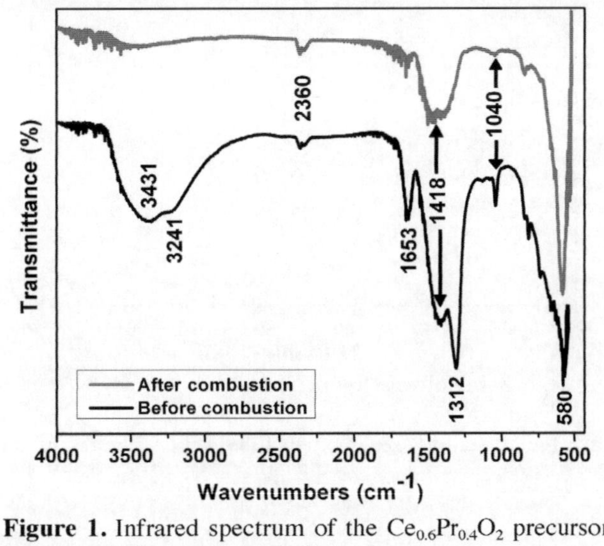

Figure 1. Infrared spectrum of the $Ce_{0.6}Pr_{0.4}O_2$ precursor before and after the combustion process.

The spectrum of precursor before combustion process, shown characteristic absorption bands associated with a polymeric citrate complex, which confirms the formation of coordination compounds between metals and citrate species as reported previously by Wahba [14]. The intense absorption band at 3431cm^{-1} is attributed for (O-H) vibration modes of the water molecules coordinated to the complex. The band stretching at 3241cm^{-1} is assigned for vibrational mode of (N-H) bond, which is associated with the addition of ammonia to dissolution for pH adjustment. The weak band observed at 2360cm^{-1} is associated with the presence of CO_2, which it is occluded in the materials during the synthesis [15]. The band located at 1653cm^{-1} is related with the asymmetric deformation mode of carboxyl group of mono-dentate coordination compounds and the bands attributed the symmetric stretching at 1418 and 1312cm^{-1} for the respective (COOH) groups, corresponds to a strong stretch mode associated to bidentate coordination compounds [16,17].

The absorption bands observed at 1041cm^{-1} corresponds the stretching modes of the (CO) bonding associated with acetals and primary alcohols, as well as bends out of plane of (C-H) bond [18]. The absorption peaks in the region 900–500cm^{-1} are associated with the bond of the metal cations with oxygen atoms. The characteristic band at around 580cm^{-1} is assigned for the symmetric stretching vibration of Ce-O-Ce, Ce-O-Pr or Pr-O-Pr [19–21]. While, there is still presence of absorption bands related to symmetrical and asymmetrical deformations of the groups (COOH) and a band to lower frequencies corresponding to the vibrational modes of the links C-O associated acetals and primary aliphatic alcohols. So, the intense band at 580cm^{-1}, indicate that the cerium and praseodymium cations are tightly bound with oxygen [22], which implies that synthesis process has a significant decrease in terms of time and efficiency.

Structural identification of $Ce_{0.6}Pr_{0.4}O_2$ system was determined using XRD as shown in Figure 2. XRD patterns reveal the formation of a single phase with cubic structure type fluorite and space group Fm-3m (225) with a lattice constant a=0.5411nm, according with the reference compound of CeO_2, ICDD code 00-034-0394.

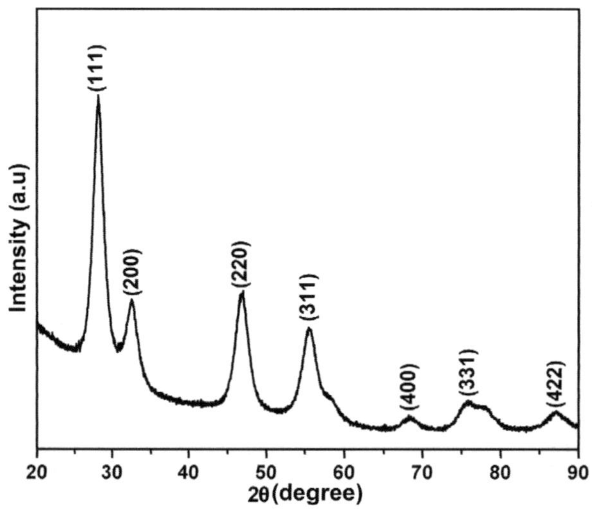

Figure 2. X-ray diffraction pattern of $Ce_{0.6}Pr_{0.4}O_2$ system.

According to XRD data, the crystallite size (D) was to be about 10 nm and was calculated using the Scherrer formula:

$$D = \frac{0.89\lambda}{\beta cos\theta} \tag{1}$$

Where λ was the X-ray wavelength, β was the width of the peak highest intensity considered in the calculation and θ was the Bragg diffraction angle of the main reflection (1 1 1) [23].

The broadening of the peaks is due to the incorporation of Pr to cerium oxide structure, which leads to an increase in the solids density and heterogeneity in the structural composition at nanoscopic level [24]. These phenomena are due to the ionic radius of Pr^{3+} (1.126Å), which is much larger than Ce^{4+} and Ce^{3+} (1.034Å) cations. However, the Pr^{4+} (0.96Å) radius is slightly smaller than that for Ce^{4+} (0.97Å)

[25–27], therefore the difference of ionic radius cause a change in the Ce–O bond length and in the crystal size of the sample [28].

Figure 3 shows the experimental data of the magnetization (M), as a function of applied magnetic field (H) for $Ce_{0.6}Pr_{0.4}O_2$ system measured at 100 and 300K. The sample exhibits a perfect paramagnetic behaviour at 100 and 300 K since the magnetization is reversible and has a linear curve.

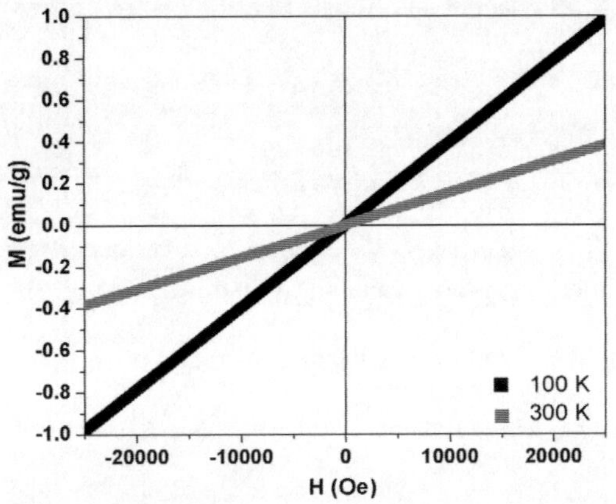

Figure 3. Magnetization (M) of $Ce_{0.6}Pr_{0.4}O_2$ system as a function of the applied magnetic field (H) at 100 and 300K.

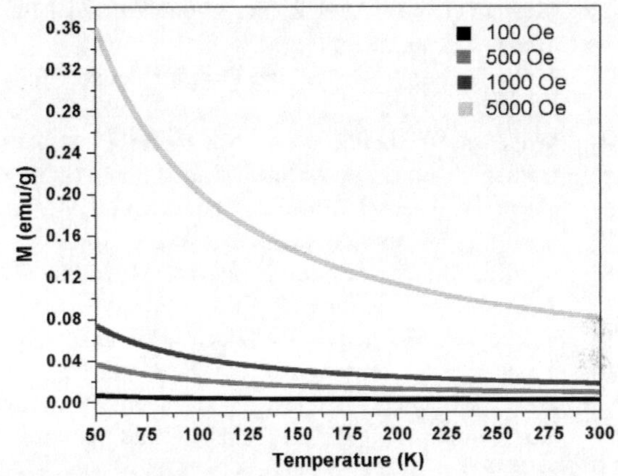

Figure 4. Magnetic susceptibility as a function of temperature for $Ce_{0.6}Pr_{0.4}O_2$ in the magnetic fields 100, 500, 1000 and 5000Oe.

The temperature dependence of the magnetic susceptibility for $Ce_{0.6}Pr_{0.4}O_2$ was performed at four magnetic fields (100, 500, 1000 and 5000Oe) as shown in Figure 4. All ZFC curves present a paramagnetic behaviour, with increased susceptibility when a temperature decreases. The behaviour that presents the sample is due to the incorporation of praseodymium in the fluorite structure of cerium oxide. The Pr^{4+} and Pr^{3+} ions have net magnetic moments that make them paramagnetic [29]. In the same way, the paramagnetic component also comes from the Ce^{3+} ions due to the one unpaired electron in a $4f$ orbital [30]. Furthermore, the diamagnetic behaviour of Ce^{4+} ions not present in measures magnetization vs applied field and the ZFC curves, so is at low proportion [31].

4. Conclusions

The $Ce_{0.6}Pr_{0.4}O_2$ system, was prepared by the combustion method, obtaining material at the nanometric order. The FT-IR analysis of metal-organic precursors, allowed evaluate the main absorption bands related with the formation of polymeric complexes between metal cations and citrate species. The structural characterization evaluated by X-ray diffraction, confirms obtaining an oxide with structure fluorite type according to the reference compound. The crystal size of sample was 10nm, using the Debye–Scherrer equation. The magnetic response produced by $Ce_{0.6}Pr_{0.4}O_2$ system was attributed to paramagnetic behaviour due to Ce^{3+}, Pr^{3+} y Pr^{4+} ions have a net magnetic moment.

References

[1] Nakagawa K, Ohshima T, Tezuka Y, Katayama M, Katoh M and Sugiyama S 2015 *Catal. Today* **246** 67

[2] Contreras-García M E, García-Benjume M L, Macías-Andrés V I, Barajas-Ledesma E, Medina-Flores A and Espitia-Cabrera M I 2014 *Mater. Sci. Eng. B* **183** 78

[3] Chen C Y, Chang K H, Chiang H Y and Shih S J 2014 *Sensors Actuators B Chem.* **204** 31

[4] Kolli M B, Manne N D P K, Para R, Nalabotu S K, Nandyala G, Shokuhfar T, He K, Hamlekhan A, Ma J Y, Wehner P S, Dornon L, Arvapalli R, Rice K M and Blough E R 2014 *Biomaterials* **35** 9951

[5] Taguchi H, Chiba R, Komatsu T, Orui H, Watanabe K and Hayashi K 2013 *J. Power Sources* **241** 768

[6] Stojadinović B, Dohčević-Mitrović Z, Paunović N, Ilić N, Tasić N, Petronijević I, Popović D and

Stojanović B 2016 *J. Alloys Compd.* **657** 866
[7] Chen B, Rao G, Wang S, Lan Y, Pan L and Zhang X 2015 *Mater. Lett.* **154** 160
[8] Deganello F, Marcì G and Deganello G 2009 *J. Eur. Ceram. Soc.* **29** 439
[9] Rajpure K Y 2015 *Superlattices Microstruct.* **77** 181
[10] Gómez-Cuaspud J A and Valencia-Ríos J S 2009 *Rev. Colomb. Química* **38** 289
[11] Rodrigues C P and Schmal M 2011 *Int. J. Hydrogen Energy* **36** 10709
[12] Mustafa G, Islam M U, Zhang W, Jamil Y, Asif Iqbal M, Hussain M and Ahmad M 2015 *J. Magn. Magn. Mater.* **378** 409
[13] Mejía-Gómez J A, Supelano-García I, Palacio C A and Parra-Vargas C A 2015 *J. Phys. Conf. Ser.* **614** 012003
[14] Wahba A M, Aboulfotoh-Ali N and Eltabey M M 2014 *Mater. Chem. Phys.* **146** 224
[15] Gómez-Cuaspud J A, Valencia-Ríos J S and Carda-Castelló J B 2010 *J. Chil. Chem. Soc.* **55** 445
[16] Chen Y and Xiao J Z 2013 *Ceram. Int.* **39** 9599
[17] Shaikh P A, Kambale R C, Rao A V and Kolekar Y D 2010 *J. Alloys Compd.* **492** 590
[18] da Silva M F P, de Souza-Carvalho F M, da Silva-Martins T, de Abreu-Fantini M C and Isolani P C 2010 *J. Therm. Anal. Calorim.* **99** 385
[19] Vanhoyland G, Pagnaer J, D'Haen J, Mullens S and Mullens J 2005 *J. Solid State Chem.* **178** 166
[20] Yang W D, Chang Y H and Huang S H 2005 *J. Eur. Ceram. Soc.* **25** 3611
[21] Esther Jeyanthi C, Siddheswaran R, Kumar P, Siva Shankar V and Rajarajan K 2014 *Ceram. Int.* **40** 8599
[22] Ho C, Jimmy Y, Kwong T, Mak A C and Sukyin L 2005 *Chem. Mater.* **17** 4514
[23] Li X, Feng Z, Lu J, Wang F, Xue M and Shao G 2012 *Ceram. Int.* **38** 3203
[24] Gerasimov E Y, Isupova L A and Tsybulya S V 2015 *Mater. Res. Bull.* **70** 291
[25] Matovic B, Boskovic S, Logar M, Radovic M, Dohcevic-Mitrovic Z, Popovic Z V and Aldinger F 2010 *J. Alloys Compd.* **505** 235
[26] Shannon R D 1976 *Acta Crystallogr. Sect. A* **32** 751
[27] Chiba R, Taguchi H, Komatsu T, Orui H, Nozawa K and Arai H 2011 *Solid State Ionics* **197** 42
[28] Phokha S, Prabhakaran D, Boothroyd A, Pinitsoontorn S and Maensiri S 2014 *Microelectron. Eng.* **126** 93
[29] Kern S 1964 *J. Chem. Phys.* **40** 208
[30] Niu G, Hildebrandt E, Schubert M A, Boscherini F, Zoellner M H, Alff L, Walczyk D, Zaumseil P, Costina I, Wilkens H and Schroeder T 2014 *ACS Appl. Mater. Interfaces* **6** 17496
[31] Bernardi M I B, Mesquita A, Béron F, Pirota K R, de Zevallos A O, Doriguetto A C and de Carvalho H B 2015 *Phys. Chem. Chem. Phys.* **17** 307

Effect of modification substrate on the microstructure of hydroxyapatite coating

J Realpe-Jaramillo[1], J A Morales-Morales[1], J A González-Sánchez[2], R Cabanzo[3], E Mejía-Ospino[3] and J Rodríguez–Pereira[3]
[1] Universidad Santiago de Cali, Cali, Colombia
[2] Universidad Autónoma de Campeche, Campeche, México
[3] Universidad Industrial de Santander, Bucaramanga, Colombia

E mail: jim.ale.mor@gmail.com

Abstract. Bioactive hydroxyapatite (HA) coatings were fabricated by a precipitation, sol-gel and dip-coating method. The effects of the aging time and the base used to adjust pH and substrate materials on the phases and microstructures of HA coatings were studied by field emission scanning electron microscopy FESEM, energy dispersive spectroscopy EDS, X-ray photoelectron spectroscopy XPS, and the vibrations of the phosphate groups were determined by Raman spectroscopy. The results showed that all the films were composed of the phases of TiO_2 and HA. With coated titanium substrate with TiO_2, the crystallinity of the HA coating increases, the structure became more compact and the Ca/P ratio increased because of the loss of P in the films. The addition of sodium hydroxide (adjusting the pH level to about 10) can increase the HA content in the coating. XPS and EDS results for steel substrate and titanium showed poor calcium content as obtained with a Ca/P ratio of 1.38 and 1.58, respectively, composition is similar to that of natural apatite. However, spectroscopic results suggest the presence of a mixture of hydroxyapatite and octacalcium phosphate. The different substrate materials have a high influence on the microstructure of the separated double films. However, hydroxyapatite nanopowders coatings were obtained using a simple method, with potential biomedical applications.

1. Introduction

Bioactive materials based on Hydroxyapatite (HA) have been widely used in hard tissue applications, such as implant coatings [1,2] and bone replacements [3,4], due to its stability in vivo conditions and its osteoconductivity. HA $[Ca_{10}(PO_4)_6(OH)_2]$ is a biocompatible and bioactive ceramic material capable of interacting and chemical bond formation with surrounding bone. However, it can not be directly used for implant devices because of its brittleness and strength limitations and hence a great deal of research is dedicated to the development of HA coatings and composites [5,6].

Growing research for development of HA coatings has led to the introduction of some strategy as the use of titanium oxide coatings, layered coatings HA/TiO_2 and TiO_2 nanotubes obtained by anodic oxidation on Ti substrates. In this study, the effects of substrate modification on the microstructure of HA coatings were investigated.

2. Material and methods

2.1. Synthesis of coatings hydroxyapatite nanopowders

It began with surface preparation of rectangular plates of 304 stainless steel and Ti respectively, with dimensions of 2.0 and 4.0cm by 0.5mm thick. The polished plates were degreased, washed and dried at room temperature. Then an inorganic coating synthesis of TiO_2 by the sol-gel method was performed. For it is required a 100mL vessel which was added 16mL of ethanol (CH_3CH_2OH, EtOH, reagent grade, Sigma-Aldrich) and 1mL of acetyl acetone (Acach) as complexing agent. It was allowed to stir the solution after 15 minutes. 3.5mL of titanium butoxide ($(Ti(O(CH_2)_3CH_3))_4$), BuTi, Sigma-Aldrich, 97% by weight) which was stirred for 1 hour was added. Finally, it was added 0.7mL of deionized water as hydrolyzing agent that was 2 hours stirred. The resulting mixture was aged 7 days. The deposit of the films was performed by dipping or dip-coating, with immersion times of 1 minute and dried at 100°C during 1 hour. Plates were heated at 450°C at 3°C/minute and 4 hours in order to obtain a mixture of anatase and rutile phases. The synthesis of hydroxyapatite nanopowders was performed by controlled precipitation. Stoichiometric volume of 0.1M calcium nitrate tetrahydrate ($Ca(NO_3)_2{\cdot}4H_2O$, Sigma-Aldrich 99.2%) aqueous solution was added dropwise to 0.06M diammonium hydrogen phosphate ($(NH_4)_2HPO_4$, (Sigma-Aldrich 99.2%) aqueous solution. The solution was electromagnetically stirred, for continuous stirring, of the dispersed medium. During precipitation, the pH was monitored and adjusted at 10±0.1 by adding NaOH (Sigma-Aldrich). Meanwhile, sodium hydroxide (Sigma-Aldrich) was added immediately to adjust the pH reaction mixture. The pH value was kept constant throughout the experiment. After the addition, stirring was continued for 2 more hours, to homogenize the solution. The precipitate was aged for 3-9 days at room temperature, at rest. Finally, the coatings hydroxyapatite nanopowders were performed by dip-coating technique dipping the 304 stainless steel and Ti plates with TiO_2 coating solution within hydroxyapatite, after aging. The substrate was placed horizontally into the solution of HA for 2 hours, then extracted vertically and dried at room temperature for 12 hours, after that they were dried at 100°C for one hour and muffle calcined at 450°C/ for 4h at 2.0°C/min with a dwell time of 2h.

2.2. Coated Powder Characterization

The X-ray diffraction was performed on a D8 ADVANCE diffractometer (Bruker) with DAVINCI design, geometry theta - theta using monochromatic radiation of Cu tube and detector LINXEYE silicon. Preliminary tests were conducted to define appropriate difractional registration and field conditions. The analysis conditions were using a monochromator 0.02 NiO, collimator 0.2mm, step time 4s, 02 increases, and 15-90° difractional field. Identifying diffractograms was performed using the software DIFFRAC. EVA version 3. The morphologies of the samples were obtained using a field emission electron scanning microscope (FESEM) brand FEI QUANTA FEG model 650, with an acceleration voltage 10kV. To improve visualization of the carbon coating samples was performed on a computer model QUORUM Q R 15R. A detector for punctual EDS analysis was used to perform elemental chemical analysis. The analysis by micro-RAMAN spectroscopy was carried out with a team HORIBA Scientific LabRAM HR Evolution.

3. Results and discussion

3.1. Microstructural analysis

The peaks obtained by wet chemical route for HA particles are shown in Figure 1. From Figure 1 HA, XRD pattern showed higher consistency with standard references in JCPDS file available in software for hydroxyapatite (09-0432). XRD pattern indicates that HA crystals were produced on stainless steel and Ti surface by wet chemical processes when taking no account of the stainless steel and Ti peaks. Some characteristic peaks of HA can also be observed at the angles of 26° and 30°–35°, which belong to the (002), (211), (112), and (300) diffractions. It is clear that there is an orientation growth of (002) plane in HA, which is consistent with the observation by other bioactive coatings [5].

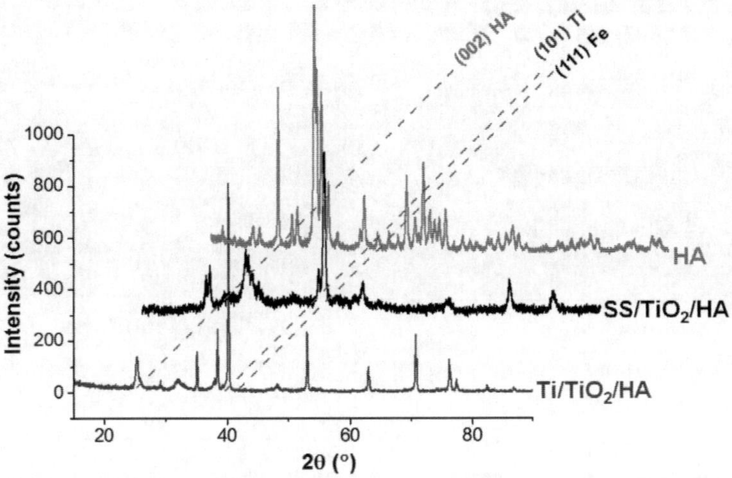

Figure 1. Growth hydroxyapatite (HA) coating on metal substrates (Stainless steel=SS and Titanium=Ti) with surface modified TiO_2 observed by XRD. The more intense peaks from HA (coating), Ti (substrate) and stainless steel (substrate) are highlighted in the graphic.

It is also found from Figure 1 pattern HA, that the diffraction peaks of (211), (112), and (300) planes are intensified going from titanium to steel. In pattern Ti/TiO_2/HA sample, the appearance of Ti reflections indicates that the coating deposited on the Ti substrate should be thin. Coatings showed a mixture of phases, the predominant phase in both cases, the reflections of higher intensity were anatase at the angles 26.9°, 39.3° and 49.7° which correspond to the (101), (112) and (202), respectively, in agreement with the crystallographic letter JCPDF 21-1272 and to a lesser extent is present in the rutile phase at the angles 28.7° and 37.6° corresponding to the (101) and (103) respectively, which coincide with the crystallographic letter JCPDF 29-1360. In the DRX pattern SS/TiO_2/HA sample, the presence of a peak (111) (2θ=43.1°) plane was detected and the signal at (200) (2θ=63.3°) indicates the presence of Fe in the γ-phase [6]. All the HA peaks were detected and indexed along with the iron peaks.

SEM images of HA coatings obtained on modified TiO_2 steel substrates (SS/TiO_2/HA) are presented in Figure 2. The images reveal considerable agglomeration among the particles, which increases the size of the crystalline powders. It also shows the morphology of the coatings of HA/TiO_2 deposited on steel at room temperature characterized by high density of microcracks in the coating of TiO_2, which could affect the uniform coating of HA growth. HA coating does not completely cover the steel substrate. In Figure 2, the TiO_2 coating areas with microarrays, porosity and formation of agglomerates were observed at the micrometer size. This morphology is well documented for TiO_2 thin coatings obtained by sol-gel [7].

Micrographic images of HA coatings obtained on modified TiO_2 titanium substrates (Ti/TiO_2/HA) are exhibited in Figure 3. The images reveal considerable agglomeration among the particles, which increases the size of the crystalline powders. It also shows the morphology of the HA/TiO_2 coatings deposited on titanium at room temperature characterized by high density of microarrays in HA coating, which may be related to microarrays in the coating of TiO_2 prepared by sol-gel which could affect the uniformity of HA growth coating. HA coating does completely cover the TiO_2 substrate coating areas with microarrays, porosity and formation of agglomerates were observed at the micrometer size.

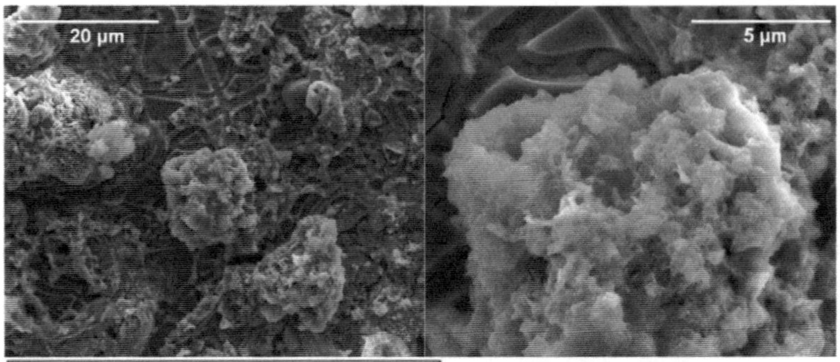

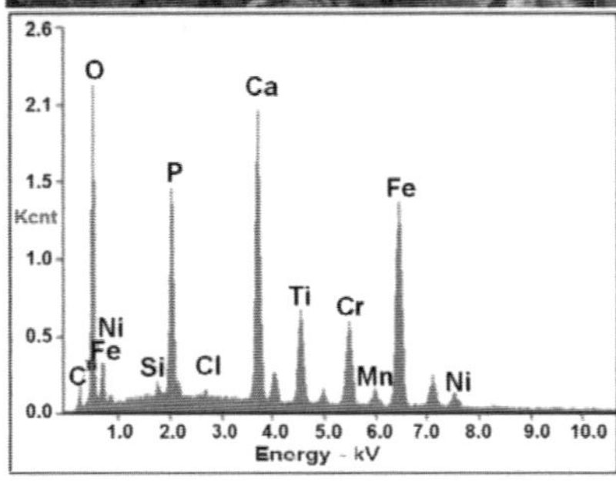

Element	*Wt%*	*At%*
CK	02.48	06.18
OK	25.69	47.99
SiK	00.55	00.58
PK	07.68	07.41
ClK	00.23	00.20
CaK	13.70	10.22
TiK	06.63	04.14
CrK	08.62	04.95
MnK	00.78	00.42
FeK	30.36	16.25
NiK	03.27	01.66
Matrix	Correction	ZAF

Figure 2. SEM images of hydroxyapatite (HA) coating on metal substrate steel=SS with surface modified TiO_2. EDS analysis of hydroxyapatite coated. Table shows EDS analysis of the atomic % of Ca, P and O and other elements present in the powders Ca/P=1.38).

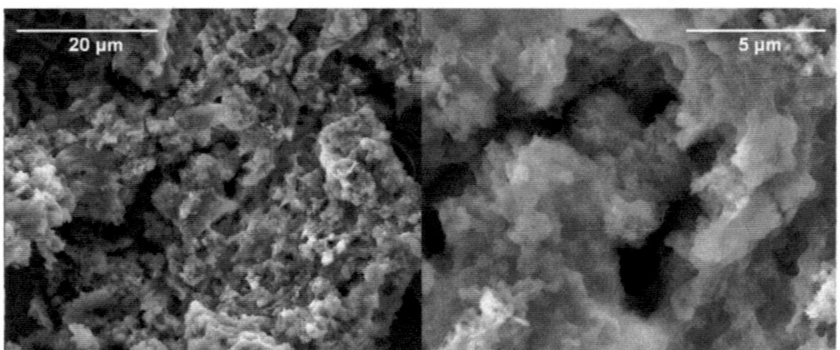

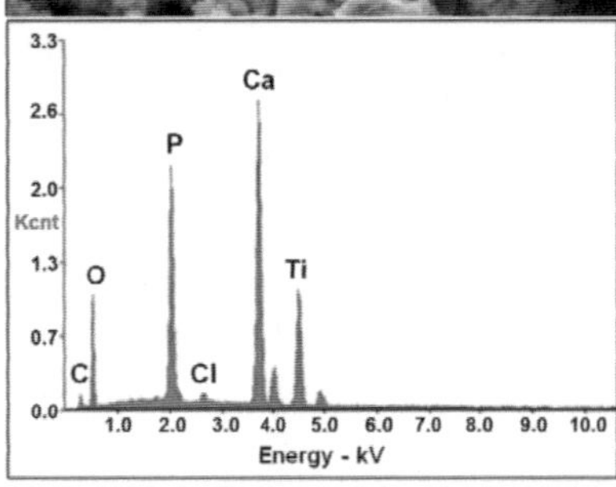

Element	*Wt%*	*At%*
CK	03.24	06.78
OK	33.63	52.77
PK	14.55	11.79
ClK	00.95	00.67
CaK	29.67	18.58
TiK	17.96	09.41
Matrix	Correction	ZAF

Figure 3. SEM images of hydroxyapatite (HA) coating on metal substrate Titanium=Ti with surface modified TiO_2. EDS analysis of hydroxyapatite coated. Table shows EDS analysis of the atomic % of Ca, P and O and other elements present in the powders. (Ca/P=1.58).

Coating the titanium substrate with TiO_2, the crystallinity of the HA coating increased, the structure became more compact and the Ca/P ratio increased due to the loss of P in the films. The addition of sodium hydroxide (to adjust the pH to 10) can increase the HA content in the coating. XPS and EDS outcomes for steel substrate and titanium showed poor calcium content as measured by a Ca/P ratio of 1.38 and 1.58, respectively, this composition is similar to that of natural apatite.

The Raman spectrum of HA coating is shown in Figure 4. The spectrum shows four distinguishable groups of spectral bands are evident. The first group consists of two bands at ~427.32 and ~443.15cm^{-1}. These bands correspond to the factor group splitting of the doubly degenerated bending mode v_2 of the PO_4 group (O-P-O bond). The bands at ~575.62, 586.76 and 604.57cm^{-1} belong to the v_4 triply degenerated bending mode of the PO_4 group. The phosphate symmetric-stretch peak was observed in the range of 958.41cm^{-1} which is characteristic for hydroxyapatite [8]. The very intense band at 958.41cm^{-1} arises from the symmetric stretching modes and is designated as v_1 fundamental vibrational mode. The bands comprising the fourth group arise from the v_3 and are due to the asymmetric stretching vibrations of the P–O bonds these bands are seen at ~1026.80, 1043.84 and 1073.56cm^{-1}. In Figure 4, it can be seen that Ti/TiO_2/HA has a lower signal intensity at 958.41cm^{-1} compared with the result in SS/TiO_2/HA at 956.26cm^{-1}, which could be due to a low adhesion material on the substrate modified TiO_2. In addition, anatase phase signals at 137.99, 391.04, 512.97 and 633.44cm^{-1} were evident. The slight displacement of the signals obtained with respect to literature, indicates the presence of other calcium phosphate phases in the microstructure of the synthesized coating.

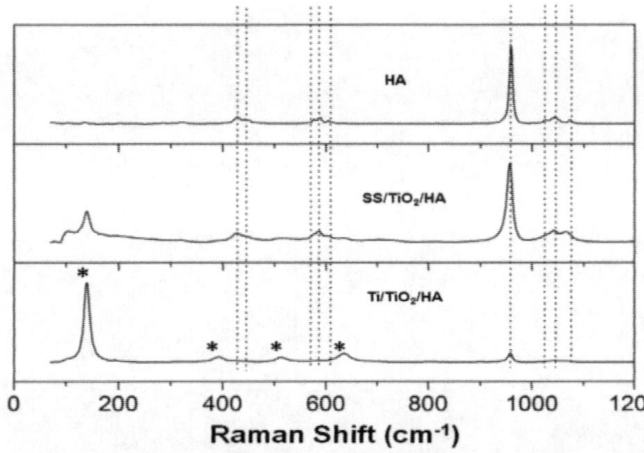

Figure 4 Raman shift spectra of hydroxyapatite (HA) coating on metal substrates (Stainless Steel=SS and Titanium=Ti) with surface modified TiO_2. The dashed line shows the signals corresponding to hydroxyapatite powder. The asterisk indicates signals corresponding to coated titanium.

4. Conclusions

HA bioceramic was synthesized using ammonium dihydrogen phosphate and calcium nitrate by simple precipitation, sol-gel, and dip-coating methods approach. XRD, XPS, FESEM-EDS and Raman results suggest the presence of crystalline phases in HA particles after calcination at 450°C. The microstructure of hydroxyapatite coating depends mainly of the aging time and the base used to adjust the pH, and substrate material parameters. The ability to generate crystalline HA coatings combined with the low temperature used, render the precipitation, sol-gel, and dip-coating approach favorable for generating HA coatings on different metallic substrates modified with sol-gel TiO_2 coatings for biomedical applications.

Acknowledgments
Authors would like to thank the University Santiago de Cali for the financial support given to the Project DGI-COCEIN-No. 939-621115-N21. Authors also thank the Center for Corrosion Research (CICORR) of the Autonomous University of Campeche and the Center for Materials and Nanoscience (CMN) for providing necessary facilities.

References

[1] Jaffe W L, Scott D F 1996 *J Bone Joint Surg Am* **78(12)** 1918
[2] Furlong R J, Osborn J F 1991 *J Bone Joint Surg B* **73(5)**:741
[3] LeGeros R Z 1991 *Monogr Oral Sci* **15** 1
[4] Wang C, Karlis G A, Anderson G I, Dunstan C R, Carbone A, Berger G, Ploska U, Zreiqat H 2009 *J. Biomed. Mater. Res. Part A* **90** 419
[5] Cameron H U, Macnab I, Pilliar R M 1977 *J Biomed Mater Res.* **11(2)** 179
[6] Zhang L, Li H, Li K, Zhang S, Fu Q, Zhang Y, Liu Shoujie 2015 *Ceramics International* **41(1)** 427
[7] Nik N, Che I, Yusairie M and Norjanah Y 2012 *APCBEE Procedia* **3** 46
[8] Torres F J, Panyayong W, Roger W, Velasquez-Plata D, Oshida Y, Moore B K 1998 *Biomed Mater Eng* **8** 25
[9] Szalkowska E, Gluszek J 2003 *Material Science* **21(4)** 367
[10] Smeulders D E, Wilson M A, Armstrong L 2001 *Ind Eng Chem Res* **40** 10

Productions of sunflower oil biodiesel and used cooking oil through heterogeneous catalysts compared to conventional homogeneous catalysts

C A Gutiérrez-Zapata[1], D Blanco Martínez[2], C A Collazos[1], H E Castellanos Acuña[1], J A Cuervo[3] and C P Fernandez[1]
[1] Universidad Manuela Beltrán, Bogotá D.C., Colombia
[2] Universidad Pedagógica Nacional, Bogotá D.C., Colombia
[3] Universidad Nacional de Colombia, Bogotá D.C., Colombia

E-mail: cesar.gutierrez@docentes.umb.edu.co

Abstract: This document compares homogeneous and heterogeneous catalysts used by production of biodiesel of sunflower oil and cooking oil used in frying. For this, NaOH was used as a catalyst homogeneous, and K_2CO_3 and Na_2CO_3 supported in gamma-alumina (K_2CO_3/γ Al_2O_3 y Na_2CO_3 /γ-Al_2O_3) were synthesized as heterogeneous catalysts, which were characterized by X-ray diffraction. The transesterification tests were carried out for the sunflower oil and used cooking oil, in a reflux system, to different molar relations methanol/oil, depending on the type of oil and characterization of the same. The reflux system is performed at a temperature of 55-60°C for one hour. Finally, biofuel was characterized and the yield of the reaction was calculated.

1. Introduction

The growing concern for the damage caused by emissions of massive consumption of fossil fuels enhances the need to seek alternatives of new sources of energy production: renewable energy sources which are characterized to be of low cost, high performance and high availability. From this, the use of vegetable oils chemically modified arises with great force, where it generates the conversion of the triglycerides present in the oils in their corresponding ester acids, through a process known as transesterification. Biodiesel is a liquid synthetic biofuel composed by monoalkyl esters, characterized because in the combustion process, is between 40% and 80% less polluting than the petrodiesel [1].

The production of biodiesel is usually performed by homogeneous basic catalysts as KOH, NaOH, CH_3ONa, and CH_3OK, with reaction times and temperatures between 1h and 60°C-70°C, with the disadvantage of needing stages of neutralization, washing and drying in the removal of the catalytic converter and the glycerol, which increase the product costs and contamination problems by the generated effluents [2].

An alternative that has been studied to counter the problems of the use of traditional homogeneous catalysts, is the use of basic catalysts heterogeneous, because with these the risk of secondary reactions and the catalytic converter can be removed by physical methods after the reaction, which decreases subsequent processes of washing and drying, and of washing and drying, and especially the generation of polluting effluents. In addition, with the use of heterogeneous catalysts is obtained high purity glycerine [2-5].

For this reason, our research has focused on the study of the process of transesterification of sunflower oil and used cooking oil, evaluating homogeneous and heterogeneous catalysts in the production of biodiesel, under conditions of reaction of 55-60°C at atmospheric pressure and at different Molar relations methanol oil, depending on the type of oil and characterization of the same. The heterogeneous catalysts were prepared in K_2CO_3 and Na_2CO_3 supported in gamma alumina (γ-Al_2O_3), from the wet impregnation method, which were characterized by X-ray diffraction. Finally, the biofuel was characterized and the yield of the reaction was calculated.

2. Experimental details

2.1. Preparation and characterization of heterogeneous catalysts

Catalysts preparation was supported on γ-Al_2O_3, was worked by the wet impregnation method, which consist in adding to the support a dissolution of the active phase desired, in this case sodium and potassium carbonates of high purity. It took 7.2g K_2CO_3 and 4.2g Na_2CO_3 supported in 15g γ-Al_2O_3. It was added 100mL of water and stirred for 9 hours. Finally, the water is removed, heating mixtures at 100°C. The catalysts prepared were dried at 120°C for 12 hours and were finally calcined at 500°C for 3 hours [1-3]. The catalysts were subjected to X-ray diffraction and atomic absorption spectrometry tests.

2.2. Transesterification by homogeneous and heterogeneous catalysis

Transesterification process began with a characterization of the raw material, at which testing was made for acidity index, refractive index, saponification index and iodine index. Table 1 represents the results obtained for each one of the oils.

Table 1. Characteristics of sunflower oil and used oil.

Features	Sunflower oil	Vegetable oil used
Linoleic 18:02	64	45.50
Oleic 18:01	14	10.80
Palmitic 16:00	4	2.20
Stearic 18:00	7	540
Acidity index (mg KOH/g sample)	2.06	10.17
Saponification value (mg KOH/g sample)	181.10	174.27
Iodine index (method Hanus (g I/10g oil)	123.60	108.23

The amount of catalyst homogeneous and heterogeneous used for the transesterification of sunflower oil and oil used was calculated by taking into account the value obtained of the acidity index [6-7]. Table 2 shows the percentages of catalyst and the molar relations methanol/oil for each of the oils. From each reaction 50 grams of oil to work were taken and everything was carried out in a reflux system, 55-60°C for one hour [6-8].

Table 2. Percentages of catalyst and molars relations methanol oil.

Oil	Catalytic Converter	% Catalytic Converter	Methanol/Oil (mol/mol)
Sunflower (AG)	NaOH	0.38	5.5/1.0
Used vegetable (AVU)	NaOH	0.50	5.5/1.0
Sunflower (AG)	Na_2CO_3 /γ-Al_2O_3	9.00	5.5/10
	K_2CO_3/γ Al_2O_3	9.20	5.5/10
Used vegetable (AVU)	H_2SO_4(98%)	2.00	3.0/10
	Na_2CO_3 /γ-Al_2O_3	9.00	5.5/10
	K_2CO_3/γ Al_2O_3	9.20	5.5/10

For oil used during testing of heterogeneous catalysts, a variation or pre-treatment was made due to its high acidity known as esterification, in where H_2SO_4 98% was used as catalyst, at 2 per cent in catalyst/oil relation and a molar relation methanol/Oil of 3/1, under the same conditions of temperature and time [7].

The product obtained is passed to a separating funnel to separate the biodiesel from the glycerin, after it was washed with distilled water until the pH 7 and dried for one hour to 105°C. Characterization was made to the biodiesel obtained, where its density and viscosity was measured and the performance for each of the reactions was determined, taking into account the following Equation [8-9].

$$\text{Yield} = \frac{\text{Grams of biodiesel}}{\text{Grams of biodiesel theoric}} \text{ X } 100 \tag{1}$$

The theoretical biodiesel grams are equivalent to the biodiesel grams that will arise from the grams of initial oil assuming a complete reaction and the grams of biodiesel correspond to the grams of biodiesel obtained at the end of each reaction.

3. Results and discussion

3.1. Preparation and characterization of heterogeneous catalysts

After the preparation of the two catalysts supported on γ-alumina by wet impregnation of carbonates tests of x-ray diffraction were carried out. Figure 1 represents the XRD patterns made for carbonates calcined, which show us the following results.

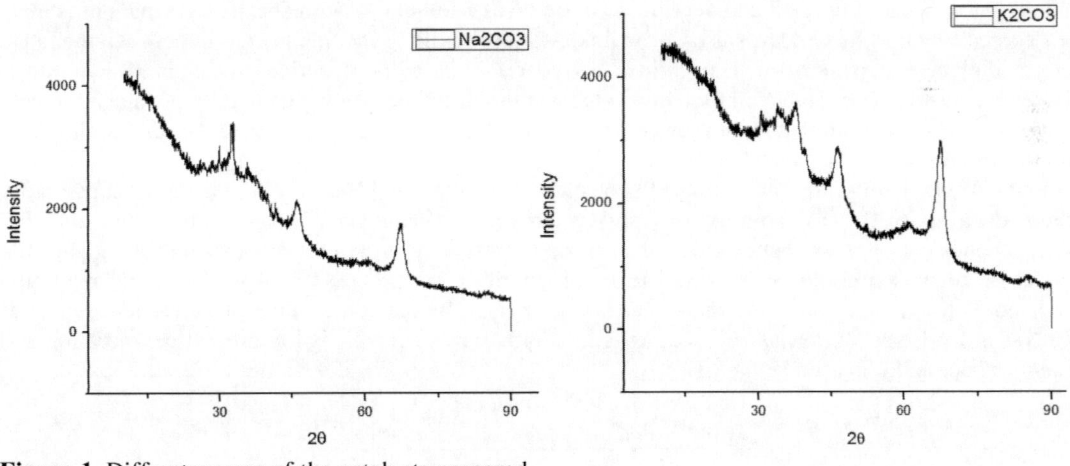

Figure 1. Diffractograms of the catalysts prepared.

There are four peaks-base some with greater intensity than the other in the shaft 2θ=32°, 37°, 46°, 67°. Likewise angles, that present the specific peaks of the range alumina, our catalytic support. Show after peaks additional to those of the alumina range, in the angles 2θ=21°, 25°, 27°, 28°, 32°, 35°, 37°, 41°, 46°, 51° and 55° reveals the presence of a new phase of the compound that contains elements of alumina and potassium, alumina and sodium. These results are consistent with different results already reported in the literature [1-4,6,10]. The new phases can be caused by the formation of species of K_2O, Na_2O and groups Al-OR-Na, Al-OR-K, which serve as active sites for the transesterification of oil with methanol.

In Table 3 presents the results for the heterogeneous catalysts of K_2CO_3 and Na_2CO_3 supported in gamma alumina (γ-Al_2O_3), respectively. It notes the concentration of the amount of metal fixed in the holder.

Table 3. Concentration of sodium and potassium impregnated in γ-alumina.

Ion	Total mg/15g γ-alumina	% Catalytic Converter
Na^+	1860	9.00
K^+	1923	9.20

Taking into account the total mg of metal fixed in the holder, was calculated as explained before the amount of heterogeneous catalyst to use for testing of transesterification. In Table 2 and 3 can observe the results obtained.

3.2. Transesterification by homogeneous and heterogeneous catalysis

Within the processes of transesterification by homogenous and heterogeneous catalysis it should be clarified, how mentioned above, that the amount of catalyst used for each of the oils was determined taking into account the valuation of acidity of the two oils because it determines the amount of fatty acids present in each one and the higher is the acidity of the oil, the lower the conversion, causing an increase in the amount and type of catalyst to use it to avoid undesirable reactions as the saponification. Assuming that by literature by each litre of vegetable oil without using are 3.5g of catalyst [5-6], it was estimated the amount of catalyst to be used for the 50mL of oil. Within the values obtained for the index of acidity for the sunflower oil and vegetable oil used was 2.06 and 10.17 respectively, which indicates the degree of hydrolysis that has suffered the link ester, by breaking the link is present between the fatty acids and glycerol present in the triglycerides, diglycerides and monoglycerides, which is evident in the greater proportion of vegetable oil used due to the heating of this, causing the quantity of catalyst to use for this oil is greater than that of sunflower oil [5-8]. In Table 1 and 2 are observed the results obtained and the specific amounts for each one of the oils.

In the Table 4 are represented the yields obtained for each one biodiesel, which shows conversion levels of triglycerides by homogenous and heterogeneous catalysis, demonstrating that for the homogeneous catalysts are better yields than those represented by heterogeneous catalysis, given the acidity of oil, which could be improved to make esterification process of oil to rise molar relations methanol/Oil and catalyst/oil. At the same time it was evaluated some kinde of heterogeneous acid catalyst instead the homogeneous used, to reduce issues by removal, pollution and acidification on reaction process if it is used in excess.

Table 4. Performance of the reaction.

Biodiesel	% Yield
AGNa	97.60
AVUNa	95.40
AG ($Na_2CO_3/\gamma Al_2O_3$)	80.00
AG ($K_2CO_3/\gamma Al_2O_3$)	89.80
AVU ($Na_2CO_3/\gamma Al_2O_3$)	80.90
AVU ($K_2CO_3/\gamma Al_2O_3$)	82.90

Then compare the characteristics of biodiesel, with the typical features of biodiesel to verify if the product obtained follows the regulations. Table 5 presents the tests carried out.

It is noted that the transesterification process if had significant changes in the conversion of triglycerides to methyl esters presenting in the biodiesel getting acceptable physical properties within the standards for the use as a mixture with diesel.

Table 5. Characterization of sunflower oil, vegetable oil used and biodiesel.

Compound	Viscosity Mm2/s	Density (g/cm3)
Sunflower oil	30.33	0.953
Used oil	29.56	0.910
AGNa	3.99	0.883
AVUNa	4.03	0.884
AG (Na_2CO_3/γ-Al_2O_3)	3.89	0.882
AG ($K_2CO_3/\gamma Al_2O_3$)	4.13	0.885
AVU (Na_2CO_3/γ-Al_2O_3)	4.66	0.879
AVU ($K_2CO_3/\gamma Al_2O_3$)	4.98	0.883
Biodiesel (Res. 18 2087 2007)	1.60-6.00	0.860-0.900

4. Conclusions

Taking into account the analyses and tests carried out on the basis of the different catalysts homogeneous and heterogeneous it could be observed that there is an effective process since yields of conversion are high. The study for the heterogeneous catalyst revealed the presence of new phases present in the catalytic converter as elements of alumina and potassium, alumina and sodium, due also to the calcination made to these. Finally, the biodiesel obtained during the transesterification process is part of the quality parameters allowed by national standard (Law 1205 of 2008) and international (European standard UNE-EN 14214), which is evident in the decrease of the oil viscosity and density.

Acknowledgements
Authors acknowledge to direction of laboratories, Universidad Manuela Beltran.

References

[1] Ríos L, Castrillon A y Zuleta E 2009 *Energética* **42** 45-52
[2] Arzamendi G, Campo I, Arguiñarena E and et al 2007 *Chemical Engineering Journal* **134** 123-130
[3] Noiroj K, Intarapong P, Luengnaruemitchai A 2009 *Renewable Energy* **34** 1145-1150
[4] da Costa J P, Araujo A, di Souza L, Gondim A 2016 *Renewable and Sustainable Energy Reviews* **59** 887–894
[5] Banerjee A, Chakraborty R 2009 *Resour Conserv Recy* **53** 490–497
[6] Castella G, Angulo E R, Cardozo B M 2014 *Prospect* **12(2)** 90-104
[7] Veljkovic V B, Lakicevic S H, Stamenkovic O S, Todorovic Z B, Lazic M L 2006 *Fuel* **85** 2671-2675
[8] Rojas Gozales A F, Giron Gallego F, Torres Castañeda H G 2009 *Ing. Investig.* **29(3)** 17-22
[9] Yin X, Ma H, You Q, Wang Z, Chang J 2012 *Appl Energy* **91** 320–5
[10] Xie W U, Li H 2006 *Journal of Molecular Catalysis A: Chemical* **255** 1–9

Obtaining and characterization of $La_{0.8}Sr_{0.2}CrO_3$ perovskite by the combustion method

A M Morales Rivera[1], J A Gómez Cuaspud[1] and E Vera López[1]
[1] Universidad Pedagógica y Tecnológica de Colombia, Tunja, Boyacá, Colombia

E-mail: angela22moralesrivera@gmail.com

Abstract: This research is focused on the synthesis and characterization of a perovskite oxide based on $La_{0.8}Sr_{0.2}CrO_3$ system by the combustion method. The material was obtained in order to contribute to analyse the effect of synthesis route in the obtaining of advanced anodic materials for solid oxide fuel cells (SOFC). The obtaining of solid was achieved starting from corresponding nitrate dissolutions, which were polymerized by temperature effect in presence of citric acid. The solid precursor as a foam citrate was characterized by infrared (FTIR) and ultraviolet (UV) spectroscopy, confirming the effectiveness in synthesis process. The solid was calcined in oxygen atmosphere at 800°C and characterized by X-ray diffraction (XRD), scanning electron microscopy (SEM) and energy dispersive of X-ray spectroscopy (EDX) and solid state impedance spectroscopy (IS). Results confirm the obtaining of an orthorhombic solid with space group *Pnma* (62) and cell parameters a=5.4590Å, b=7.7310Å and c=5.5050Å. At morphological level the solid showed a heterogeneous distribution with an optimal correspondence with proposed and obtained stoichiometry. The electrical characterization, confirm a semiconductor behaviour with a value of 2.14eV Band-gap according with previous works.

1. Introduction

Since its discovery 70 years ago, the solid oxide fuel cells (SOFC) have taken great importance in the clean production of energy, because these devices can produce electricity from a continuous feed of hydrogen and oxygen, without generation of greenhouse gases with a high efficiency or limitations established by the Carnot cycle [1,2]. In this regard, the high temperature solid oxide fuel cells (HT-SOFC), allow to use different raw materials to provide the hydrogen fuel at anodic level, however in many cases, the purity of feed gases need to be improved to avoid problems related with the carbon deposition in anodic component, represented by the current state of art of Ni-YSZ cermet. Therefore, the need arises to find more efficient ceramic systems that do not allow the formation of secondary phases and carbonaceous deposits [3-6]. Among the most promising materials to replace the conventional cermet anodes, are included some mixed oxides derived from a perovskite structure particularly concentrated in lanthanum and strontium chromites, which exhibit good levels of conductivity in air and fuel hydrogen rich atmospheres, additionally are resistant to graphite deposition [7], due to chromium cation that promote the catalytic oxidation of light hydrocarbons like methane and propane in the reforming process with steam [8].

Under above considerations in order to find new and simple methods to obtain innovative materials at low cost in design of SOFC anodes, this paper focused on the implementation of a synthesis route, which enables the obtaining of homogeneous ceramic materials as opposed to conventional methods of solid stet reaction [9-16]. For this purpose we developed a comparative study highlighting the main advantages of this methodology in terms of the main techniques of chemical, structural and electrical characterization.

2. Experimental

The ceramic oxide was synthesized starting from corresponding nitrates $La(NO_3)_3.6H_2O$, $Sr(NO_3)_2$, $Cr(NO_3)_3.9H_2O$ in 1.0mol L^{-1}. The dissolutions of nitrates were dosed in stoichiometric form in a glass vessel with magnetic stirring at 80°C. The addition of citric acid was performed in a 1:1 relation with respect to total metal cations concentration. The obtaining of a dense sol was achieved after thermal treatment for 3 hours after that, the temperature was increased at 250°C to form a viscous material that which is then entered into spontaneous combustion, providing a solid carbonaceous precursor. The precursor was characterized by infrared (FTIR) and ultraviolet (UV) spectroscopy in a Perkin-Elmer FTIR-1000 computer and HP 8453 UV-Vis respectively, in order to confirm the presence of citrate species. The solid was calcined at 800°C under oxygen flow (50mL min^{-1}) for 2 hours to eliminate remnant carbon from previous stage.

The obtained oxide was macerated and characterized by X-ray diffraction (XRD), to identify the nature of the crystalline phase and purity. The analysis were performed on a PANalytical X'Pert PRO MPD-, equipped with X'Celerator detector in Bragg Brentano configuration, using the Cu K_α radiation (λ=1.54186Å) between 20 and 80°. The X-ray data were analysed using the X'Pert High-Score software and the crystallite size was evaluated by the Debye-Sherrer equation. The SEM analysis was performed in a LEO 440 microscope (Leica-Zeis), equipped with an electron gun and microprobe analysis for EDX. The solid state electrical impedance analysis (IS), was performed on an AUTOLAB potentiostat-galvanostat using modified cell at room temperature and a $La_{0.8}Sr_{0.2}CrO_3$ blank to validate the obtained material. The blank and the oxide was compressed in a uniaxial system at 5.0MPa pressure and sintered at 900°C for 1 hour in similar way to avoid measurements errors.

3. Results and discussion

The UV analysis of precursor, showed two major electronic transitions located at 340 and 470nm (Figure 1), in the range of 200 to 700nm, where the citrate ligand provides a strong signal of electronic transition ($\pi \rightarrow \pi^*$), while the $n \rightarrow \pi^*$ transitions are considerably minor. Such transitions are characteristic and are related with the presence of unsaturated compounds with metal heteroatoms in correspondence with Pino *et al.* [17]. From UV results was possible to identify the value of Band-gap around 2.14eV for the obtained material, which is consistent with studies of Strehlow *et al.* [18].

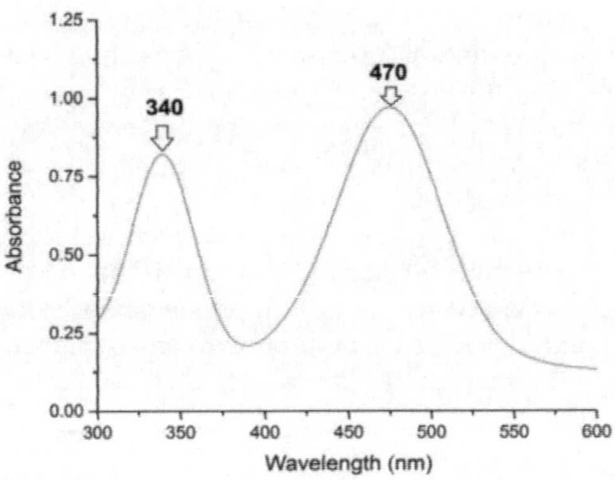

Figure 1. UV spectrum of metalorganic precursor La0.8Sr0.2CrO3 between 300 and 600nm.

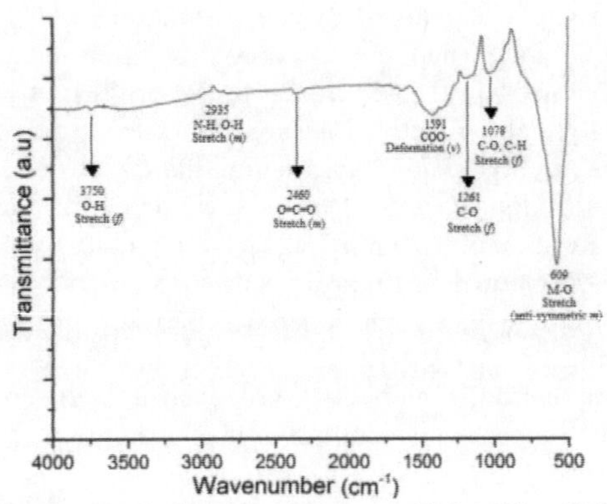

Figure 2. Infrared spectrum of $La_{0.8}Sr_{0.2}CrO_3$ precursor in KBr pellet obtained at room temperature with its main transmittance signals.

The FTIR results shown in Figure 2 and analysed comparatively with the SDBS database, allow obtaining the relationship between the transmittance signals with the main vibration bands. The presence of these signals is justified by the thermal decomposition process of the gels at 250°C and the signs located below $1000cm^{-1}$ are related with the vibrational movements associated with the different metal cations La^{3+}, Sr^{2+} and Cr^{3+} with hydroxyl groups and oxygen atoms, which confirm the bond between citrate ligands and the metal cations as has been previously confirmed for similar perovskite compositions [16].

The XRD analysis of the calcined oxide using the X'Pert High Score software (Figure 3), permit to classify the oxide with the $La_{0.85}Sr_{0.15}CrO_3$ reference compound, with a ICSD 051197 code, space group *Pnma* (62), orthorhombic system and cell parameters a=5.4590Å, b=7.7310Å and c=5.5050Å, with a preferential crystal orientation along (1 2 1) facet as shown in Figure 4. The crystallite size evaluated with the highest diffraction signals permit to identify nanometric sizes of sample around 4.4nm.

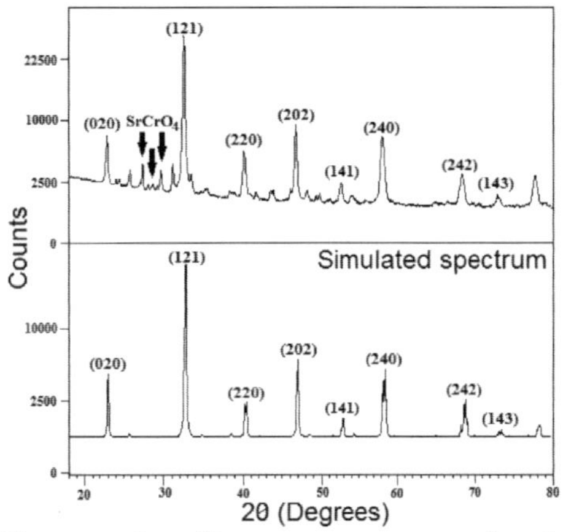

Figure 3. X-ray spectrum of the La0.8Sr0.2CrO3 sample and its corresponding Rietveld refined spectrum.

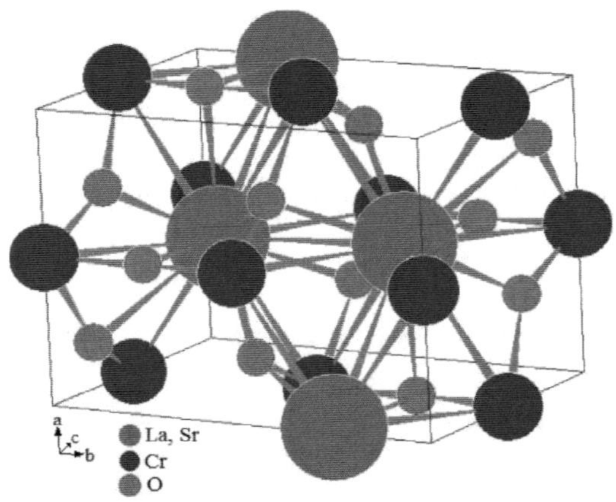

Figure 4. Unit cell of La0.8Sr0.2CrO3 obtained using the PCW software.

The Rietveld refinement analysis, allowed to identify that the signals located about 24 and 31° 2θ are associated with the presence of strontium chromate ($SrCrO_4$), as secondary phase of system. The above explained by the formation of oxygen vacancies, allow the mobility of ions at high temperatures and the formation of intermediate low concentration species, as has been shown previously [19]. According to Waernhus et al. [20] the $SrCrO_4$ has a high resistivity and is catalytically inactive, because the strontium cation consent the stabilization of the perovskite structure by a load balancing, therefore deficiencies at site A of the structure, promote such substitutions causing that B site elements form different kinds of binary oxides.

Results of scanning electron microscopy (SEM) and energy dispersive X-ray spectroscopy (EDX) shown in Figure 5, allow to corroborate the obtaining of a heterogeneous distribution of particles with a high coherence between proposed and obtained compositions, articulate with the synthesis process [21]. In Table 1, is shown the difference between proposed and obtained composition, which does not exceed a difference by more than 0.10% (w/w), confirming the effectiveness in the control of used stoichiometry.

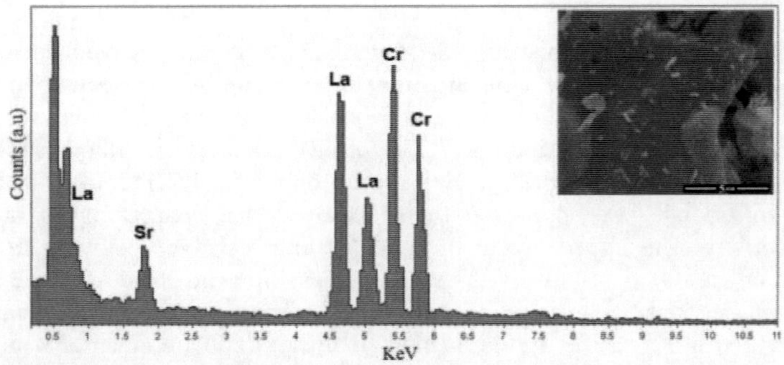

Figure 5. X-ray energy dispersive spectrum and scanning electron microscopy of $La_{0.8}Sr_{0.2}CrO_3$ oxide.

Table 1. Composition results (%w/w) of La0.8Sr0.2CrO3 sample obtained by EDX technique.

Percentage values by weight (%)	La	Sr	Cr	O	Total
Experimental	48.63	7.722	22.65	21.00	100
Theoretical	48.60	7.664	22.75	20.99	100
Difference	0.03	0.058	0.10	0.01	

Figure 6, shown the solid state impedance spectrum, where the two curves shown the impedance versus frequency and phase versus frequency change. The incidence of chromium in the structure relates to the development of charge carriers, permit to understanding the interactions between electrons and atoms in the solid material, so, the obtained resistivity value (6131 ohm), is result of $SrCrO_4$ in the $La_{0.8}Sr_{0.2}CrO_3$ oxide, requiring its elimination through thermal treatments as claimed by Gómez *et al.* [22,23]. According to these results, the solid state impedance spectroscopy was a useful tool to validate the contamination of main crystalline phase, since the Bode impedance plot for blank did not show the same behaviour.

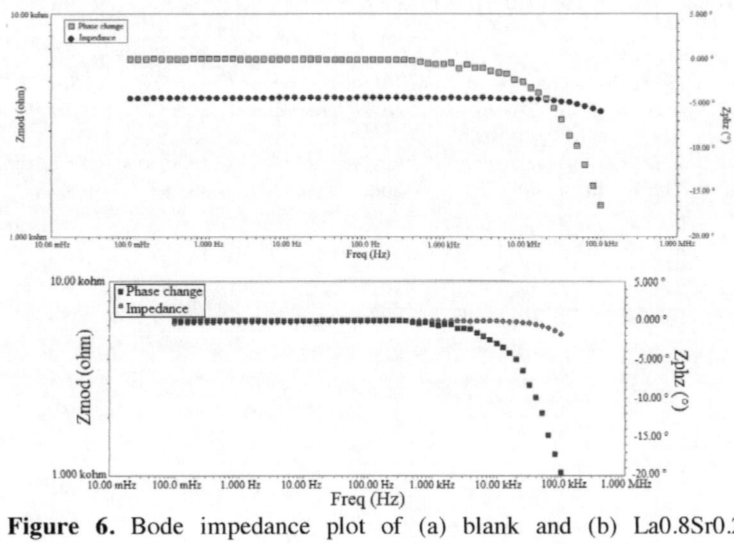

Figure 6. Bode impedance plot of (a) blank and (b) La0.8Sr0.2CrO3 samples.

4. Conclusion

The UV and FTIR analysis, corroborate the obtaining of citrate species corresponding with synthesis method, which favour the consolidation of a crystalline phase with morphology and surface characteristics to be explored in design of SOFC anodes. The XRD analyses confirmed that a highly crystalline material was obtained, with *Pnma* (62) space group, orthorhombic structure and crystallite sizes around 4.4 nm with preferential orientation along (1 2 1) facet. The SEM and EDX analysis, allow corroborate several morphological and surface characteristics of relevance for electrocatalytic activity. The impedance analysis, validate a resistive behaviour of sample due to secondary phases of $SrCrO_4$ in contrast with the blank measurements, leading to the conclusion that the main crystal phase needs to be consolidated to promote integration of the remaining oxides to the main structure. Nevertheless, the synthesis method allows to obtain material with morphological and surface properties of relevance for the design of SOFC anodes.

References

[1] Yamamoto O 2000 *Electrochim* **45** 2423
[2] Soldati A L, Baqué L, Troiani H, Cataro C, Schreiber A, Caneiro A and Serguis A 2011 *Journal of Hydrogen Energy* **36** 9180
[3] Gómez J A and Carda J B 2014 *Ciencia en Desarrollo* **5** 55
[4] Alvarado J and Ávalos J 2013 *Revista Mexicana de Física* **59** 66
[5] Plint S, Connor P, Tao S and Irvine J *Solid State Ionics* **177** 19
[6] Williams M C, Strakey J P and Surdoval W A 2005 *Int. J. Appl. Ceram. Technol.* **2** 295
[7] Lamas D G 2011 *Hifusen* **3** 105
[8] Stambouli B and Traversa A 2002 *Renew. Sustain. Energy* **6** 433
[9] Priyatham T and Bauri R 2010 *Mater. Charact.* **61** 54
[10] Irvine J T 2010 *Solid State Ionics* **180** 1683
[11] Lee J J, Moon H, Park H G, Yoon D I and Hyun S H 2010 *Int. J. Hydrogen. Energ.* **35** 738
[12] Kao W K, Lee M C, Lin T N, Wang C H and Chang Y C 2010 *J. Power Sources* **195** 2220
[13] Knapp M C and Woodward P M 2006 J *Solid State Chem.* **179** 1076
[14] Deganello F, Marci G and Deganello *J. Eur. Ceram. Soc,* **29** 439
[15] Valdés T, Marbán G and Fuertes A B 2005 *Chem. Mater.* **17** 1919
[16] Goméz J A and Valencia J S 2009 *Rev. Colombiana de Química* **38** 289
[17] F Pino Pérez, María Dolores Pérez Bendito 1983 *Análisis de elementos-traza por espectrofotometría de absorción molecular Uv-visible* (España: Universidad de Córdoba)
[18] Strehlow W H and Cook E L 1973 *J. Phys. Chem.* **2** 44
[19] Sakai N, Yamaji K, Horita T and Yokokawa H 2003 *Solid State Ion* **152** 517
[20] Waernhus I, Sakai N, Yokokawa H, Grande T, Einarsrud M A and Wiik K *Solid State Ionics* **178** 907
[21] Ye X, Wang S, Wang Z, Hu Q, Sun X and Wen T 2008 *Journal of Power Sources* **183** 512
[22] Várez A 1993 *Conductores de litio basados en superconductores de alta temperatura* (Madrid: Universidad Complutense de Madrid) p 90
[23] Gómez J A 2010 *Síntesis y caracterización del sistema LaSrCrFeO soportado sobre óxidos de cerio dopados con elementos de transición interna* (Bogotá: Universidad Nacional de Colombia)

IOP Conf. Series: Journal of Physics: Conf. Series **786** (2017) 012027 doi:10.1088/1742-6596/786/1/012027

Synthesis and characterization of the Cu_2ZnSnS_4 system for photovoltaic applications

D L Sánchez Pinzón[1], G Y Soracá Perez[1], J A Gómez Cuaspud[1] and E Vera López[1]
[1] Universidad Pedagógica y Tecnológica de Colombia (UPTC), Tunja, Colombia

E-mail: dayana.sanchez@uptc.edu.co

Abstract. This paper focuses on the synthesis and characterization of a ceramic material based on the Cu_2ZnSnS_4 system, through the implementation of a hydrothermal route. For this purpose, we started from nitrate dissolutions in a 1.0mol L^{-1} concentration, which were mixed and treated in a teflon lined vessel steel at 280°C for 48h. The Physicochemical characterization of the solid was evaluated by means of ultraviolet visible spectroscopy (UV-VIS), X- ray diffraction (XRD), Raman spectroscopy, scanning and transmission electron microscopy (SEM-TEM) and solid state impedance spectroscopy (IS). The initial characterization through UV measurements confirms a Band-gap around 1.46eV obtained by the Kubelka-Munk method, which demonstrates the effectiveness of the synthesis method in the obtaining of a semiconductor material. The XRD results confirm the obtaining of a crystalline material of pure phase with tetragonal geometry and I-42m space group. The preferential crystalline orientation was achieved along (2 2 0) facet, with crystallite sizes of nanometric order (6.0nm). The morphological aspects evaluated by means electron microscopy, confirmed the homogeneity of the material, showing specifically a series of textural and surface properties of relevant importance. Finally, the electrical characterizations allow to validate the semiconductor behaviour of CZTS system for development of photovoltaic technologies.

1. Introduction

The research in the field of renewable energies is oriented to the search for new and enhanced systems that not only meet the growing energy demand that the current world requires, but also mitigate the environmental impact caused by the excessive use of fossil fuels used mainly for the generation of conventional energy. In this way, photovoltaic technology becomes an important alternative, since it promises a decreased environmental impact while supplying security and lowers the cost of the energy requirements of modern world [1]. However, photovoltaic technology has not achieved sufficient maturity to allow it to contribute significant energy contributions at the global level. Therefore, it is of the utmost importance to development of new materials that can eventually contribute to the research and development of this technology [2]. Traditionally the manufacture of photovoltaic devices involves the use of materials with high purity, structural perfection and high levels of toxicity. These restrictions have considerably limited the utilization of photovoltaic solar energy. The kesterite (Cu_2ZnSnS_4) or sulphur of copper, zinc and tin is one of the most interesting materials in the field of renewable energies, mainly due to its interesting photovoltaic properties. The CZTS offers optical and structural characteristics of significant importance for applications in thin film technology. Likewise, it is composed of elements that are not toxic and are readily available in the terrestrial crust [3]. The

Cu_2ZnSnS_4 is a quaternary type p semiconductor that has recorded values of Band-gap 1.40 to 1.55eV [4], coefficients of absorption in the visible region of $10^4 cm^{-1}$ [5] and efficiencies of around 31 to 33% calculated theoretically [6]. However, the solar cells based on CZTS have managed to reach values of energy conversion efficiencies close to 9.66% in solar panels [7].

In this regard, they have used a great variety of techniques for the development of thin film solar cells based on this material. However, a great part of them involve the use of equipment and routes of complex and costly synthesis that dramatically raise the cost associated with the manufacture of this type of material [8-9]. The hydrothermal synthesis method is an alternative technique to synthesize these materials at low temperatures and low residence times [10-11] that give rise to aggregates with important physical, chemical, and electrical properties [12]. Therefore, the present investigation is focused on the synthesis and characterization of a ceramic material based on Cu_2ZnSnS_4 system for application in thin film solar technology.

2. Experimental

The Cu_2ZnSnS_4 material was synthesized by a hydrothermal route in a single stage of reaction, starting from precursor solutions of Cu, Zn and Sn in 1.0mol L^{-1} concentration, together with the addition of thiourea as source of sulphur. The dosage of the reagents is carried out in a stainless steel autoclave, where it were treated under stirring conditions at 280°C for 48h. Once completed the synthesis process, it was obtained a dark precipitate, which was washed repeatedly with absolute ethanol. Subsequently the solid was dried at 100°C (60min) to get the ceramic material.

The characterization by UV-VIS was conducted in a MAPADA UV-VIS 1800PC equipment, between 200 and 1100nm. The morphological characteristics were evaluated by X ray diffraction, in a PANAlytical X´pert PRO MPD diffractometer, using the Cu Kα (λ=1,54Å) radiation between 10 and 80° 2θ with steps of 0.02°. The X-ray pattern was analysed using the X'Pert High Score software, while the size of crystallite was estimated using the highest diffraction signal by the Debye-Scherrer equation. The evaluation of the purity of the solid and the formation of the crystalline phase was evaluated using Raman spectroscopy in a HR-UV 800 infinity microprobe (Jobin-Yvon) apparatus, equipped with a CCD detector (-70°C) and a laser of 10.7mW. The Raman spectrum of the solid was taken between 100 and 800cm^{-1} projecting a continuous radiation laser of He-Ne, which supplying the excitation line at 632nm. The analysis by scanning electron microscopy (SEM) was performed in a Leica-Zeiss LEO 440 system with an electron gun and an accelerating voltage of 30kV, with analogic and digital image of EDX. The analysis by transmission electron microscopy (TEM), was conducted in a JEOL 2100 microscope with a thermionic gun of LaB_6 and an accelerating voltage of 200kV, allowing the obtaining of high-resolution images. Finally, the impedance analyses were performed with an AUTOLAB potentiostat-galvanostat at room temperature, to which the material was compressed in an uniaxial system at 5.0MPa pressure.

3. Results and discussion

The material obtained was evaluated by UV-VIS spectroscopy between 190-1100nm (see Figure 1). The Band-gap values were calculated from the inflection points, using the more characteristic wavelength providing a value around 1.46eV, which allow concluding that the dominant phase in the material corresponds to Cu_2ZnSnS_4 phase. The obtained values of Band-gap, permit to compare current result with the optimal value around 1.5eV and evaluate the effectiveness in the synthesis process. The slight decrease in assignation of wavelength and therefore in the calculation energy values could be attributed to the presence of Cu_2SnS_3 phase in the material, according to previous studies Malerba et al., this phase can be corrected by light thermal treatments not greater than 500°C [13].

The X-ray analysis, confirm the obtention of the Cu_2ZnSnS_4 system in a kesterite phase, showing a clear tetragonal geometry and a preferential crystal orientation along (2 2 0) facet. The estimation of the crystallite size was made by the Debye-Scherrer equation using the highest signal of the spectrum, resulting in a value around 5.46nm. In the Figure 2 is observed the main signals associated to the

CZTS system. These representative signals of material in kesterite phase show a great similarity with the works published recently by Vanalakar et al. [14]. The signals located at 34.3°, 37.1° 62.2° 2θ are directly related with the formation of covalite (CuS), this secondary phase is associated with the low temperatures in the synthesis process, however it can be eventually corrected with subsequent thermal treatments as has been previously documented by Agawane et al. [15].

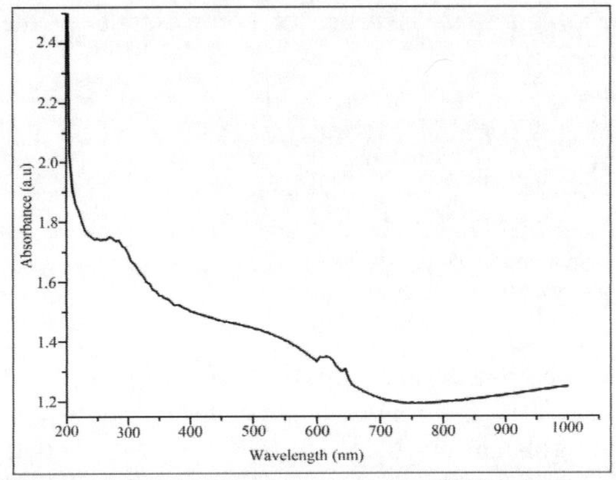

Figure 1. UV-VIS spectrum for CZTS between 200-1000nm. Spectrum.

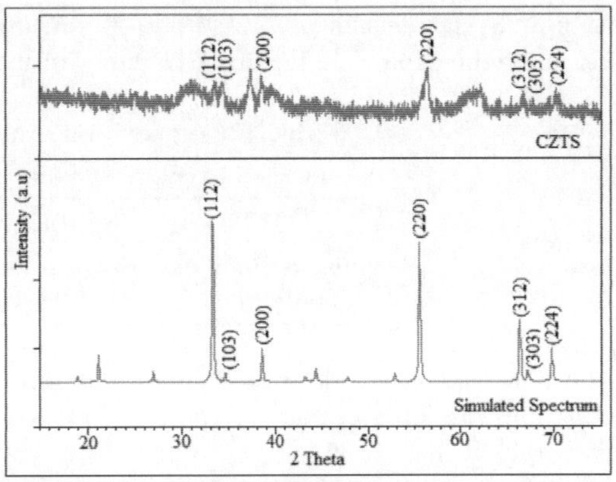

Figure 2. X-ray diffraction pattern with simulated material of CZTS system.

The characterization by Raman spectroscopy shown in the Figure 3, allow to highlight a strong signal around 289cm^{-1} associated to the main mode vibrational asymmetric mode of the CZTS system in the kesterite phase [16]. Of equal way, this technique confirms the presence of other signals with lower intensity located at 225cm^{-1} (SnS), 326cm^{-1} (Cu_2SnS_3), 346cm^{-1} (ZnS) and 475cm^{-1} (CuS_2), which are linked to the formation of intermediate compounds. The existence of a low peak located at 326cm^{-1}, is associated with the tetragonal phase of Cu_2SnS_3, the formation of this phase in the structure is probably caused by the strong presence of Cu^{2+} cation in the material [17]. Of this form, the consolidation of secondary phases is a consequence of the incomplete conversions of sulphides during the synthesis process [18]; therefore, a progressive increase in the synthesis temperature could improves the morphological, structural and optical characteristics of CZTS material [15].

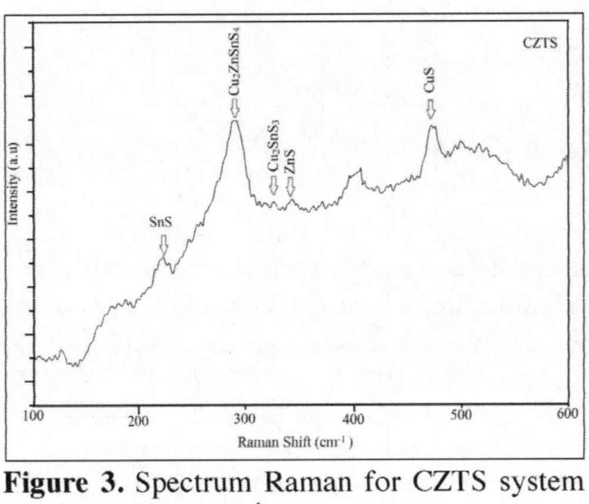

Figure 3. Spectrum Raman for CZTS system between 100-600cm^{-1}.

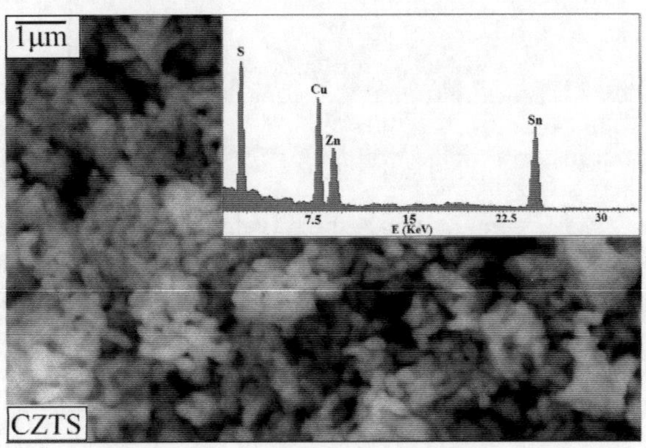

Figure 4. Scanning electron micrograph with X-ray energy dispersive of the system CZTS.

The analysis by scanning electron microscopy (Figure 4), show the obtaining of a material conformed by conglomerates distributed homogeneously. The morphology of the material allows to identify a solid formed by aggregates with irregular spherical geometry. According to Wang et al. [19], the morphology of the material is improved with the increases in the synthesis temperature since is a key factor in the consolidation of appropriate kinetic mechanisms in the formation of crystalline phases [19,20]. In addition, the chemical composition of CZTS nanoparticles using the energy dispersive of X-rays spectroscopy (EDX), permit to validate an excellent relationship between proposed and obtained composition (Table 1). The data show an optimal correlation between Cu/Zn/Sn/S elements in a 2:1:1:4 proportion, this corroborates the absence of contaminants in the solid, allowing appreciate the grade of purity of the material.

Table 1. Composition results (% w/w) of CZTS sample obtained by EDX technique.

Percentage values by weight (%)	Cu	Zn	Sn	S	Total
Experimental	28.65	15.69	26.89	28.77	100
Theoretical	28.91	14.87	27.03	29.18	99.99

On the other hand, the morphology and surface characteristics of the material were investigated using transmission electron microscopy (TEM), as shown in the Figure 5. The results show that nanoparticles have a diameter around 5-6nm with interplanar distances of 0.70nm, current data confirm the results obtained from previous size analysis performed by the Debye-Scherrer equation.

The electrical response of the material was evaluated using solid-state impedance spectroscopy, the results confirm that the material presents a semiconductor behaviour, all resistivity values are values near to 50000Ω. The semicircle in Figure 6, show a region controlled by charge transfer processes, which show the way in which the charge carriers react during the potential change and its relation with the generation of electric power [21].

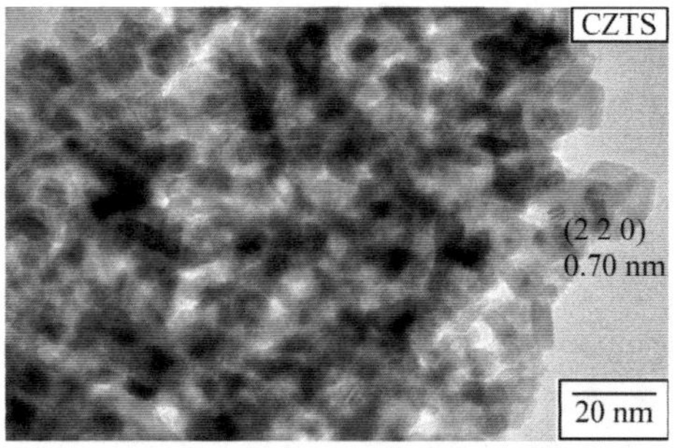

Figure 5. Transmission electron microscopy images for the CZTS system.

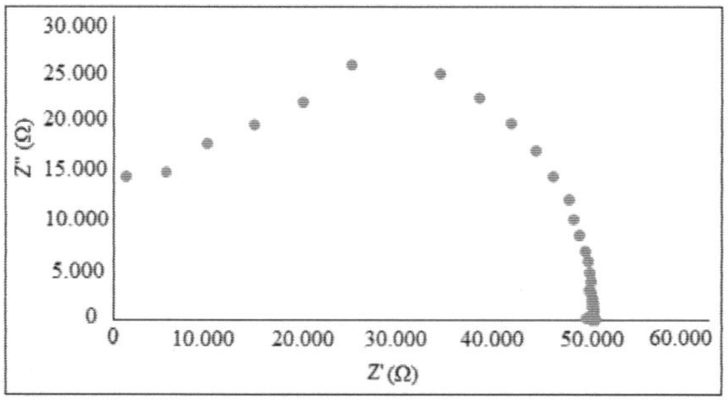

Figure 6. Impedance diagram for the CZTS system.

The Figure 7 show the equivalent circuit, modelled for the CZTS system taking as reference the impedance shown in Figure 6, which is consistent with a constant phase element (CPE). In this case, the experimental values conform a value of n=1.1. In this way, the use of the CPE allows make an excellent fit of the experimental data because it takes into account the independent gap of frequency between the alternate potential and its response in terms of current. Therefore, the results derived from the electrical circuit demonstrated that the material presents a superficial heterogeneity, resulting in a superficial roughness. This physical phenomenon is probably related to the presence of the minority of covalite phase in the structure of the material, responsible for a notable decline in the process of transfer of load on the material. These data are consistent with the results previously obtained through the analytical techniques described above [22, 23].

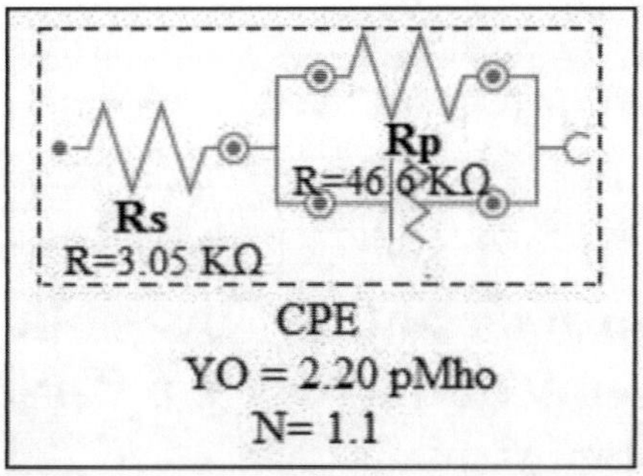

Figure 7. Equivalent circuit for the CZTS system.

4. Conclusions

The CZTS material obtained by means a hydrothermal route, showed physicochemical features of marked importance in the field of photovoltaic applications. The results of X-ray diffraction showed that the material are composed of nanometric particles (5-6nm), which agree with the transmission electron microscopy results. The obtained Band-gap value (1.46eV), allow to classify the solid as a semiconductor material in accordance with previous works. In this form, the synthesis method permits the obtaining of a material with a characteristic morphology textural and optical properties, for potential use as absorbent layer in photovoltaic devices.

References

[1] S A Kalogirou 2004 *Manage* **45** 3075
[2] Energy US 2005 *US Department of Energy* **9** 276
[3] M P Suryawanshi, G L Agawane, S M Bhosale, S W Shin et al 2013 *Technol* **28** 98
[4] S Huang, W Luo and Z Zou 2013 *J Phys D Appl Phys* **46** 235108
[5] J S Seol, S Y Lee, J C Lee, H D Nam and K H Kim 2003 *Energy Mater Sol Cells* **75** 155
[6] Q Guo, H W Hillhouse and R Agrawal 2009 *J Am Chem Soc* **131** 11672
[7] Teodor K Todorov, Kathleen B Reuter and B Mitzi 2010 *Adv Mater* **20** 156
[8] F Liu, Y Li, K Zhang, B Wang, C Yan, Y Lai, Z Zhang, et al 2010 *Sol Energy Mater Sol Cells* **94** 2431
[9] N Kamoun, H Bouzouita and B Rezig 2007 *Thin Solid Films* **515** 5949
[10] Y L Zhou, W H Zhou, M Li, Y F Du and S X Wu 2011 *J Phys Chem C* **115** 19632
[11] J Xu, X Yang, Q D Yang, T L Wong and C S Lee 2012 *J Phys Chem C* **116** 19718
[12] Mali S S, Shim C S and Hong C K 2014 *Mater Res Bull* **59** 249
[13] C Malerba, F Biccari, C L A Ricardo, M Valentini, R Chierchia, M Müller, et al 2014 *J Alloy Compd* **582** 528
[14] Vanalakar S A, Shin S W, Agawane G L, Suryawanshi M P, Gurav K V, et al 2014 *Ceram Int* **40** 15097
[15] Agawane G L, Shin S W, Vanalakar S A, et al 2015 *J Mater Sci Mater Electron* **26** 1900
[16] A Khare, B Himmetoglu, M Johnson, D J Norris, M Cococcioniet, et al 2012 *J Appl Phys* **111** 083707
[17] P A Fernandes, P M P Salomé, A F da Cunha 2011 *J Alloy Compd* **509** 7600
[18] G Rajesh, N Muthukumarasamy, E P Subramaniam, S Agilan, et al 2013 *J Sol-Gel Sci* Technol **66** 288
[19] Wang Y, Zhao X, Liu F, Zhang X, Chen H, Bao F, et al 2014 *RSC Adv* **4** 16022
[20] Huang W, Tseng C, Chang S, Tuan H, Chiang C Lyu, L et al 2012 *Langmuir* **28** 8496
[21] Vasquéz L G 2007 *Use of the technique of spectroscopy of impedances electrochemical characterization of biomaterials Application to a biomedical alloy of Co-Cr-Mo* (España: UPV)
[22] A Igual Muñoz, J García Antón, J L Guiñón, V Pérez Herranz 2006 *Corrosión Science* **48(12)** 4127
[23] Norlin A, Pan J, Leygraf C 2002 *Biomolecular Engineering* **19** 67

Synthesis and characterization by solid-state impedance spectroscopy of semiconductor Cu_2ZnSnS_4 material for photovoltaic technologies

M Muñoz, E Vera-López[1], J A Gómez-Cuaspud[1] and Y Pineda-Triana[1]
[1] Universidad Pedagógica y Tecnológica de Colombia Tunja, Colombia

E-mail: ingenero.mauricio.munoz@gmail.com

Abstract: Current work is focused on the synthesis and characterization of a Cu_2ZnSnS_4 material (Abbreviated CZTS), identified as a potential candidate for the manufacture of photovoltaic cells. The material was obtained by means of a hydrothermal route which permits a simple and economical alternative to synthesize advanced materials for photovoltaic applications. The synthesis of a solid started from corresponding metal nitrates of $Cu(NO_3)_2.6H_2O$, $Zn(NO_3)_2$, $Sn(NO_3)_4.6H_2O$ and thiourea as S source, which were dissolved in deionized water until complete a 1.0mol L^{-1} concentration. The solution was kept in a Teflon lined steel vessel with magnetic stirring (150 rpm) and treated at 300°C for 12 hours to form the crystalline phase. The initial characterization of solid was done using UV spectroscopy to validate the chemical process and identify the corresponding Band-gap around (1.43eV). The structural characterization by X-ray diffraction, confirmed the presence of nanometric solids (140-260nm). The morphological characterization by SEM analysis evidenced a homogeneous material in the form of micrometric aggregates, by a related synthesis method. Finally, the electrical characterization by means of solid state impedance spectroscopy demonstrated a semiconductor behaviour which evidenced the transport phenomena associated with a Warburg resistance.

1. Introduction

Recently, research related with photovoltaic materials has been focused on the search of new and sustainable energy sources that include non-toxic and abundant materials, with aim of reducing contamination by heavy metals and diminishing greenhouse emissions for the preservation of the atmosphere for future generations [1]. In this context, new materials based on kesterite and stannite structures have been developed for use in new photovoltaic cells. Such quaternary materials based on Cu_2ZnSnS_4 composites (CZTS), represent a real alternative to increase production and expand the photovoltaic technology to other fields of technology [2].

Based on the above, for several years, both the physics and chemistry of materials have focused attention on the progress of new synthesis routes which enable the economic production of CZTS materials in, allowing increased production and diversification of these technologies in the development of unconventional systems and previously un-explored solar energy systems [2]. Current interest is focused on the use of hydrothermal routes of synthesis in one or two-step reaction stages, that preserve and provide morphological characteristics of relevance in design of solar cell materials [4]. Therefore, present work is focused on the synthesis and characterization of a Cu_2ZnSnS_4 material, using a two-step hydrothermal route, allowing the study of the physicochemical properties of solids, to identify key chemical aspects of the synthesis process that allow the modelling of the best conditions to generate the highest level of photovoltaic performance. Accordingly, a series of characterization techniques which support the identification of these key aspects opens a wide field of exploration in the design and construction of devices for the generation of clean and renewable energy with low environmental impact [2].

2. Materials and methods

The precursors used to prepare the CZTS system were $Cu(NO_3)_3.3H_2O$, $Zn(CH_3COO)_2$, $SnCl_2$ and SH_4N_2S; all from Merck. The initial stage corresponded with the amount of each metal cation to reach a 1.0mol L^{-1} concentration of precursors in deionized water, using KOH as mineralizing agent. The solutions were kept in a Teflon lined steel vessels for 30 minutes with magnetic stirring (250rpm) at room temperature to provide a homogenous reaction medium in each case [3].

The solutions obtained were treated at 300°C for 12 hours to obtain the CZTS powders. The resultant materials present as a dark powder, which was washed in absolute ethanol several times to eliminate by products of the reaction. The solid was dried at 100°C in an electric oven for 3 hours, at the end of which the solid was calcined at 400°C in a tubular furnace for 2 hours using a helium flow (50mL min^{-1}), to promote the crystalline formation of desired material and avoid the possible oxidation of S in the material. The initial characterization of solid was done by means UV measurements in a HP 8453 UV-VIS spectroscope. The characterization by means X-ray diffraction (XRD), was done on a PANalytical X' per PRO MPD difractrometer provided with an ultra-fast X'Celerator detector in Bragg-Brentano configuration, using the Cu K_α radiation (λ=1.54Å). The diffractograms were taken between 10 and 90° 2θ and the results were analysed using the X'Pert High Score software and the Rietveld refinement.

The morphological characterization was performed by scanning electron microscopy (SEM), in a Leica-Zeiss LEO 440 electron gun equipment with an accelerating voltage of 30kV and the electrical and conductivity properties were evaluated using the solid state impedance spectroscopy technique (IS), in a GAMRY potentiostat-galvanostat instrument between 1 and 10MHz over 10mm pellets compacted in uniaxial form at 5.0Mpa pressure with a 10mm diameter and 1mm standard thickness, to avoid variations in measurements.

3. Results and discussion

The initially UV measurements provided information regarding the main electron transitions related with the CZTS composition as shown in Figure 1.

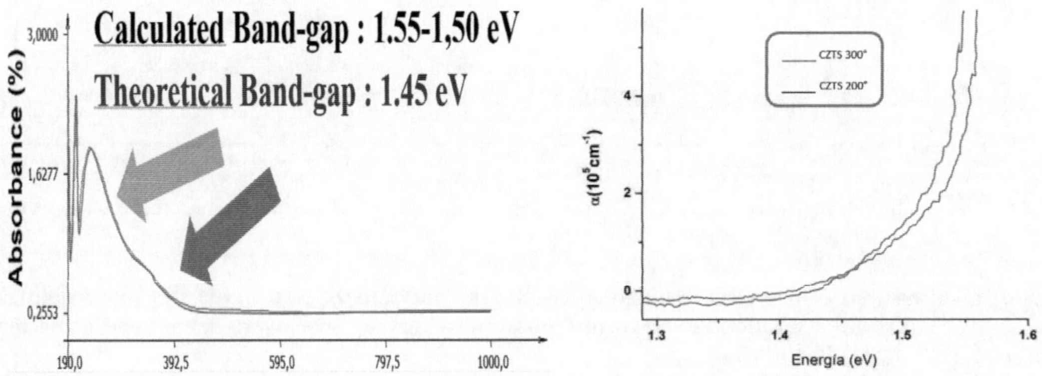

Figure 1. UV spectrums with corresponding Band-gap energy values obtained from general scan between 190-1100nm,

The Band-gap values were obtained using the Kubelka Munk technique, providing a value of 1.55eV, in accordance with a semiconductor material. Additionally, evident was a strong absorption phenomena around the 220nm zone in which the solid has the capacity to absorb the most energetic portion of UV radiation. The structural characterization of solid by X-ray diffraction confirmed the presence of the main diffraction signals of CZTS in a tetragonal (Figure 2) system oriented along (0 1 1), (0 0 2) and (0 2 0) facets, as the most intense signals (Figure 3). A rise in temperature in the second thermal treatment of material provided more definitive evidence of a corresponding elimination of secondary phases associated with villaminite and wursite as shown in Figure 3.

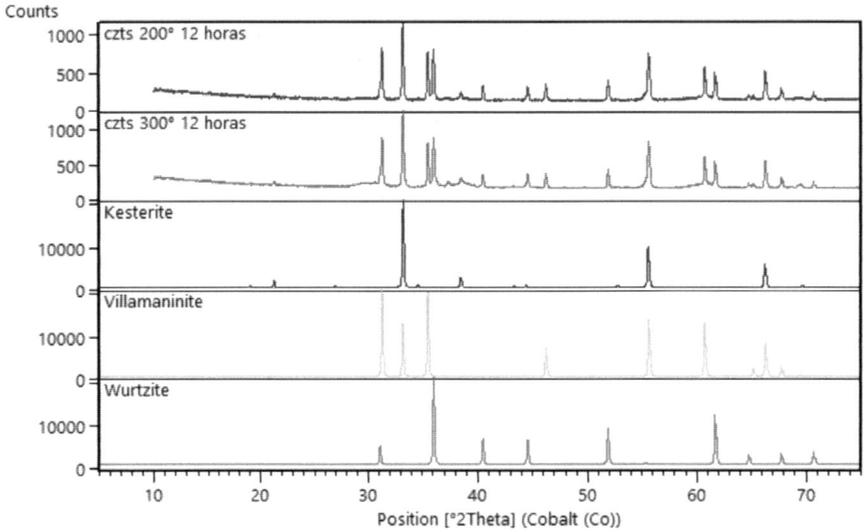

Figure 2. Comparison CZTS X-ray patterns obtained with the simulated patterns of secondary phases.

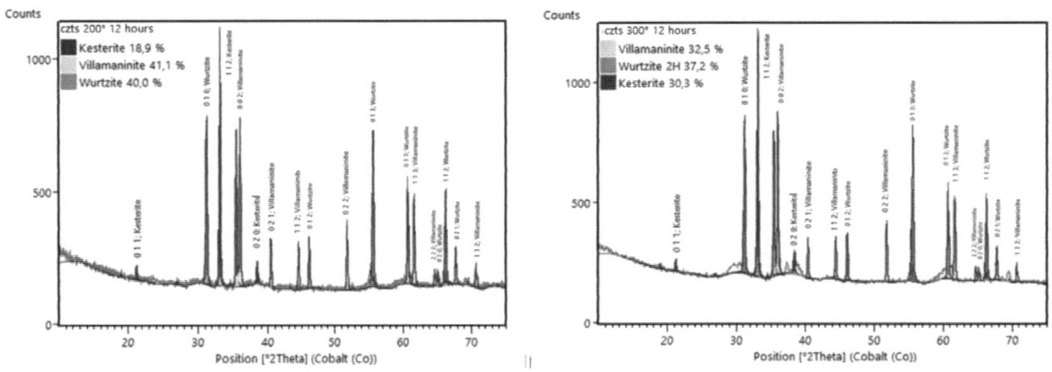

Figure 3. Quantitative analysis of CZTS powders obtained by Rietveld refinement, show the incidence of villaminite and wursite.

A detailed analysis of the XRD results demonstrate that differences exist in the size of crystallites after the second stage of reaction as shown in Figure 4. The crystal sizes were determined using the Scherrer equation as follows:

$$D = K\lambda/\beta\cos\theta \qquad (1)$$

Where λ is the wavelength generated by the copper source supplying radiation to the sample, K is a constant close to 1 and θ is the angle corresponding to the main diffraction signal, resulting in a crystallite size around 140nm to CZTS sample before thermal treatment and 260nm for the sample annealed at 400°C. These results could be important in determination of electric properties, since these results mean that resistance values due to grain boundaries in the material annealed at 400°C exhibit a better electrical response in terms of the effectiveness of carriers transfer [12].

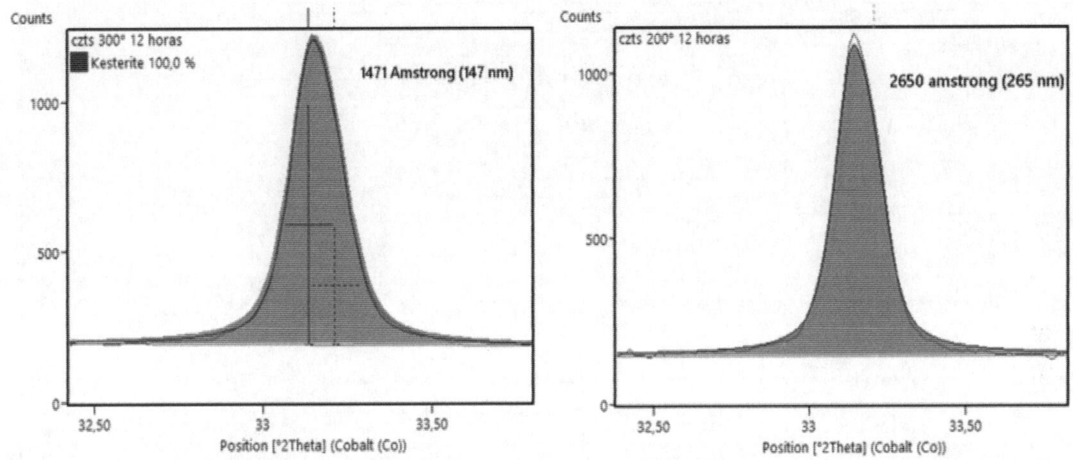

Figure 4. Crystallite size of CZTS materials obtained at 300 and 400°C.

Figure 5 (a) and (b), shows the images obtained by the scanning electron microscopy (SEM) at identical magnification, with clear evidence of the development of homogeneous powders with a typical morphology related with the synthesis process [4]. The diameter of CZTS semispherical particles changes from 2.8 to 3.2μm with increasing temperature from 300 to 400°C, indicating that thermal treatment in CZTS solid strongly influenced the crystal growth. Therefore, the morphology was found to be dependent upon the synthesis conditions as well as the reaction time.

Figure 5. Morphology of the CZTS sample obtained in (a) one-step and (b) two-step reaction stages.

Finally, the electrical characteristics of CZTS powders were evaluated by solid-state impedance in a cell adapted and modified for this purpose. This characterization enabled the establishment of a semiconducting behaviour with conductivity levels around 5.4x10^{-4}S. The interpretation of the data needs the study of the Nyquist plots, which was adjusted to a transfer function (impedance), where in addition to the classic elements such as resistors or elements of constant phase, it was necessary to define an electronic equivalent circuit to fit the electrochemical response of material as shown in Figure 6. The identification of a Warburg (W) resistance in the system obtained at low frequencies, confirmed that the materials display a transfer phenomenon attributed mainly to the presence of heterogeneous surfaces and polymorphisms, which increase the tortuosity of electric charge carriers as has been established in preliminary works [10,11].

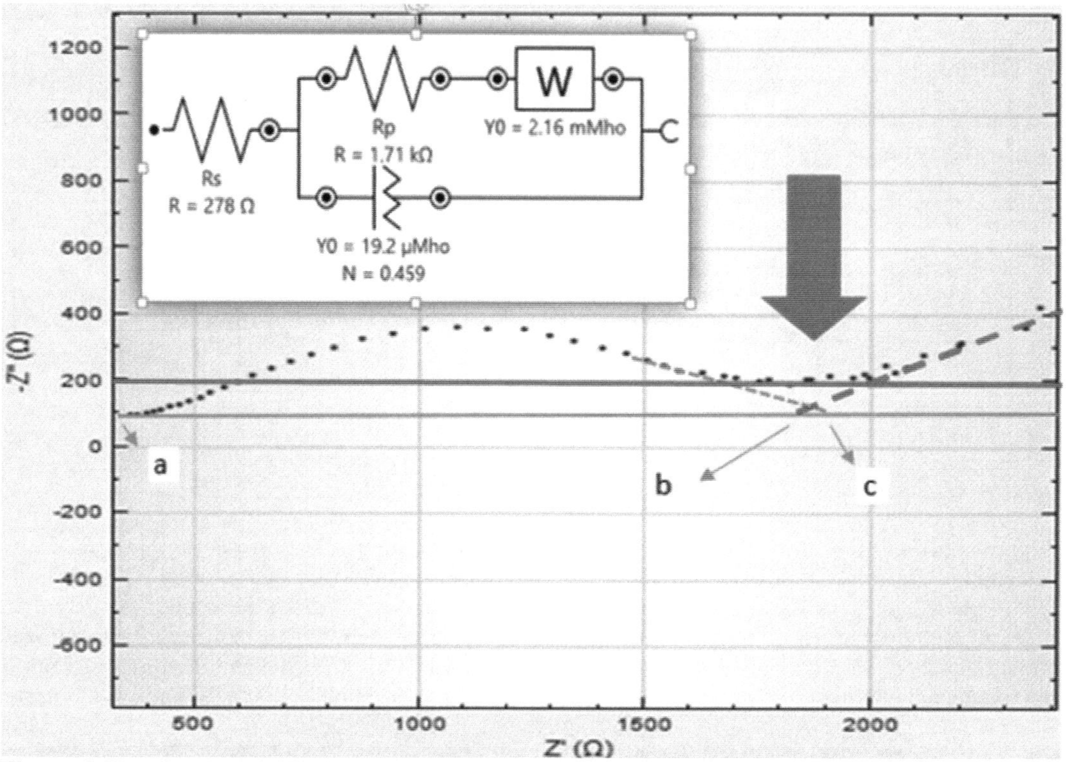

Figure 6. Nyquist plot of CZTS material obtained at 400 °C.

4. Conclusions

The current work confirms how the development of a second thermal treatment in the hydrothermal synthesis of CZTS powders, provides a more efficient tool for the design of more effective photovoltaic materials. The UV measurements, evidence the obtaining of a material with an optimal Bang-gap (1.55eV), which permit the ideal absorption in the UV region. The XRD characterizations confirm the obtaining of a crystalline material in tetragonal configuration with presence of secondary phases of villaminite and wursite, which could be eliminated by a thermal treatment in the second stage of reaction. The morphological results demonstrate that temperature promotes the obtaining of solids with a crystal size around 260nm, reflecting a strong dependence with the electrical behaviour. The implementation of the solid-state impedance spectroscopy to validate the electrical characteristics of CZTS powders was effective enabling the study of semiconductor materials for photovoltaic applications.

Thanks to

The authors are grateful for the support of the instituto de investigación e innovación en ciencia y tecnología de los materiales and to investigation unit of universidad pedagógica y tecnológica de Colombia, who financed the funds and equipment necessary for the successful development of the project.

References

[1] García Verduch 1994 Cerámica tradicional y no tradicional *Bol. Soc. Esp. Ceram. Vidrio* **33(6)** 309313

[2] L Sánchez Muñoz, I T Marinova, G F de la Fuente, I Nuñez, M A Rodríguez, J Sanz, J B Carda 2010 I+D para las baldosas cerámicas del siglo XXI: Competencia, diversidad y funcionalidad *Técnica Cerámica* **385** 468-477

[3] S M Camara, L Wang and X Zhang 2013 Easy hydrothermal preparation of Cu2ZnSnS4 (CZTS) nanoparticles for solar cell application *Nanotechnology* **24(49)** 495401

[4] Tuan Truong Mau 2012 Synthesis of Cu2ZnSnS4 thin film absorbers by sulfurizing dip-coated precursors *Journal of Ceramic Processing Research* **13** 301-304

[5] A Moreno Berto 2007 Ceramic tiles: above and beyond traditional applications *J. Eur. Ceram. Soc.* **27** 16071613

[6] D Iencinella, E Centurioni, M G Busana 2009 Thin film solar cell on comercial ceramic tiles *Solar. Energy Mat. Solar Cells* **93** 206210

[7] Y N Xia, P D Yang, Y G Sun, Y Y Wu, B Mayers, B Gates, Y D Yin, F Kim, H Q Yan 2003 One dimensional nanostructures: synthesis, characterization, and applications *Adv Mater.* **15** 353389

[8] Y F Zhu, D H Fan, W Z Shen 2008 Chemical conversion synthesis and optical properties of metal sulfide hollow microspheres *Langmuir* **24** 1113111136

[9] J Mendoza Florez, R Duran Romero 2012 Espectroscopia de impedancia electroquímica *Corrosion* **24** 5-24

[10] U Piratoba Morales, A Mariño Camargo 2009 Impedancia electroquímica-interpretación de diagramas típicos con circuitos equivalentes *Dyna rev.* **77(164)** 0012-7353

[11] N Diaz Abril, J Bárcera Ruiz, E Fernández Ruiz 2008 *Espectrofometría: Espectros de absorción y cuantificación colorimétrica de biomoléculas* (Córdoba: Universidad de Córdoba) pp 1-8

[12] S C Riha, B A Parkinson and A L Prieto 2009 Solution based synthesis and characterization of Cu2ZnSnS4 nanocrystals *J. Am. Chem. Soc.* **131(34)** 12054–12055

IOP Conf. Series: Journal of Physics: Conf. Series **786** (2017) 012029 doi:10.1088/1742-6596/786/1/012029

Structural and electrical study of $LaCrO_3$ modified with Fe and Co

W Rativa-Parada[1], J A Gómez-Cuaspud[1], E Vera-López[1] and J B Carda[2]
[1] Universidad Pedagógica y Tecnológica de Colombia, Tunja, Colombia
[2] Universitat Jaume I, Castellón de la Plana, España

E-mail: wilson.rativa@uptc.edu.co

Abstract: The present work shows the effect of a modification on $LaCrO_3$ oxide using Fe and Co cations over the structural and electrical properties in order to evaluate its response to applications as electrode in solid oxide fuel cells (SOFC). The $LaCr_{0.8}Fe_{0.2}O_3$ and $LaCr_{0.6}Co_{0.2}Fe_{0.2}O_3$ oxides were synthetized by polymerization-combustion method using citric acid as complexing agent. The characterizations by means of X-ray diffraction (XRD), energy dispersive X-ray (EDX), and Raman spectroscopy revealed the conformation of a $LaCrO_3$ single phase with an orthorhombic structure (Pnma 62) in both ceramics and average crystal size of 131nm. The images of scanning electron microscopy (SEM) and transmission electron microscopy (TEM) confirmed the conformation of materials with relevant morphological characteristics obtained because of synthesis method used. Analysis by Impedance spectroscopy (IS) from room temperature to 800°C, indicated that both oxides were semiconductor type in accordance to thermal activation process.

1. Introduction

Solid oxide fuel cells (SOFC) are an important technological alternative for the optimal use of hydrogen as fuel on the production of sustainable energy. These devices are a key factor in power generation with low greenhouse emissions and a high efficiency [1]. The current study on these systems confirm that the use of modified perovskites as electrode components, mainly focused on development on anodic components based on $LaCrO_3$ systems, could resolve the problems related with the low catalytic efficiency and the stability of solid at high temperatures when the oxidation fuel process take place [2]. However, some drawbacks related with the obtainment of impure phases and the necessity of diminish the calcination periods in the synthesis by the conventional solid state reaction that affect the structure and the electrical response, have allowed the development and adaptation of new synthesis methods that provide a major control on aspects related with particle size, chemical and structural stability, electrical behaviour, sinterability and reduction of chromium volatilization [3-4]. In this context, this investigation is focused in the synthesis by the polymerization-combustion method and corresponding characterization of two oxides based on $LaCr_{0.8}Fe_{0.2}O_3$ and $LaCr_{0.6}Co_{0.2}Fe_{0.2}O_3$ systems, in order to identify the effect of doping in the B site of perovskite over the structural and electrical properties.

2. Experimental

For the synthesis of materials, stoichiometric amounts of $La(NO_3)_3.6H_2O$, $Fe(NO_3)_3.9H_2O$, $Cr(NO_3)_3.9H_2O$ and $Co(NO_3)_3.9H_2O$ (99.99%) were dissolved and dossed in glass reactors with temperature control, magnetic stirring and reflux. Each solution was heated to 80°C during 120min.

Then a quantity of citric acid 2.00M was added in a 0.5:1 molar ratio. The solutions were heated at 350°C for 24 hours obtaining the solvent evaporation and the consolidation of dense gels. The gels were calcinated at 850°C during 12 hours to obtain the corresponding powders.

The X-ray diffraction measurements were performed in a Panalytical X´pert PRO-MPD equipment, with a Bragg-Brentano configuration using Cu radiation (Kα=1.54056Å), between 10 and 80° (2Θ). The Debye-Scherrer equation permitted to evaluate the crystal size of samples, using a constant of 0.89 in all cases. The surface morphology was observed using scanning electron microscopy and transmission electron microscopy using a Leica-Zeiss LEO 440 microscope and a JEOL 2100 equipment using a LaB_6 thermionic gun operated with an acceleration voltage of 200kV, equipped with a CCD imaging system, respectively. The elementary chemical composition was confirmed by energy dispersive X-ray analysis. Raman spectroscopy measurements were carried out in a HR-UV 800 infinity microprobe (Jobin-Yvon) spectrometer equipped with a CCD detector (−70◦ C) and a laser power of 10.7mW. The Raman spectrums of the solids were collected between 100 and 800cm^{-1}, projecting a continuous wave laser of He-Ne. The electrical characterization was performed using a Agilent 4294A spectrometer from 25°C to 800°C, with solid pellets compacted in an uniaxial form at 5Mpa. The electrical results were collected and corrected using a solid state reference cell in each determination.

3. Results and discussion

The XRD results are shown in Figure 1, and reveal the formation of a single perovskite phase in both oxides. No secondary phases were detected. The detailed analysis of the samples allows to establish that $LaCr_{0.8}Fe_{0.2}O_3$ and $LaCr_{0.6}Co_{0.2}Fe_{0.2}O_3$ oxides have a high similarity with the reference pattern of $LaCrO_3$ (JCPDS=01-089-0478), an orthorhombic structure Pnma (62), cell parameters a=5.4790Å, b=7.7562Å, c=5.5161Å and a cell volume of 234.41Å^3, in accordance with similar compositions [5,6]. In all cases the crystal size was calculated using the highest diffraction signals along (1 1 2) facet, confirming the obtainment of crystals around 131nm.

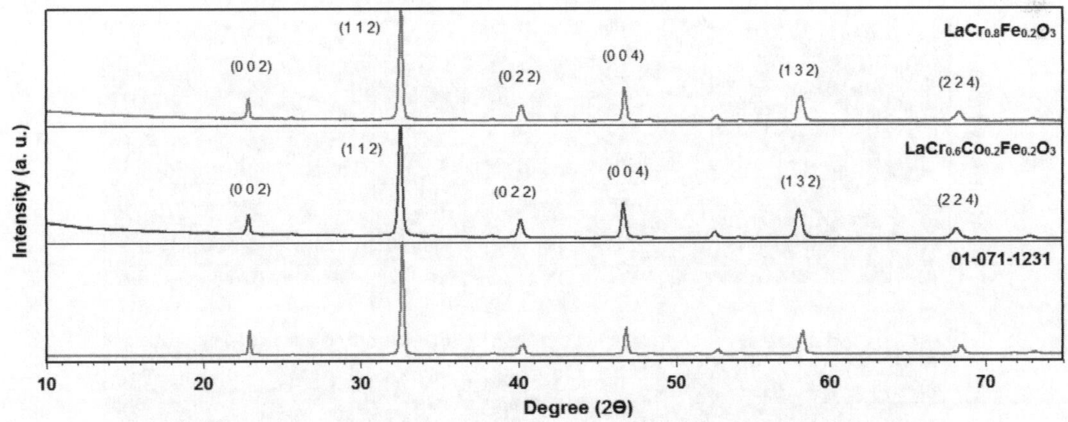

Figure 1. XRD patterns of $LaCr_{0.8}Fe_{0.2}O_3$ and $LaCr_{0.6}Co_{0.2}Fe_{0.2}O_3$ oxides.

The Figure 2 show the SEM images of $LaCr_{0.8}Fe_{0.2}O_3$ and $LaCr_{0.6}Co_{0.2}Fe_{0.2}O_3$ oxides. In both materials, particles present semispherical and regular morphology with a distribution according to synthesis process. Additionally, it is observed interconnect irregular porosity originated by decomposition of volatile compounds in the combustion and calcination of samples [7,8]. The EDX analysis of the Figure 3 reveal the presence of Cr, Fe, Co and O, validating the proposed stoichiometry in terms of a high purity and chemical stability of oxides.

IOP Conf. Series: Journal of Physics: Conf. Series **786** (2017) 012029 doi:10.1088/1742-6596/786/1/012029

(a)

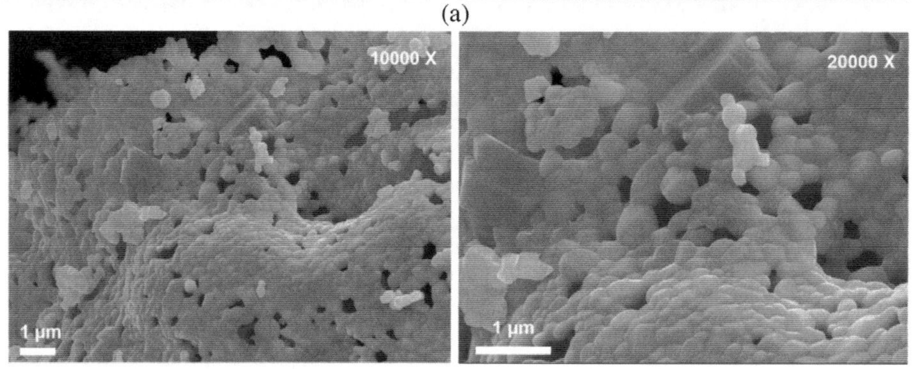

(b)

Figure 2. SEM images of (a) $LaCr_{0.8}Fe_{0.2}O_3$ and (b) $LaCr_{0.6}Co_{0.2}Fe_{0.2}O_3$ to different magnifications.

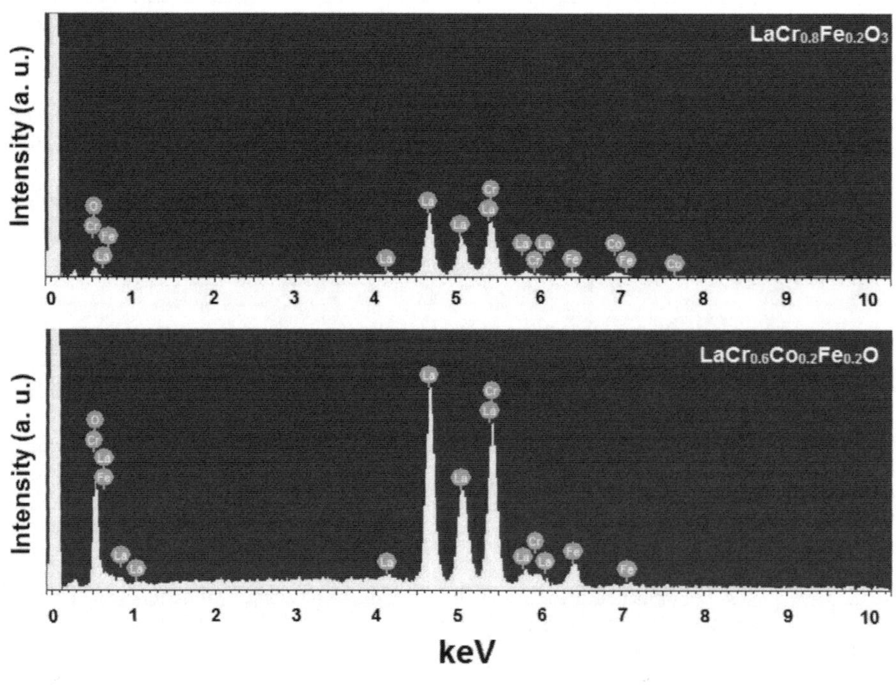

Figure 3. EDX analysis of $LaCr_{0.8}Fe_{0.2}O_3$ and $LaCr_{0.6}Co_{0.2}Fe_{0.2}O_3$.

Figure 4 with the TEM images allows to observe the crystalline nature of the particles, showing a homogeneous distribution of particles with an average size around 168nm, and interplanar distances of 0.86nm corresponding to the (1 1 2) facets in accordance with XRD results [9].

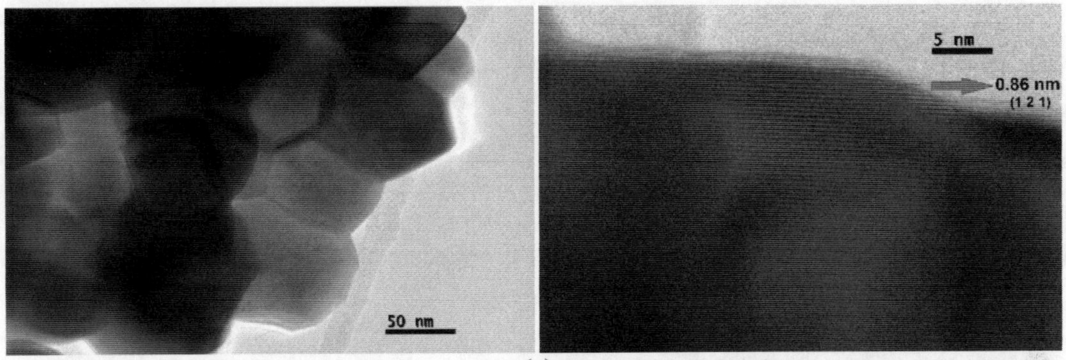

(a)

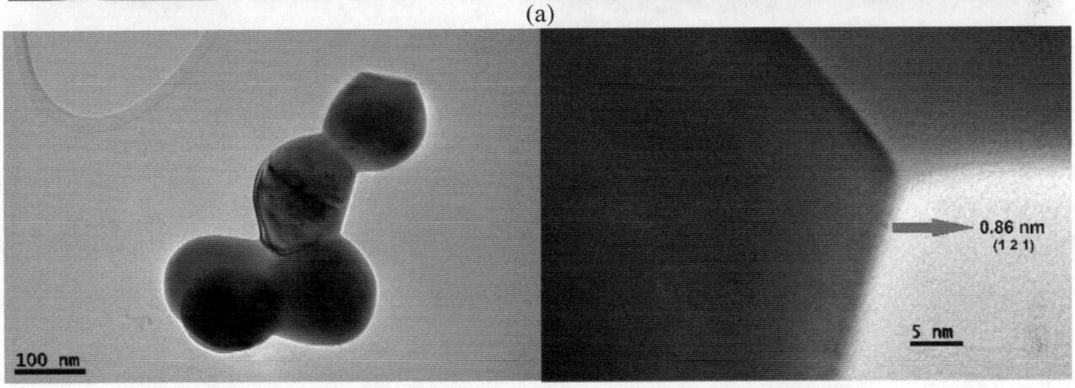

(b)

Figure 4. TEM images of interplanar distance and average particle size of (a) $LaCr_{0.8}Fe_{0.2}O_3$ and (b) $LaCr_{0.6}Co_{0.2}Fe_{0.2}O_3$.

The Raman results of the Figure 5 confirm the presence of vibrational active modes $\Gamma=A_g+7B_{1g}+5B_{2g}+5B_{3g}+8A_u+10B_{1u}+8B_{2u}+10B_{3u}$, with 24 Raman actives to $\Gamma=7A_g+7B_{1g}+5B_{2g}+5B_{3g}$ According to above, the signals located at 432 and 467cm^{-1} are related with A_g modes, and the signals at 524 and 701cm^{-1} corresponding with B_{2g} modes [10]. The absence of another additional peaks suggests the consolidation of a pure and single structures related with $LaCr_{0.8}Fe_{0.2}O_3$ and $LaCr_{0.6}Co_{0.2}Fe_{0.2}O_3$ oxides.

Finally, the electrical characterization shown in Figure 6, confirm that conductivity of $LaCr_{0.8}Fe_{0.2}O_3$ and $LaCr_{0.6}Co_{0.2}Fe_{0.2}O_3$ samples increases with temperature in a thermal activation process, which provoke a major mobility of the charge carriers through B-O-B bonds [11]. The modification of oxides with a high concentration of Co promotes an increase in the conductivity, allowing to calculate the activation energy from corresponding Arrhenius plots. The low values of activation energy indicate that conductivity is presented by means of electronic mechanisms more than ionics [8,12]. Furthermore, the incorporation of Co in perovskite oxides produce an increasing in density of charge carriers which are stimulated by thermal activation process related with the conductivity.

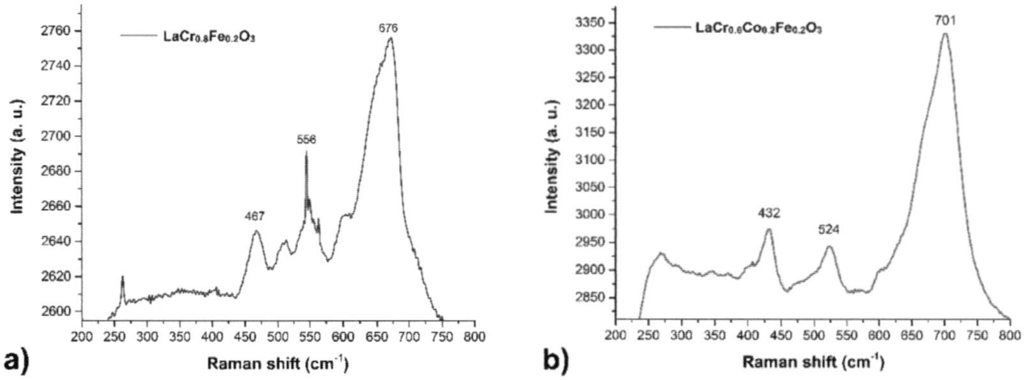

Figure 5. Raman shift of (a) $LaCr_{0.8}Fe_{0.2}O_3$ and (b) $LaCr_{0.6}Co_{0.2}Fe_{0.2}O_3$.

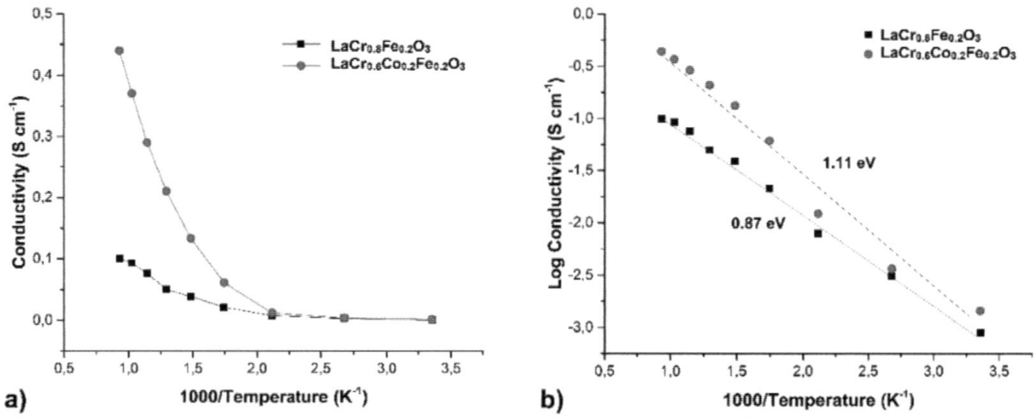

Figure 6. (a) Plot of conductivity and (b) Arrhenius plot from 25 °C to 800 °C of $LaCr_{0.8}Fe_{0.2}O_3$ and $LaCr_{0.6}Co_{0.2}Fe_{0.2}O_3$.

4. Conclusions

The $LaCr_{0.8}Fe_{0.2}O_3$ and $LaCr_{0.6}Co_{0.2}Fe_{0.2}O_3$ oxides were obtained by means of polymerization-combustion method with crystal sizes of nanometric scale and morphological properties of relevance. The XRD analysis allowed to verify the formation of a pure phase with an orthorhombic structure, Pnma (62) space group and crystal sizes around 131nm in both perovskites. SEM and TEM characterization techniques confirmed that the synthesis method facilitates the obtaining of materials with homogeneous surfaces, high porosity, and regular morphology. Impedance spectroscopy analysis showed as the addition of Co reduced the resistivity to high temperatures of both semiconductor type ceramics for potential applications in design of advanced electrodes for solid oxide fuel cells.

References

[1] R D Purohit, Sathi R Nair, Deep Prakash, P K Sinha, B P Sharma, K P Sreekumar, P V Ananthapadmanabhan, A K Das, L M Gantayet 2010 Development of Ca-doped $LaCrO_3$ feed material and its plasma coating for SOFC applications *Journal of Physics: Conference Series* **208** 012125

[2] Romualdo S Silva Jr, Petrucio Barrozo, N O Moreno, J Albino Aguiar 2016 Structural and magnetic properties of $LaCrO_3$ half-doped with Al *Ceramics International* **42** 14499–14504

[3] V A Kolotygin, E V Tsipis, M F Lü, Y V Pivak, S N Yarmolenko, S I Bredikhin, V V Kharton 2013 Functional properties of SOFC anode materials based on $LaCrO_3$, La(Ti,Mn)O_3 and Sr(Nb,Mn)O_3 perovskites: A comparative analysis *Solid State Ionics* **251** 28–33

[4] Nitin Labhasetwar, Govindachetty Saravanan, Suresh Kumar Megarajan, Nilesh Manwar, Rohini Khobragade, Pradeep Doggali, Fabien Grasset 2015 Perovskite-type catalytic materials for environmental applications *Sci. Technol. Adv. Mater.* **16** 036002

[5] Ribedi Bora, S Ravi 2014 Effect of Ce doping on the magnetic properties of $LaCrO_3$ *Physic B* **448** 233–236

[6] W Acchara, C R C Sousa, S R H Mello-Castanho 2012 Mechanical performance of $LaCrO_3$ doped with strontium and cobalt for SOFC Interconnect *Materials Science and Engineering A* **550** 76–79

[7] H P S Corrêa and C. O. Paiva-Santos 2008 Crystal structure refinement of Co-doped lanthanum chromites *Powder Diffraction Suppl.* **23(2)** S18-S22

[8] San Ping Jiang, Li Liua, Khuong P Ong, Ping Wu, Jian Li, Jian Puc 2008 Electrical conductivity and performance of doped $LaCrO_3$, perovskite oxides for solid oxide fuel cells *Journal of Power Sources* **176** 82–89

[9] Luke M Daniels, Mads C Weber, Martin R Lees, Mael Guennou, Reza J Kashtiban, Jeremy Sloan, Jens Kreisel, Richard I Walton 2013 Structures and Magnetism of the Rare-Earth Orthochromite Perovskite Solid Solution $La_xSm_{1-x}CrO_3$ *Inorg. Chem.* **52** 12161–12169

[10] V D Nithya, R Jacob Immanuel, S T Senthilkumar, C Sanjeeviraja, I Perelshtein, D Zitoun, R Kalai Selvan 2012 Studies on the structural, electrical and magnetic properties of $LaCrO_3$, $LaCr_{0.5}Cu_{0.5}O_3$ and $LaCr_{0.5}Fe_{0.5}O_3$ by sol–gel method *Materials Research Bulletin* **47** 1861–1868

[11] Handong Jiao, Junxiang Wang, Jianbang Ge, Long Zhang, Hongmin Zhu, Shuqiang Jiao 2016 Fabrication, characterization and electrical conductivity of Ru-doped $LaCrO_3$ dense perovskites *Solid State Communications* **231-232** 53–56

[12] Hui Qi, Yongxin Luan, Songtian Che, Ling Zuo, Xudong Zhao, Changmin Hou 2016 Preparation, characterization and electrical properties of Ca and Sr doped $LaCrO_3$ *Inorganic Chemistry Communications* **66** 33–35

Effect of substrate nature on the electrochemical deposition of calcium-deficient hydroxyapatites

A F Gualdrón-Reyes[1], V Domínguez-Vélez[2], J A Morales-Morales[2], R Cabanzo[1] and A M Meléndez[1]
[1] Universidad Industrial de Santander, CMN, Piedecuesta, Santander, Colombia
[2] Universidad Santiago de Cali, Cali, Colombia

E-mail: amelende@uis.edu.co , angelemet@gmail.com

Abstract. Calcium phosphates were obtained by reducing nitrate ions to produce hydroxide ions on TiO_2/stainless steel and TiO_2/titanium electrodes. TiO_2 coatings on metallic substrates were prepared by sol-gel dip-coating method. The morphology of deposits was observed by FESEM. Chemical nature of calcium phosphate deposits was identified by Raman micro-spectroscopy and FESEM/EDS microanalysis. Electrochemical behavior of nitrate and nitrite reduction on stainless steel and titanium electrodes was studied by linear sweep voltammetry. In addition, voltammetric study of the calcium phosphate electrodeposition on both electrodes was performed. From these measurements was selected the potential to form a calcium phosphate. A catalytic current associated to nitrate reduction reaction was obtained for stainless steel electrode, leading to significant deposition of calcium phosphate. Ca/P ratio for both substrates was less than 1.67. The formation of calcium deficient hydroxyapatite was confirmed by Raman spectroscopy.

1. Introduction

Calcium phosphates (CP) are ceramics used as coatings to improve the biocompatibility between an implant and the damaged tissue during the surgery recovering, because its composition is similar to osseous tissues, besides has been demonstrated that CP promote the tissue growth [1]. Depending on pH, temperature of synthesis and mainly Ca/P ratio, different CP phases can be obtained, such as dicalcium phosphate dehydrate (DCPD), octacalcium phosphate (OCP), hydroxyapatite (HAp) and calcium-deficient hydroxiapatite (CDHAp), being this latter one the most studied in biomedicine. Although HAps are widely known by their high bioactivity in reparation or substitution of bond [2], other CP exhibits a more extended applicability. Thus, CP as DCPD and CDHAp have a high commercial interest since they have been used as bioinorganic material in tooth calcification [3], and as drugs nanocarriers for periodontitis treatment [4].

The CP are usually obtained as coatings on metallic substrates for biomedical applications, employing conventional techniques as plasma spraying [5], chemical vapor deposition [6] and sputtering [7], which requires simultaneously high vacuum and control of temperature, high power laser or an electron beam to perform the process. In addition, CP powders are used as starting material, thereby elevating the cost of synthesis. Electrochemical deposition of CP has been considered as an attractive alternative to obtain coatings on diverse sizes or shapes of metallic substrates. Modification either of the applied current or potential, or the chemical composition of the deposition bath, can lead to morphological change. Besides depending on time and temperature, different coating thicknesses are achieved [1]. Different electrolytes have been used to prepare CP coatings on titanium [2] and stainless steel [3] substrates. Nitrates are preferred over other electrolytes because they can be electrochemically reduced generating hydroxide ions (OH^-), which increases the interfacial pH, and causes the precipitation of CP [2,3].

Recently, stainless steel surfaces have been coated with a TiO_2 sol-gel film to improve their corrosion resistance and the adhesion of CP during its deposition [8], but as far as it knows, there is no reports on the influence of substrate nature on the CP electrodeposition.

2. Experimental

2.1. Preparation of electrodes

Commercial AISI 304 stainless steel (SS) and titanium (Ti, 99.5% of purity) plates were polished with SiC emery papers of different grades. Samples were degreased with ethanol, acetone, and rinsed with deionized water for 5min in each solvent. TiO_2 films were prepared by sol-gel method and deposited on SS and Ti substrates by dip-coating technique, details of the method can be found elsewhere [9]. Coatings were dried at 100°C and annealed at 400°C during 2h at a heating rate of 3°C min^{-1}.

2.2. Physicochemical and electrochemical characterization

Electrochemical measurements were performed in a three-electrode cell with a jacket for temperature control. SS or Ti disk electrodes (diameter: 6.0mm) or TiO_2/SS or TiO_2/Ti plates were used as working electrode. A graphite rod (AGKSP grade) was employed as counter electrode while an Ag/AgCl (3M KCl) was used as reference electrode. The composition of electrolyte solution to deposit electrochemically CP was 0.1M $Ca(NO_3)_2{\cdot}4H_2O$ and 0.06M $NH_4H_2PO_4$ at pH 5.0, adjusted with 1M NaOH. In order to determine the condition to perform the CP electrodeposition, voltammetry measurements for SS and Ti electrodes without TiO_2 coating were carried out in negative direction from open circuit potential at 298K, 313K and 323K. The aforementioned electrodes were also characterized in a) 0.1M $NaNO_2$, b) 0.1M $NaNO_3$ and c) 1M $NaNO_3$ with and without stirring. The deposition of CP on TiO_2/SS and TiO_2/Ti electrodes was performed applying a bias potential of -1.41V at 323K during 30min. All measurements were performed using an Autolab PGSTAT 302N potentiostat and Nova software. The surface of TiO_2/SS and TiO_2/Ti electrodes coated with CP was observed with a FESEM (Quanta FEG 650), and their chemical composition was analyzed with an EDAX Apollo X. Raman spectra were obtained using a confocal high-resolution Raman spectrometer with a 473nm blue laser, integration time of 2s and an average of 100 scans.

3. Results and discussion

3.1. Electrochemical characterization

Cracking of TiO_2 films has frequently been reported in the literature [9,10]. When a sol-gel film is immersed in a solution, inter-crack spacing causes exposure of the substrate to solution. For this reason, initial electrochemical study was performed on metallic substrates. Figure 1 shows the electrochemical behavior of nitrate and nitrite ions on stainless steel in the potential range from ~-0.3V to -1.2V. In both cases, a reduction peak appears at -0.61V and -0.65V under quiet condition. These peaks can be associated either to nitrate or nitrite reduction [11], or to iron oxides reduction from passive layer of SS [12]. In order to know more about the reduction process, solution was stirred (Figures 1(a'), (b')). An increase of the current was registered for nitrate and nitrite solutions in disturbed condition, indicating that process depends on ion concentration at interface. However, increasing the nitrate concentration 10 times and maintaining the solution stirred, the magnitude of the current stayed the same (Figures 1(b'), (c)); hence, the reduction peak at ~-0.6V corresponds to Fe(III) reduction from passive layer of SS [12].

Figure 2 shows the electrochemical behavior of nitrate and nitrite ions on titanium, a significant increase of current around -1.1V is observed, which is also registered for SS. In addition, current of C2 and C3 peaks increases as the temperature of system increases. According to literature [11], these processes are associated to nitrate and nitrite reduction, Equations (1) and (2).

$$NO_3^- + H_2O + 2e^- \rightarrow NO_2^- + 2OH^- \quad (1)$$

$$NO_2^- + 5H_2O + 6e^- \rightarrow NH_3 + 7OH^- \quad (2)$$

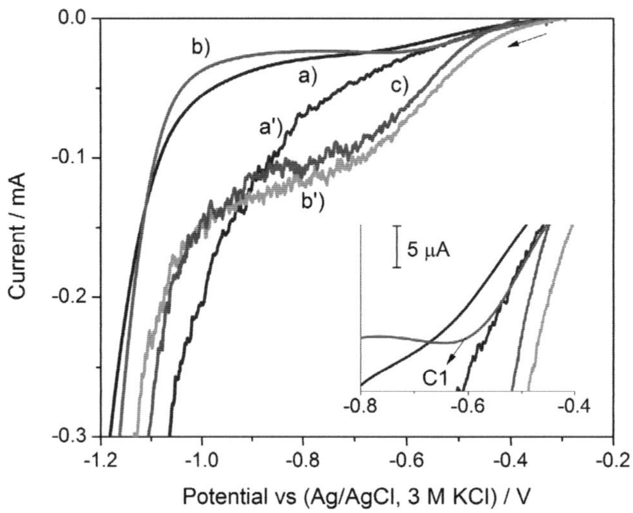

Figure 1. Voltammetric responses of (a) 0.1M $NaNO_2$, (b) 0.1M $NaNO_3$ and (c) 1.0M $NaNO_3$ under quiet (a), (b) and disturbed (a'), (b'), (c) conditions.

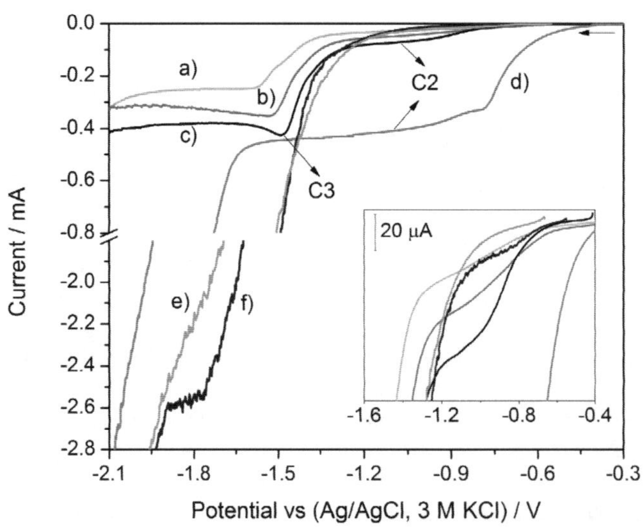

Figure 2. Nitrate reduction on (a)-(c) titanium and (d) stainless steel in a phosphate containing electrolyte at (a) 298K, (b) 313K and (c) 323K. Reduction of (e) nitrite and (f) nitrate ions on stainless steel at 298 K under disturbed condition.

The nitrate and nitrite reduction processes occurs at C2 and C3 peaks, respectively. On the other hand, the current generated by SS at 323K was higher than that of Ti. It has been reported that Cu or Ni has a catalytic activity on nitrate reduction [13]; hence, nickel present in the passive layer of SS could catalyze the reduction of nitrate. According to Equations (1) and (2), the potential selected for carrying out the electrochemical deposition of CP was -1.41V (Figure 3). In order to prevent the depletion of electroactive species at interface, a temperature of 323K was chosen to compact the thickness of the diffusion layer, and thereby increase the flux of electroactive species diffusing to the interface.

During the nitrate and nitrite reduction, hydroxide ions are formed and react with the monobasic phosphate to produce dibasic phosphate and then produce phosphate ions (Equations (3) and (4)), which lead to the formation of CP [1].

$$H_2PO_4^{-}+OH^{-} \rightarrow HPO_4^{2-}+H_2O \quad (3)$$

$$HPO_4^{2-}+OH^{-} \rightarrow PO_4^{3-}+H_2O \quad (4)$$

The high current generated by TiO_2/SS in comparison with TiO_2/Ti electrode indicates indirectly the formation of a high amount of phosphate ions (Figures 3 and 4), as result of a high production of hydroxide ions at interface.

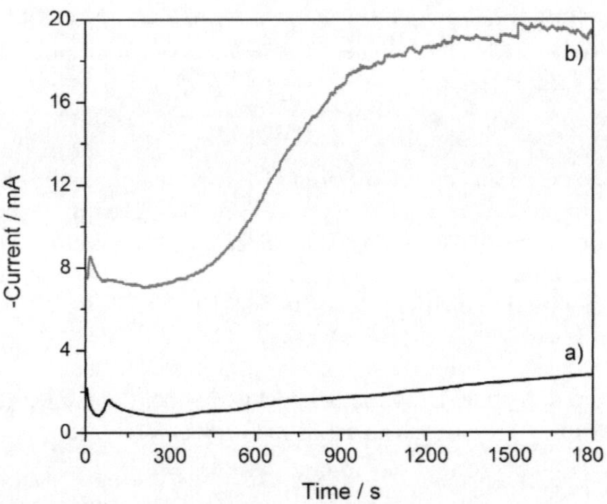

Figure 3. Transient current generated upon the CP deposition on (a) titanium and (b) stainless steel by holding the potential at -1.41V vs Ag/AgCl. Temperature: 323K.

3.2. Characterization of calcium phosphate deposits

Figure 4 shows FESEM images of electrodeposited CP coatings on TiO_2/SS and TiO_2/Ti. The CP deposit is coat more evenly on SS substrate (Figure 4(a), (a')) than that on Ti substrate (Figure 4(b), (b')). In order to determine the chemical nature of electrodeposited CP, Ca/P ratio was estimated from EDS analysis. A Ca/P ratio of 1.51 and 1.57 was determined for TiO_2/SS and TiO_2/Ti, respectively. This suggests the formation of CDHAp because Ca/P<1.67 [4]. Indeed, Ca/P values determined from EDS corresponds to an average of chemical composition, and more than one CP species could be present. Hence, other technique is required to confirm the presence of CDHAp. Figure 4(c) shows the Raman spectrum of CP deposited on TiO_2/SS. Typical peaks of anatase phase from TiO_2 film appears at 140.3cm^{-1}, 191.4cm^{-1}, 393.3cm^{-1}, 512.9cm^{-1} and 635.7cm^{-1}. Peak centered at 958.4cm^{-1} is representative of CP deposit on TiO_2/SS and TiO_2/Ti, and it has been commonly attributed to HAp [14]. Take into account EDS results and the structural similarity between HAp and CDHAp [4], CP deposit corresponds to CDHAp.

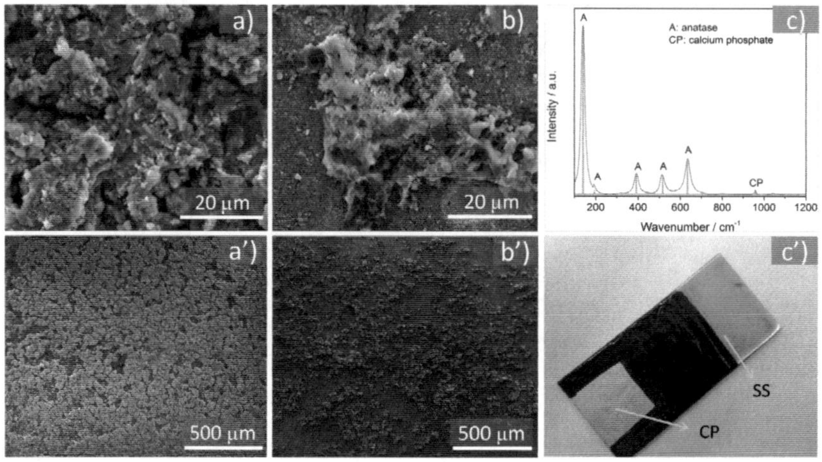

Figure 4. FESEM images of CP electrodeposited on TiO_2/SS (a), (a') and TiO_2/Ti substrates (b), (b´). Typical Raman spectrum and photo of CP deposited on SS (c), (c´).

4. Conclusions

It has been found an electrochemical condition to deposit calcium-deficient hydroxyapatites on stainless steel and titanium substrates coated with a TiO_2 sol-gel film. Nickel from stainless steel catalyze the production of hydroxide ions on TiO_2/SS electrode, which increases the production of calcium-deficient hydroxyapatite and thus enabling a homogeneous coating. The condition chosen in this study can further be modified to obtain others CP.

Acknowledgements

This research was supported by the Universidad Industrial de Santander and was funded by the Vicerrectoría de Investigación y Extensión as part of the project VIE-UIS1336.

References

[1] Montero-Ocampo C, Villegas D and Veleva L 2008 *Surface Eng* **24** 18
[2] Liu D, Savino K and Yates M 2011 *Surf Coat Tech* **205** 3795
[3] Prasad B and Kamath P 2013 *Bull Mater Sci* **36** 475
[4] Madhumathi K and Kumar T S 2014 *Biomed Mater* **9** 035002
[5] Ellies L, Nelson D and Featherstone J D 1992 *Biomaterials* **3** 313
[6] Sato M, Tu R and Goto T 2007 *Mater Trans* **12** 3149
[7] Boyd A R, O'Kane C and Meenan B J 2013 *Surf Coat Tech* **233** 131
[8] Sidane D, Chico D, Yala S, Ziani S, Khireddine H, Iost A and Decoopman X 2015 *Thin Solid Films* **593** 71
[9] Gualdrón-Reyes A F, Meléndez A M, Niño-Gómez M E, Rodríguez-González V and Carreño-Lizcano M I 2015 *Rev Acad Colomb Cienc Ex Fis Nat* **39** 77
[10] Kappert E, Pavlenko D, Malzbender J, Nijmeijer A, Benes N and Tsai P 2015 *Soft Matter* **11** 882
[11] Ghodbane O, Sarrazin M, Roué L and Bélanger D 2008 *J Electrochem Soc* **155** F117
[12] Gopi D, Indira J, Kavitha L and Ferreira J M F 2013 *J Appl Electrochem* **43** 331
[13] Mattarozzia L, Cattarin S, Comisso N, Guerriero P, Musiania M, Vázquez-Gómez L and Verlato E 2013 *Electrochim Acta* **89** 488
[14] Cukrowski I, Popović L, Barnard W, Paul S, Rooyen P and Liles D 2007 *Bone* **41** 668

Journal of Physics: Conference Series **786** (2017) 012031 doi:10.1088/1742-6596/786/1/012031

Enriched alloy layer on an Al-Cu alloy studied by cyclic voltammetry

S J García Vergara[1], C E Blanco Pinzon[2] and P Skeldon[3]
[1] Universidad Industrial de Santander, Bucaramanga, Colombia
[2] Instituto Colombiano del Petróleo, Piedecuesta, Colombia
[3] University of Manchester, Manchester, United Kingdom

E-mail: sagarver@uis.edu.co

Abstract. The behaviour of enriched Al-0.7at.%Cu alloy is investigated using cyclic voltammetry. Enriched alloy layers at the interface between the alloy/oxide film were developed by alkaline etching at $5mAcm^{-2}$ in 0.1M sodium hydroxide solution at 298K, with the time of etching determining the extent of enrichment. Cyclic voltammograms were recorded at a scan rate of $10mV\ s^{-1}$ in naturally aerated 0.1M ammonium pentaborate solution at 298K. The current overshoot of the enriched alloys was different from that for non-enriched alloy. The latter material revealed the usual single peaks, which are very similar. In contrast, the overshoot comprised two or more components for the enriched alloys. The behaviour is suggested to be associated with the atomic bonding of aluminium in copper-rich and aluminium-rich regions of the enriched alloy layer, with influence on the activation potentials for oxidation of aluminium.

1. Introduction

Oxidation of many aluminium alloys by anodizing leads to formation of a thin layer of alloy, which is highly enriched in the alloying element with respect to the bulk alloy [1,2]. However, enrichment can be generated also by surface pre-treatments, such as chemical polishing [3], electropolishing [4] and alkaline etching [5], making the enrichments of general relevance to preparation of alloy surfaces and their subsequence performance. In particular, the corrosion potential of Al-Cu alloys is affected by the presence of the enriched alloy layer [6,7].

In the present study, the effect of the copper enriched layer on the transient behaviour of an Al-0.7at.%Cu alloy is investigated using cyclic voltammetry. Dilute aluminium alloys containing relatively noble alloying elements offer the ability to change the composition of the alloy at the metal/oxide interface, without affecting the composition of the oxide. For a single alloy composition, a wide range of local alloy compositions can be developed by anodic pre-treatment of the alloy. The possibility of control of the local alloy composition arises because copper is not oxidized until it is sufficiently enriched in the alloy. In the case of the alloy of interest, enrichment to about 40at.% Cu in an enriched alloy layer of 2nm thickness is necessary [8]. The copper enriches in the initial stages of oxidation of the alloy, during which only aluminium atoms are oxidized, forming an alumina film. When copper later begins to oxidize, it enters the oxide as Cu^{2+} ions that migrate through the oxide faster than the Al^{3+} ions, by about a factor of 2 [9]. At this stage of the oxidation process, the average composition and thickness of the enriched alloy layer do not change significantly with time, indicating

Journal of Physics: Conference Series **786** (2017) 012031 doi:10.1088/1742-6596/786/1/012031

that copper and aluminium atoms enter the oxide in proportion to their concentrations in the bulk alloy rather than their concentrations in the enriched layer [10].

2. Experimental details

The Al-0.7at.%Cu alloy was deposited onto electropolished high purity aluminium sheet by magnetron sputtering. Targets were composed of 99.999% aluminium and 99.99% copper. The deposited layers were about 400-500nm thick. In order to develop enrichments of the alloy, anodic etching was carried out at $5mAcm^{-2}$ in 0.1M NaOH at 293K for different times, followed by an immersion in nitric acid. Cyclic voltammograms were then recorded at a scan rate of $10mV\ s^{-1}$ in naturally aerated 0.1M ammonium pentaborate electrolyte at 293K. The cell comprised a platinum counter electrode and a saturated calomel reference electrode (SCE). All potentials are expressed with respect to SCE. Compositions of enriched alloy layers were determined by Rutherford backscattering spectroscopy, using 2.0MeV He^+ ions produced by the Van de Graaff accelerator of the University of Paris, with spectra interpreted using the RUMP program [11].

3. Results and discussion

The enrichments of copper in the Al-0.7at.%Cu alloy following anodic etching are summarized in Table 1. The average compositions of the enriched layer are given, assuming a layer of thickness of 2nm and a weighted average of the atomic densities of aluminium and copper. The thickness of the alloy is reduced by 110nm after 41s of etching, indicating an etching rate of about $2.7nm\ s^{-1}$. Typical RBS spectra showing enrichments of Al-Cu alloys have been presented in previous work [6]. It can be seen that the copper enrichment increases as the etching time increases and that for the longer times the oxidation of copper at the alloy/oxide interface took place. The latter is evident from the differences between the experimental and expected enrichment values.

Table 1. Enrichment of copper, determining by RBS, following anodic etching at 5mA cm^{-2}, of Al-0.7at.% Cu alloy in 0.1M sodium hydroxide solution at 293K.

Time of etching (s)	Enrichment in alloy (Cu atoms cm^{-2})		Average composition of enriched layer (at.%Cu)
	RBS	Expected	
0	0	0	0
12	0.9×10^{15}	0.9×10^{15}	7.2
25	1.8×10^{15}	1.8×10^{15}	15.2
41	2.9×10^{15}	2.9×10^{15}	24.3
71	4.7×10^{15}	5.2×10^{15}	38.7
110	5.1×10^{15}	8.0×10^{15}	42.5

The first anodic cycle of the voltammograms for Al-0.7at.%Cu alloy in the as-deposited condition disclosed the transient peak in current density, centred at about -100mV, followed by a shallow dip and gradual approach to a steady value of about $33\mu Acm^{-2}$ that is maintained from about 500mV until the anodic limit of 1V (Figure 1). The final current density corresponds to the steady-state oxide growth. The current density on reversal of the anodic scan fell rapidly to a low value as oxide growth declined rapidly. The anodic current density remained low on subsequent cycles, with a small rise towards the anodic limit, when some film growth occurred.

Different behaviour is observed for the alloy after the anodic etching. The transient peak is shifted to more positive potentials as the etching time increases, achieving an increment of about 500mV after 71s of etching. The height of the transient peak is significantly reduced. In contrast, the steady current density was relatively unaffected. The transient peak appears to comprise at least two sub-peaks separated by about 200-300mV. The shape of the peak often differed in repeat measurements. However, consistently, the potential range between the initial rises in current density to the tail of the transient peak was from 500-700mV compared with about 300mV for the as-deposited alloy. In the cathodic scan only the hydrogen evolution is evident after -1.7V.

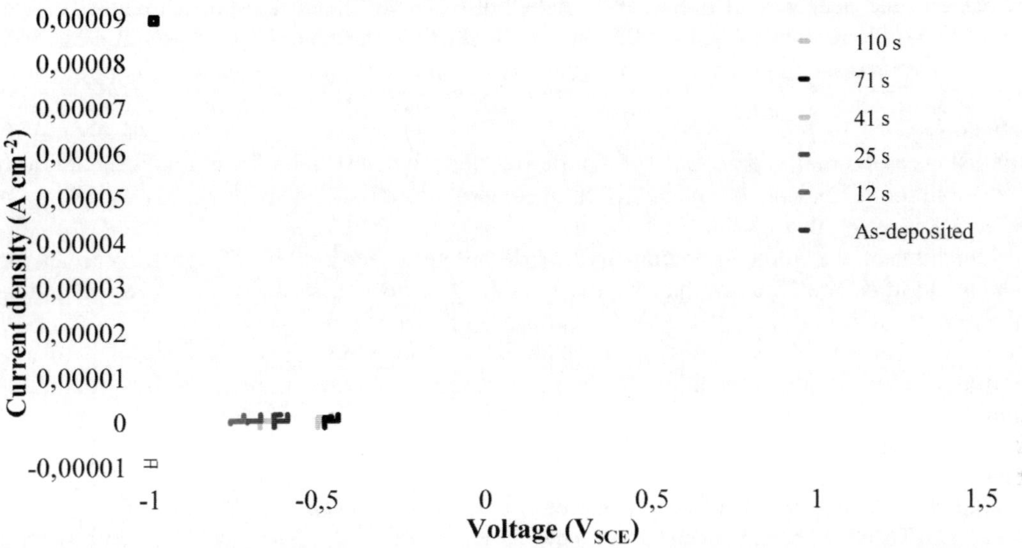

Figure 1. Cyclic voltammograms (First cycle) of Al-0.7at.%Cu alloy after anodic etching in 0.1M sodium hydroxide, in 0.1M ammonium pentaborate electrolyte at 298K.

The copper-enriched alloy layer appears to have a significant effect in the initial anodic oxidation. For an Al-0.7at.%Cu alloy, the required level of enrichment of the alloy for oxidation of copper to take place is about 6×10^{15}Cu atoms cm^{-2}, in a layer about 2nm thick. Assuming an average atomic density of the enriched layer of 7×10^{22} atoms cm^{-3}, between those of copper and aluminium of 8.4×10^{22} and 6.0×10^{22} respectively, the average concentration of copper is 43at.%. During etching, copper enriches initially at a rate of about 9×10^{13}Cu atoms cm^{-2} s^{-1}, thus achieving the maximum enrichment at about 67s. This stage follows oxidation of about 142nm of the alloy. By 12 s of etching, the enrichment corresponds to an average copper concentration of 7at.%, with levels of 15 and 24at.% after 25 and 41s, respectively. The enriched layer is composed of aluminium-rich and copper-rich region. Some aluminium atoms are bound to copper atoms, while others are bound to other aluminium atoms, and probably some copper atoms are bound to others copper atoms. The previous considerations suggest that the developing enriched alloy layer alters the local lattice structure of the metal.

The aluminium atoms in the various locations will require different activation potentials for their oxidation and incorporation into the oxide. The altered transient of the etched alloy may relate to oxidation of aluminium atoms at local sites of relatively low copper content, when a low activation potential is sufficient, and at sites of relatively high copper content, where an increased activation potential is necessary. The shape of the peak will depend upon the contribution to the current from oxidation of aluminium at the various sites. Although there is experimental evidence in some voltammograms for two peaks in the overshoot, there may be a range of types of site that contribute to the observed behaviour. The aluminium ions would be generated preferentially from the sites with low activation potential, but these would be more limited as enrichment proceeds, which in broad terms is consistent with the experimental observation of a smaller contribution to the overshoot of the lower potential contributions for longer etching times. However, the experimental data do not reveal a systematic change with time of etching, and in some cases, differences in data occurred for repeated measurements at a single etching time.

The maximum current density in the overshoots of the etched alloys moves to higher potentials with longer times of etching up to the point at which copper can oxidize, when there is no further increase. The magnitude of the potential shift for the etched alloys, about 400-500mV, is very similar to the increase in the open-circuit potential of etched alloys with increasing enrichment of copper in the alloy. The similarity suggests that the potential shift at the open-circuit potential is largely determined by the increase in potential of the anodic reaction due to the enrichment of copper.

4. Conclusions

The initial stages of anodic oxidation of sputtering-deposited Al-0.7at.%Cu alloy in ammonium pentaborate solution is associated with a current overshoot, comprising a single relatively narrow peak over a potential range of about 300mV.

Anodic etching of the alloy in sodium hydroxide solution leads to enrichment of copper in the alloy, which modifies the form of the current overshoot. The overshoot occurs over an extended potential range, approximately 500-700mV, and consists of two or more components.

The differing behaviour resulting from enrichment of copper is suggested to be due to altered bonding of aluminium atoms in the alloy, leading to a range of activation polarization for oxidation of aluminium.

References

[1] Hashimoto T, Zhou X, Skeldon P and Thompson G E 2015 *Electrochim. Acta* **179** 394-401
[2] Donatus U, Thompson G E, Elabar D, Hashimoto T and Morsch S 2015 *Surf. Coat. Tech.* **277** 91-98
[3] Feliu Jr S, Bartolomé Ma J, González J A, López V and Feliu S 2008 *Appl. Surf. Sci.* **254** 2755-62
[4] Zhou X, Habazaki H, Shimizu K, Skeldon P, Thompson G E and Wood G C 1996 *Corros. Sci.* **38** 1563-77
[5] Wu X and Herbert K 1996 *J. Electrochem. Soc.* **143** 83-91
[6] Garcia-Vergara S J, Colin F, Skeldon P, Thompson G E, Bailey P, Noakes T C Q, Habazaki H and Shimizu K 2004 *J. Electrochem. Soc.* **151** B16-21
[7] Garcia-Vergara S J, Skeldon P, Thompson G E, Williams G and McMurray N 2012 *J. Electrochem. Soc.* **159** C428-33
[8] Habazaki H, Paez M A, Shimizu K, Skeldon P, Thompson G E, Wood G C and Zhou X 1996 *Corros. Sci.* **38** 1033-42
[9] Habazaki H, Zhou X, Shimizu K, Skeldon P, Thompson G E and Wood G C 1997 *Electrochim. Acta* **42** 2627-35
[10] Garcia-Vergara S J, Skeldon P, Thompson G E, Bailey P, Noakes T C Q, Shimizu K and Habazaki H 2003 *Appl. Surf. Sci.* **205** 121-27
[11] Doolittle L R 1985 *Nucl. Instrum. Methods B* **9** 344-51

Effect of the layer of anodized 7075-T6 aluminium corrosion properties

R D Montoya Z[1], E Vera L[1], Y Pineda T[1] and M L Cedeño[1]
[1] Universidad Pedagógica y Tecnológica de Colombia, Tunja, Colombia

E-mail: enrique.vera@uptc.edu.co

Abstract. Aluminium alloys are widely used in various sectors of industry. The 7075-T6 alloy corresponding to an Al-Zn T6, is mostly used as structural component in the aviation industry, due to the good relationship between weight and mechanical properties. However, the negative point of this alloys is the resistance to corrosion, which is why they need to be coated with an anodic film. Different surface treatments, such as anodizing, are used to improve corrosion resistance. Anodizing is an electrolytic process by which a protective layer on aluminium known as "alumina " is formed, this is formed by the passage of an electric current in an acidic electrolyte. This investigation presents a study of the effect of the thickness of layers of alumina deposited by anodized method, in the corrosion resistance of 7075-T6 aluminium. This study was performed by using in a solution of tartaric acid - sulfuric acid and an inorganic salt. To evaluate the influence alumina layer thickness on the corrosion properties some tests were carried out by using the electrochemical spectroscopy impedances (EIS) technique and Tafel polarization curves. It was found that the grown of the thickness of film favourably influences in the corrosion resistance.

1. Introduction

Aluminium alloys are widely used in different industry sectors [5-10]. The negative point of these alloys is the corrosion resistance, whereby they need to be coated with an anodic film. Different types of surface treatments, such as anodized, are used to improve corrosion resistance [1-11]. The anodized is an electrolytic process, in which a protective layer of "alumina", it is generated on the surface of the aluminium through current flow, gives the aluminium excellent anticorrosive properties.

Despite the high corrosion protection obtained by the anodizing technique, one of the main current concerns is the great environmental effect resulting from the use of chromic acid [6,7], commonly used as electrolyte in the realization of these coatings [2-4]. It has been known that chromium (VI) compounds (also known as chromates) have been used for the last 60 years and are toxic and dangerous to the ecosystem [5], in addition to causing carcinogenic problems to people who are exposed and in contact with chromates.

For this reason, one of the challenges facing the coating industry is the production of cleaner and less harmful products for the environment. This regard, some electrolytic baths of lower toxicity have been investigated and capable of even improving the mechanical and corrosive properties of anodizing coatings of aluminium alloys) [8-9]. In this work, the anodized were made using a solution of tartaric acid - sulfuric acid and a molybdic sodium salt as the electrolytic bath and studied the effect of the thickness of the alumina layer in corrosion resistance of the 7075-T6 aluminium alloy.

2. Experimental development

2.1. Anodized process

A The specimens of 7075-T6 aluminium alloy (chemical composition showed in Table 1), with dimensions of 40mm, 50mm and 5mm of width, long and high respectively, were subjected to a conventional cleaning treatment, consisting of degreased with NaOH 5% to 50°C for about 3 minutes. Subsequently, the specimens were rinse in distilled water for 2 minutes, and were pickled for 5 seconds with a solution of HCl 5% to 25°C. Then this were rinse in distilled water for 2 minutes.

In the step of anodized, the specimen was completely immersed in an electrolytic cell containing an aqueous acidic solution composed of 0.40M sulfuric acid, 0.53M of tartaric acid and 0.25M sodium molybdate a Temperature of 20 °C and pH between 1 and 1,2 [11] . The aluminium parts played the role of anode, while the cathode was composed of AISI 304 stainless steel plates.

A potential difference of 14V was used and immersion times of 40, 60 and 90 minutes. After this time the pieces were extracted and were rinse in deionized water for 3 minutes. Subsequently, were sealed with boiling deionized water for 40 minutes and dried with a hot air stream.

2.2. Characterization of specimens

The thickness of the alumina layer was determined by scanning electron microscopy (SEM) with a EVO MA 10 equip operated at 20kV.

Tests of electrochemical impedance spectroscopy and potentiodynamic polarization were performed with a Gamry potentiostat model PC4. All tests were at room temperature, using a glass cell in where located working electrode (aluminium alloy) with an area of 0.079 cm^2 a reference electrode of Ag/AgCl and a platinum wire as counter electrode, immersed in a solution of NaCl 3.5%l. This solution was chosen because of simulates a marine solution, which corrodes active materials, forming chlorides over the metal [4].

Table 1. Chemical Composition of Alloy 7075-T6.

Alloy	Al	Zn	Cu	Mg	Mn
7075-T6	bal	5.1	1.152	2.163	0.037

Impedance spectroscopy test and potentiodynamic polarization were performed. An AC voltage of 10mV was use in the first case, with initial and final frequencies of 100,000Hz and 0.1Hz respectively. On the potentiodynamic polarization test were used potentials between -250mV to 250mV with respect to corrosion potential, scan rate of 2mV/min in anodic direction.

3. Results and discussion

3.1. Thickness of the anodized layer

The thickness of the anodized layer was measured from the micrographs shown in Figure 1. It is noted that the thickness of the alumina layer in specimen AA7075-T6 varies directly with the anodizing time. A 40 minutes immersion time produces a thickness of 6.14μm, while times of 60 and 90 minutes produce an increase of the oxide film of 8.4μm and 14.9μm respectively. The images of the left side of Figure 1 correspond to the results of X-ray energy scattering spectroscopy microanalysis, EDS made to the coating. It is observed that the layers are composed mainly of oxygen and aluminium, this corroborates the existence of a layer of deposited alumina.

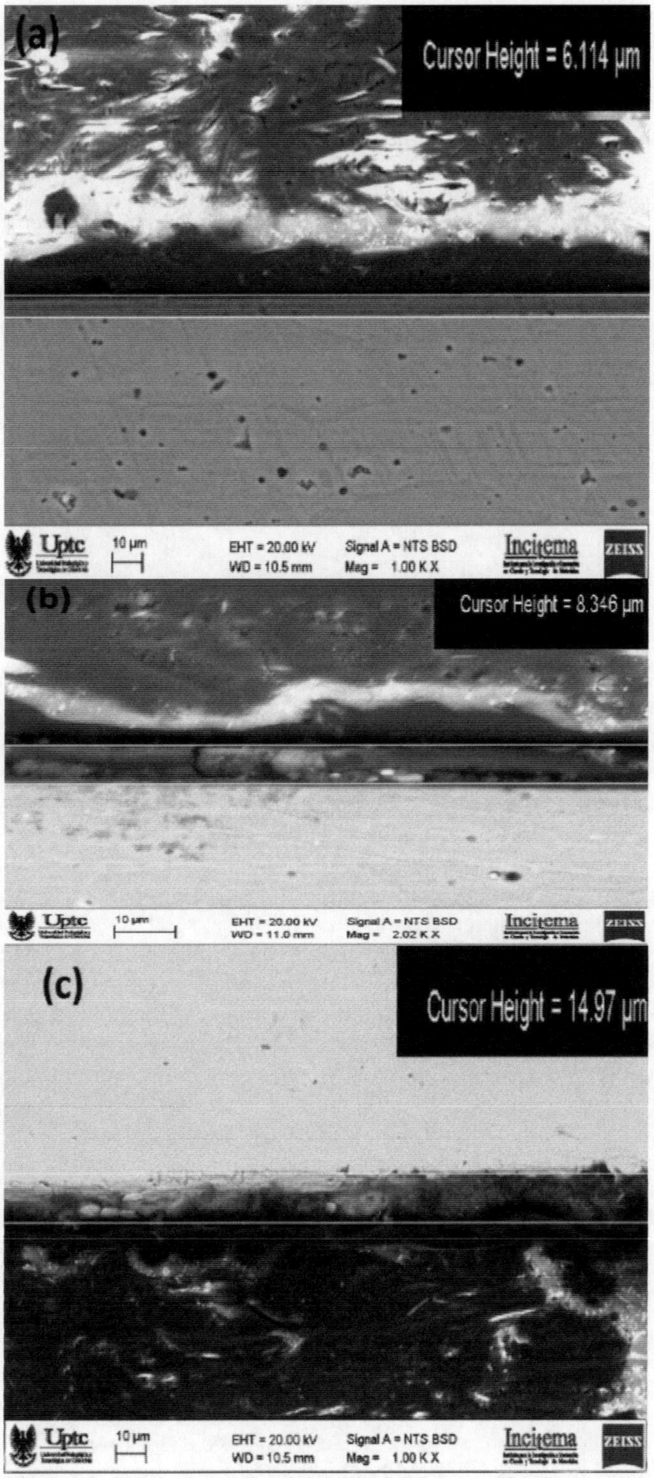

Figure 1. SEM of anodized AA7075-TSA (a) 40 minutes, (b) 60 minutes and (c) 90 minutes.

3.2. Electrochemical Impedance Spectroscopy (EIS)

The results obtained in the Nyquist diagrams (Figure 2) show an increase of the diameter of the semicircles with the increase in the thickness of the anodized layer, for thicknesses lower than 14.9µm. It is observed that the less resistive behaviour is obtained with the aluminium free of superficial treatment, this phenomenon is attributed to that the anodized samples contemplate a layer of alumina that prevents the transfer of reducible species to the base metal. The behaviour of the curve of EIS for the aluminium with an anodized thickness of 14.9µm shows a diffusion phenomenon that indicates a gradient of the concentration of ionic charges in the interface metal-solution, which evidences the protection or barrier of the coating w to the corrosion.

The obtained values of the resistance to the polarization of the electrode with the thickness of the layer were of 1,610 Mohms for a coating of 14.9µm microns and values of 1.25 Mohms, 38.05 Kohms for 8.34µm and 6.14µm respectively.

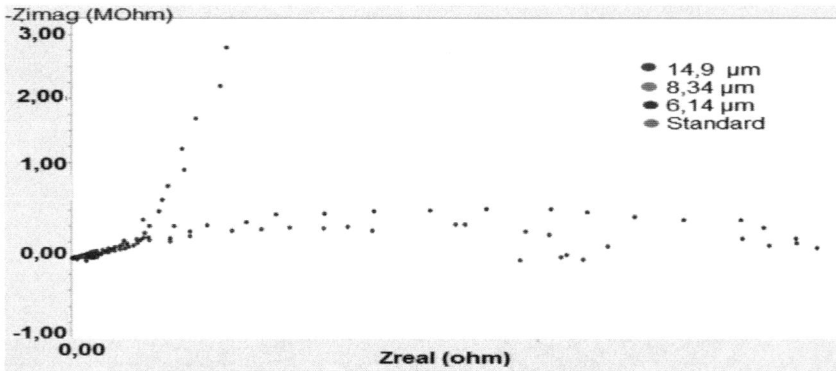

Figure 2. Nyquist for anodic layers 6.14µm, 8.34µm, 14.9µm and for the Specimen pattern.

3.3. Potentiodynamic polarization curves (Tafel)

Figure 3 shows the potentiodynamic polarizationcurves (Tafel) of the aluminium 7075-T6 with and without anodized exposed to a saline environment of 3.5% by weight of NaCl. It is observed that the increase in the thickness of the anodized layer causes displacement of the cathodic zones towards the smaller values of current, leading to the decrease of the speed of corrosion of the material. The current major of the corrosion is reached when the aluminium is found without anodized, while when coated with a layer of Al_2O_3 of 14.9µm decreases considerably. The change the corrosion potential to the highest values indicates that aluminium tends to behave as noble with respect to the other samples.

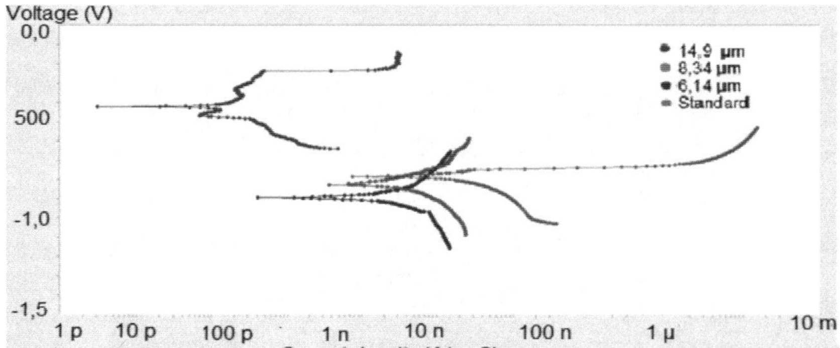

Figure 3. Polarization curves for the anodic layers 6.14µm, 8.34µm, 14.9µm and for the specimen pattern.

The Figure 4 shows the behaviour of the corrosion rate versus the thickness of the layer of anodized.

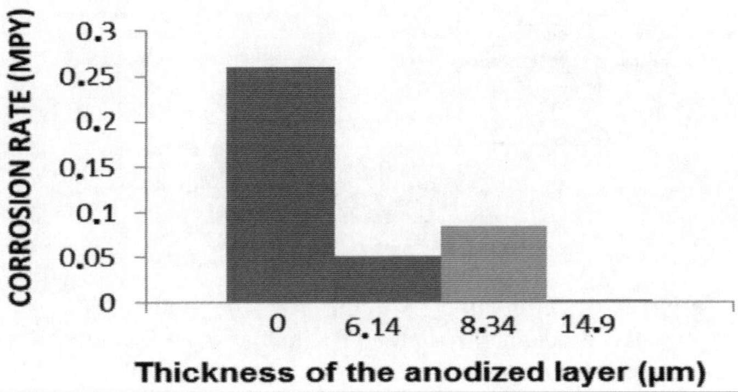

Figure 4. Corrosion rate vs. thickness of the layer of anodized.

3.4. Equivalent circuits

Equivalent circuits are used to describe the electrochemical behaviour and to calculate the parameters of interest such as electrolyte resistance (Rs), charge transfer resistance (Rct) and electrochemical double layer capacitance (Cdl). In real cells the double layer capacitor, with behaves with frequency as a constant phase element (constant phase element: CPE) and not as an ideal capacitor. The CPE element is use in the model in place of the capacitor to compensate for non-homogeneity in the system. The impedance of this element (ZCPE) is determined by the following equation:

$$ZCPE = Y_o\text{-}1(iw)^{-n} \quad (1)$$

Where Y_o is the constant CPE, w is the angular frequency and n is an exponent used to indicate the heterogeneity of the roughness surface, usually is the value of the slope and is in the range of 0.5 to 1.

In order to analyse the metal/electrolyte interface in anodic layers with thickness lower than 14.9μm, a modified Randless circuit with a constant phase element was chosen, which allowed to obtain a good fit with the experimental data. Figure 5 represents the proposed equivalent circuit.

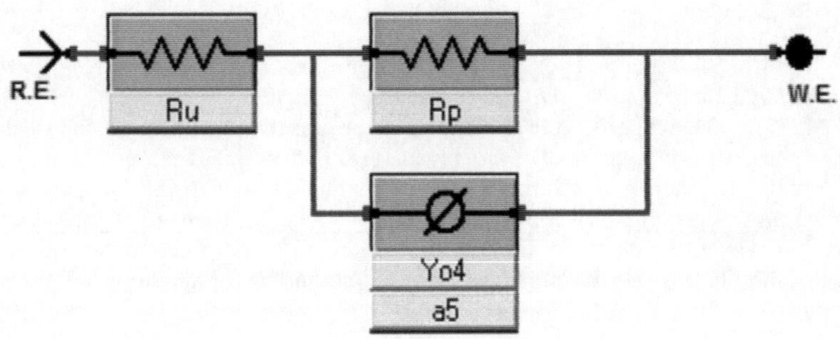

Figure 5. Equivalent Circuit for impedance diagrams obtained for low thicknesses to 14.9μm of the specimen of Al7075-T6.

For a thickness of 14.9μm, was chosen the equivalent circuit of Figure 6. In this the Randles circuit is included an impedance of Warburg (W) that represents the diffusion phenomena that take place.

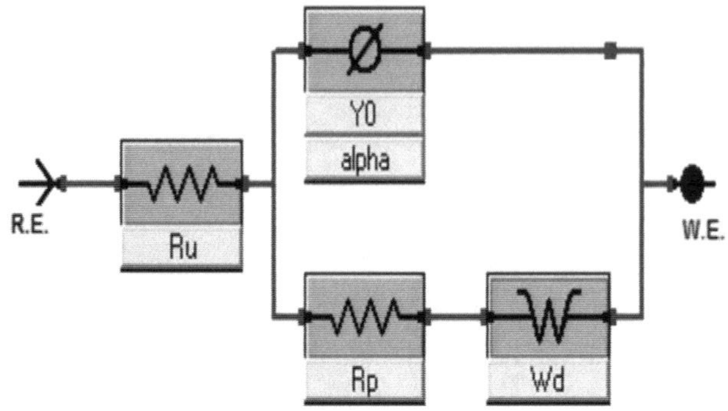

Figure 6. Equivalent Circuit for the diagrams of impedance obtained for an equal thickness to 14.9μm of the specimen of Al70754-T6.

4. Conclusions

It was determined that the thickness of the oxide film increases the corrosion resistance of the aluminium. Higher thicknesses generate lower corrosion rates. The best results were obtained with a 14.9μm layer.

References

[1] Rubio M N 2009 *Optimisation of a non-chromium-containing tartaric acid/sulphuric acid anodising bath for aluminium alloys for aerospace industry application* (España: Departamento de Tecnología de superficies Airbus) 13-16

[2] Hurtado M J 2007 The Anodic Oxidation of Aluminium: Fabrication and Characterization Portugaliae *Electrochimica Acta* **25(1)**153-162

[3] Hideki M, Fumio H, Sachiko O 1997 Self Ordering of Cell Arrangement of Anodic Porous Alumina Formed in Sulfuric Acid Solution *J. Electrochem. Soc* **144(5)** 127-130

[4] Thompson G E, Wood G C 1981 Porous anodic film formation on aluminium, *Nature* **290** 230-232

[5] Bello V, Russo E D 1992 Proceedings of Materials Development in Rail, Tire, Wing, Hulí transportation (Italia) 1.11-1.29

[6] Chun M W, Yukimori M 1990 *Method for anodizing aluminum* (United Estates: Patent US 4894127 A)

[7] Camargo A, Aperador W, Rios A, Ortiz C, Vera E 2009 Caracterización Mediante Espectroscopia de Impedancia Electroquímica de Películas Anódicas Crecidas Sobre Al 2024-T3 *Revista Colombiana de Física* **41(2)** 261

[8] Dattilo A, Tamiro S, Romano C 2002 *Anodizing process, with low environmental impact, for a woodpiece of aluminum or aluminum alloys* (United States: Patent US 20020157961 A1)

[9] Montero J M, Sarret M, Müller C 2007 Influence of the aluminum surface on the final results of a two-step anodizing *Surface and Coatings Technology* **201(14)** 6352–6357

[10] Shina J S, Songb J B, Choia S H, Kimb J T, Ohc S G, Yunb J Y 2016 Plasma Corrosion in Oxalic Acid Anodized Coatings Depending on Tartaric Acid Content *Appl. Sci. Converg. Technol* **25(1)** 15-18

[11] Rubio M N 2007 Ocón P, Font A C-, Smith R W, Curioni M, Thompson Ge, Skeldon P, Lavía A, García I, Influence of molybdate species on the tartaric acid/sulphuric acid anodic films grown on AA2024 T3 aerospace alloy *Rev. Corrosion Science* **51(9)** 2034–2038

IOP Conf. Series: Journal of Physics: Conf. Series **786** (2017) 012033 doi:10.1088/1742-6596/786/1/012033

Influence of Na^+, K^+, Mn^{2+}, Fe^{2+} and Zn^{2+} ions on the electrodeposition of Ni-Co alloys: Implications for the recycling of Ni-MH batteries

S Blanco[1], R Orta-Rodriguez[2] and P Delvasto[1]
[1] Universidad Industrial de Santander, Bucaramanga, Colombia
[2] Universidad Simón Bolívar, Caracas, Venezuela

E-mail: delvasto@uis.edu.co

Abstract. A hydrometallurgical recycling procedure for the recovery of a mixed rare earths sulfate and an electrodeposited Ni-Co alloy has been described. The latter step was found to be complex, due to the presence of several ions in the battery electrode materials. Electrochemical evaluation of the influence of the ions on the Ni-Co alloy deposition was carried out by cyclic voltammetry test. It was found that ions such as K^+, Fe^{2+} and Mn^{2+} improved the current efficiency for the Ni-Co deposition process on a copper surface. On the other hand, Na^+ and Zn^{2+} ions exhibited a deleterious behaviour, minimizing the values of the reduction current. The results were used to suggest the inclusion of additional steps in the process flow diagram of the recycling operation, in order to eliminate deleterious ions from the electroplating solution.

1. Introduction

Ni-MH batteries are mainly composed of nickel (in oxidized and metallic form), cobalt, alkaline electrolyte substances and rare earths metal alloys [1-5]. Although it is difficult to get exact numbers for the usage of Ni-MH batteries in Latin America, it has been reported that this kind of batteries account for 2.7% (by weight) of the total household spent batteries, as collected in communal recycling procedures in Venezuela [6]. This number can be regarded as representative, for most countries within the region. New legislation changes issued in the Latin American countries have encouraged the application of valorisation strategies, not only to prevent environmental damage, but also to recover metal values from this kind of electronic wastes. A feasible solution for this local problem could be the adoption of small-scale hydrometallurgical facilities, in order to process these wastes without the need of exporting them to other countries, most of which are usually located in Europe and East Asia [7]. Nickel, cobalt and rare earths are strategic materials, which are worth to be recovered from depleted Ni-MH batteries; nevertheless, the hydrometallurgical recycling of Ni-MH batteries is a challenging issue, due to its internal chemical complexity [1-8]. In a previous paper [7], a small-scale hydrometallurgical processing route was devised, comprising a chemical stage and an electrochemical stage. During the chemical stage, the active electrode materials of the Ni-MH batteries were leached using aqueous H_2SO_4. The obtained pregnant solution was then submitted to pH adjustment in a settling reactor, in which the rare earths were recovered as an insoluble complex sulphate. The remnant solution, containing nickel, cobalt and some other ions is then electrolyzed, in order to recover a Ni-Co alloy [7]. This latter stage is not yet well understood, because of the interaction of the several ions present in the solution.

Based on the above described situation, the present paper deals with the application of electrochemical techniques to study the co-deposition of nickel and cobalt from solutions containing several ions, as those obtained after leaching the internal components of Ni-MH batteries with H_2SO4. In particular, cyclic voltammetry was performed over Ni and Co sulphate synthetic solutions, so that the effect of ions such as Na^{1+}, K^{1+}, Mn^{2+}, Fe^{2+} and Zn^{2+} was individually evaluated. The objective is to draw useful conclusions, which will be used for the design of small scale recycling procedures, intended for the local processing of spent Ni-MH batteries in the Latin American countries.

2. Experimental

The synthetic electrolytes were formulated in concordance with the composition of the solutions obtained by the processing of spent Ni-MH batteries [7]. In order to evaluate the influence of each ion independently, 6 solutions were prepared (Table 1). Nickel, cobalt, iron, manganese and zinc ions were obtained by the addition of the respective sulphate, ($NiSO_4$, $CoSO_4$, $FeSO_4$, $MnSO_4$, $ZnSO_4$). In order to reproduce the solution management conditions given during the recycling of Ni-MH spent batteries [7], the synthetic solutions studied in this work were acidified to pH 1 using H_2SO_4 before being adjusted to pH 4 with NaOH or KOH, as source of Na^+ or K^+, respectively. Electrochemical evaluation of the influence of the ions on the Ni-Co alloy deposition was carried out by cyclic voltammetry test. A three-electrode electrochemical cell was used, with a Pt wire as auxiliary electrode, a 1mm diameter copper disk as working electrode and an Ag/AgCl electrode as reference. All experiments were performed using a potentiostat AUTOLAB PGSTAT 302N.

Table 1. Chemical composition (mole/liter) of the evaluated Ni-Co solutions.

Code	Ni	Co	Na	K	Fe	Mn	Zn
Ni-Co	0.4	0.04					
Ni-Co+0.015M Na	0.4	0.04	0.015				
Ni-Co+0.02M K	0.4	0.04		0.02			
Ni-Co+0.001M Fe	0.4	0.04			0.001		
Ni-Co+0.012M Mn	0.4	0.04				0.012	
Ni-Co+0.005M Zn	0.4	0.04					0.005

3. Results and discussion

Figure 1(a) shows cyclic voltammograms of a 1mm radius polycrystalline copper disk electrode in 0.4M Ni^{2+}-0.04M Co^{2+}solution. The observed behaviour is similar to results reported in the literature [9-13], showing the typical reduction and oxidation zones. The reduction zone starts at -0.45V vs Ag/AgCl with a maximum at -0.66V vs Ag/AgCl characteristic of the co-deposition of nickel and cobalt according to the following reactions:

$$Ni^{2+} + 2e^- \rightarrow Ni_{(s)} \tag{1}$$

$$Co^{2+} + 2e^- \rightarrow Co_{(s)} \tag{2}$$

The increment in the cathodic current at more negative potentials (-0.69V vs Ag/AgCl) correspond to the hydrogen evolution reaction,

$$2H^+ + 2e^- \rightarrow H_{2(g)} \tag{3}$$

These reactions is catalysed by the nickel compound, and represent the principal limitation for the electrodeposition process, the hydrogen evolution decreases the current efficiency and it is expected to affect negatively the mechanical properties of the obtained coating, due to hydrogen entrapment. The inclusion of potassium, iron and manganese ions in the solution clearly modified the electrodeposition process and the hydrogen evolution reaction. With the addition of potassium, the reduction reaction starts at the same potential than that of the solution without extra ions, and the reduction maximum is maintained near -0.66VvsAg/AgCl. The main variation is shown by the reduction current, which changed from -8.9×10-5A in the Ni-Co solution to -27.4×10^{-5}A in the Ni-Co-K solution. It is well known that the reduction potential of potassium is more negative, so the potassium co-electrodeposition can be dismissed, then, the variation in the reduction current could be related to a change in the ions mobility in the solution, and its effect over the migration current. The inclusion of manganese in the solution produced a similar effect, in which the reduction potential was maintained and the reduction current changed to -33.8×10^{-6}A. The influence of the iron ions in the solution is different, in this case the reduction current shows an increment but the curve does not show a well-defined reduction peak. The reduction potential of iron ions is in the evaluated range, and the co-deposition with nickel and cobalt is possible, a chemical analysis of the obtained deposit is needed to confirm this hypothesis and evaluate the formation of alloys or the segregation of the components in different areas of the electrode. The reduction of the hydrogen evolution current could be related to a change in the deposit composition. The catalytic effect of the nickel could be reduced by the co-deposition of iron or by an increment of the concentration of cobalt.

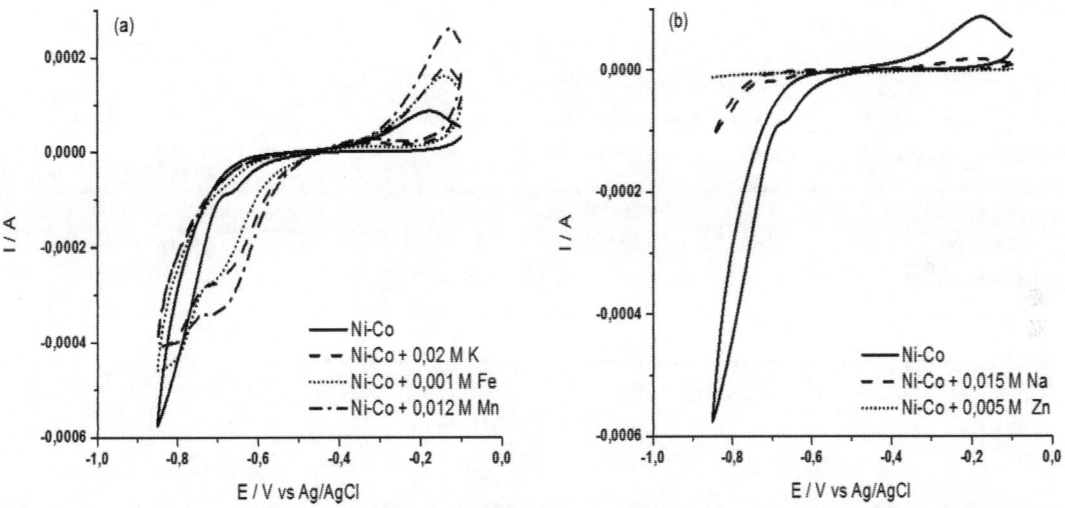

Figure 1. (a) Cyclic voltammetry of copper in 0.4M Ni^{2+}, 0.04M Co^{2+} solutions including the influence of the presence of K^{+}, Fe^{2+} and Mn^{2+} ions, pH:4.0, T:25°C, v:50mV.s^{-1}. (b) Cyclic voltammetry of copper in 0.4M Ni^{2+}, 0.04M Co^{2+} solutions including the influence of the presence of N^{a+} and Zn^{2+} ions, pH:4.0, T:25°C, v:50mV.s^{-1}.

Figure 1(b) shows the influence of the presence of sodium and zinc ions in the electrolyte on the electrochemical response. With the addition of these elements to the solution, the reduction and oxidation peaks show a significant decrease. In the case of the addition of zinc, the decrease in the current could be related to the precipitation of non-conductive zinc compounds on the electrode surface. The stability of the zinc ions could be affected by the presence of sulphate ions, promoting the formation of zinc sulphate complex on the electrode surface. These less conductive sulphates could act as a physical barrier [14]. In addition, despite the explanation for the deleterious effect of Na ions is not straightforward, it has been previously reported that Na ions decrease the current efficiency in the electrodeposition of Co powders [15] and this was attributed to a non-uniform layer of deposited metal on the surface of the electrode, that seems to facilitate hydrogen evolution. A similar phenomenon could be happening in the experiment showed in Figure 1(b).

4. Implications of these findings for the electrochemical recycling of spent Ni-MH batteries

The chemical complexity of the leachates obtained from the hydrometallurgical processing of spent Ni-MH implies important challenges that need to be addressed. Briefly, in our previous work, we suggested a route comprising: the leaching in sulfuric acid of the internal component of the batteries, a precipitation stage to generate an insoluble sulphate containing the rare earths and a final electroplating stage to recover Ni and Co in the form of an alloy coating or deposit. The results showed herein indicate that once the rare earths are segregated from leachate, further conditioning of the solution must be done, in order to get rid of deleterious ions present such as Na and Zn. Since most Na present may arise from the use of NaOH as precipitating agent, other Na-free bases should be explored if a technology based in this process is to be implemented. In addition, an intermediate process needs to be included, aiming to scavenge the Zn and to withdraw it from the solution, in order to prevent its build up on the surface of the deposition electrode and to improve the current efficiency in the recovery process of the Ni-Co alloy. A scheme of the process flow diagram for such operation sequence is shown in Figure 2.

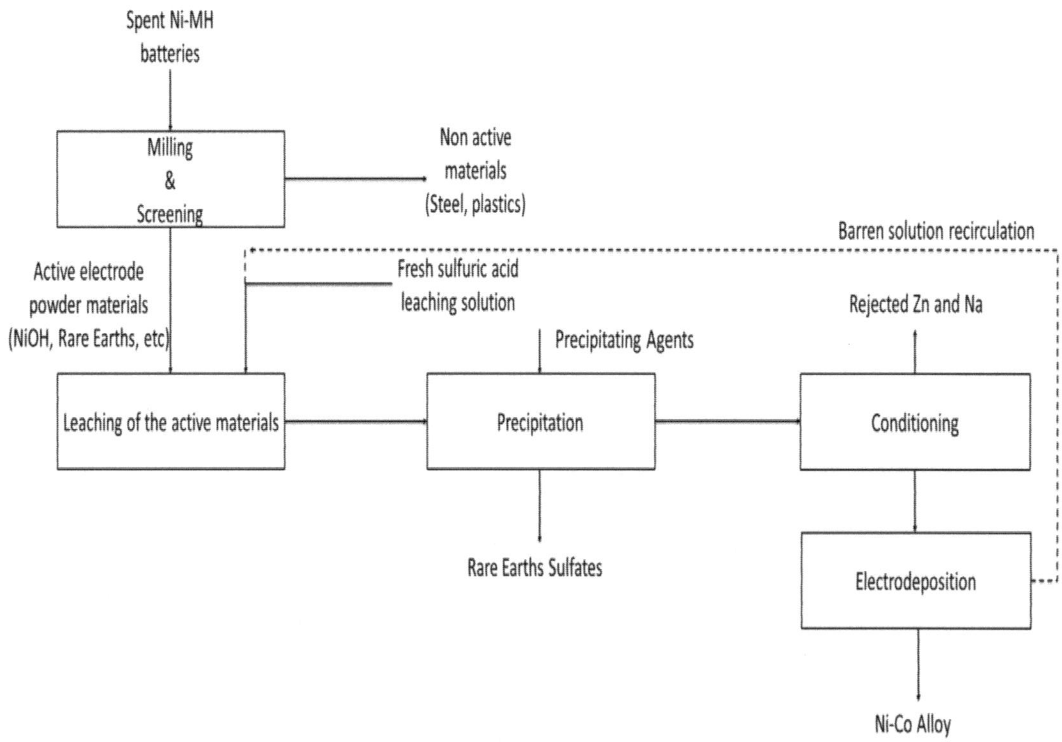

Figure 2. Process flow diagram of the proposed recycling operation for spent Ni-MH batteries.

5. Concluding remarks

The deposition of Ni-Co coatings from multi-metallic sulphate solution is a complex process. It was found that ions such as K^+, Fe^{2+} and Mn^{2+} improve the current efficiency for the Ni-Co deposition process on a copper surface. On the other hand, Na^+ and Zn^{2+} ions exhibited a deleterious behaviour, minimizing the values of the reduction current. The latter implies that a hydrometallurgical recycling procedure to recover a Ni-Co alloy from spent Ni-MH batteries needs to include additional stages for the conditioning of the solution, in order to eliminate those deleterious ions, which are present in the electroplating solution.

References

[1] Zambrano A, Romero C, Moccia A, Orta-Rodríguez R, López J and Delvasto P 2015 *Producción+Limpia* **10(1)** 51
[2] Santos V E O, Celante V G, Lelis M F F and Freitas M B J G 2012 *J. Power Sources* **218** 435
[3] Rozario A, Silva e Silva R K and Freitas M B J G 2006 *J. Power Sources* **158** 754
[4] Ito M, Kashiwaya K, Sumiya N, Furuya H, Hiroyoshi N and Tsunekawa M 2010 *Int. J. Miner. Process.* **97** 92
[5] Ni-MH Linyan Li L, Xu S, Ju Z and Wu F 2009 *Hydrometallurgy* **100** 41
[6] Moccia A, Romero C, Orta-Rodríguez R and Delvasto P 2015 *Av. Quím.* **10** 11
[7] Delvasto P, Orta-Rodríguez R and Blanco S 2016 *J. Phys.: Conf. Series* **687** 012107
[8] Yang X, Zhang J and Fang X 2014 *J. Hazard. Mater.* **279** 384
[9] Zamani M, Amadeh A and Lari Baghal S 2016 *Trans. Nonferrous Met. Soc. China* **26** 484
[10] Rahman I Z, Khaddem-Mousavi M V, Gandhi A A, Lynch T F and Rahman M A 2007 *J. Phys.: Conf. Series* **61** 523
[11] Vazquez-Arenas J, Altamirano-Garcia L, Treeratanaphitak T, Pritzker M, Luna-Sánchez R and Cabrera-Sierra R 2012 *Electrochim. Acta* **65** 234
[12] Vazquez-Arenas J, Treeratanaphitak T and Pritzker M 2012 *Electrochim. Acta* **62** 63
[13] Schweckandt D S and Aguirre M C 2015 *Procedia Mater. Sci.* **8** 91
[14] Granados Neri M, Mendoza Huizar L and Rios-Reyes C 2011 *Quim. Nova* **34** 439
[15] Zhelibo E and Andrushchenko V 1986 *Sov. Powder Metall+* **25** 531

Crevice corrosion of nickel superalloy in deaerated chloride/sulphate solution at 90°C

M Rincón Ortiz[1], R M Carranza[2] and M A Rodríguez[2]
[1] Universidad Industrial de Santander, Bucaramanga, Colombia
[2] Comisión Nacional de Energía Atómica, Instituto Sábato, Buenos Aires, Argentina

Email: mrincono@saber.uis.edu.co

Abstract. Alloy 22 is a nickel base alloy highly resistant to all forms of corrosion. It was designed to resist to most aggressive environments for industrial applications. Electrochemical studies such as Potentiodynamic-Galvanostatic-Potentiostatic (PD-GS-PD) tests and Electrochemical Impedance Spectroscopy (EIS) and Scanning Electron Microscopy (SEM) observations were performed to determine the corrosion behaviour of alloy 22 (N06022). The effect of sulphate ion in chloride containing solutions at 90°C, were studied in this work under aggressive conditions where this material might be susceptible to crevice corrosion. The electrolyte solution, which consisted of 0.1M and 1M NaCl and different sulphate concentrations. It was observed that there were complete inhibitions of crevice corrosion for $R_{CRIT}=[SO_4^{=}]/[Cl^-]=1$ in the 0.1mol/L NaCl solutions and $R_{CRIT}=2$ in the 1mol/L NaCl solutions. The corrosion rate obtained was about 0.1μm/year at 24 hours of immersion.

1. Introduction

Alloy 22 (UNS N06022) belongs to the Ni-Cr-Mo family, and it is extremely resist to general and localized corrosion [1-2]. It contains nominally 22% Chromium, 13% Molybdenum and 3% Tungsten. Alloy 22 is generally not susceptible to pitting corrosion in chloride solutions, however, it might suffer crevice corrosion under certain aggressive conditions [3-5]. The localized corrosion will only occur when the corrosion potential is equal o higher than a critical potential. The susceptibility to crevice corrosion is measured by the value of repassivation potential $E_{R,CREV}$, in each tested condition. The crevice corrosion repassivation ($E_{R,CREV}$) potential has been proposed as the critical potential [6]. Chloride is the only known ion that is able to promote the crevice corrosion of Alloy 22 [5,7-8]. The alloy susceptibility to crevice corrosion is increased by temperature and higher chloride concentrations. Species such as nitrate, carbonate, hydroxyl, phosphate, Fluoride, organic acids are also able to mitigate or inhibit the crevice corrosion [2,5-6,9-10]. In the literature, there is a critical concentration ratio R_{CRIT} =[Inhibitor]/[Cl^-] for the inhibition to be complete. The lower the R_{CRIT} value, the better the effectiveness of the inhibitor. R_{CRIT} for nitrate is 0.15 to 0.5 while R_{CRIT} for fluoride is 5 to 10, and it is 5 to 20 for organic acids [11-12]. The occurrence of crevice corrosion is due to the formation of a hydrochloric acid solution in the creviced region. Inhibitors act by eliminating the occurrence of hydrochloric acid or by hampering its actions [13]. Different standardized techniques have a been used for determining $E_{R,CREV}$ for alloy 22, such as cyclic potentiodynamic polarization (CPP) [14] technique (ASTM G6), Tsujikawa – Himasatsu Electrochemical (THE) method (ASTM G192) [14]. Non-standardized techniques, such as Potentiodynamic-Galvanostatic-Potentiodynamic PD-GS-PD [15], have been also used. The objective of this work was to evaluate the general and crevice corrosion performance of Alloy 22 in solutions containing different concentration of chloride and sulphates at 90°C, by PD-GS-PD technique. The repassivation potentials were used as the parameter for comparison of the crevice corrosion susceptibility.

2. Experimental procedure

The chemical composition of the Alloy in Weight percent was 59.56% Ni, 20.38%Cr, 13.82% Mo, 2.64% W, 2.85% Fe, 0.17%V, 0.16% Mn, 0.008% P, 0.0002% S, 0.05% Si, and 0.005% C. Electrochemical measurements were carried out using a glass cell. A platinum foil was used as a counter-electrode. A saturated calomel electrode was used as a reference (SCE) ($V_{SCE}=V_{ENH}+0.244V$). Two different types of specimens, prismatic and PCA (Prism Crevice Assembly), were used to study uniform and crevice corrosion behaviour respectively. I) Prismatic specimen for the study of corrosion in the passive state. II) PCA (Prism Crevice Assembly) specimen with crevice formers of PTFE-coated ceramic material (ASTM G 78) [14], especially designed for the study of crevice corrosion. The samples were polished with SiC sandpaper to particle size 600, rinsed with alcohol and water and then dried with a stream of air. The studied solutions included chloride 0.1M, 1M plus different concentrations of Na_2SO_4. The tests temperature was 90°C. Oxygen was eliminated by N_2 bubbling. Potentiodynamic-Galvanostatic-Potentiodynamic (PD-GS-PD) tests were performed using PCA specimen to establish crevice corrosion susceptibility. PD-GS-PD test consisted of three consecutive stages [15]: (1) A potentiodynamic polarization (at 0.167mV/s) in the anodic direction up to reaching an anodic current of 30μA. (2) Application of a constant anodic current of $I_{GS}=30\mu A$ for 2h. (3) A potentiodynamic polarization (at 0.167mV/s) in the cathodic direction, from the previous potential up to reaching alloy repassivation. General or uniform corrosion rates were obtained by means of Electrochemical Impedance Spectroscopy (EIS). The scan started from 10kHz to 1mHz. EIS measurements were performed after 1 hour and 24 hours of immersion. Simple equivalent circuits were fitted to the experimental data in order to obtain the polarization resistance (R_P) and calculate the corrosion rate (CR), using Equation 1.

$$CR(\mu m / year) = \frac{K \cdot EW}{\rho} \cdot \frac{B}{R_P} \quad (1)$$

Where K is the faradic conversion constant ($K=3.27*10^6 \mu m g A^{-1} cm^{-1} year^{-1}$), B is the Stern and Geary constant, EW is the Equivalent Weight (EW=23.28), ϱ is the alloy density (ϱ=8.69g/cm^3) and R_P is the polarization resistance. At the end of the tests, the specimens were examined in the optical and scanning electron microscopes (SEM).

3. Results and discussion

Figure 1 shows PD-GS-PD tests on Alloy 22 in 0.1mol/l $NaCl+Na_2SO_4$ solutions at different R values. The three stages as described above are observed. The PD-GS-PD technique provides a cross-over potential ($E_{R,CREV}$) which is determined at the intersection of the forward (stage 1) and reverse (stage 3) scan. When the no cross-over between the forward and the reverse scans, $E_{R,CREV}$ was determined at the intersection of the reverse scan and the extrapolation of the passive current in the forward scan at higher potentials. Eco increased as R increased. The potential drop during the galvanostatic step was negligible for R=1 suggesting crevice corrosion did not occur, which was verified by SEM. Figure 2 Shows $E_{R,CREV}$ of Alloy 22 as a function of R, for chloride plus sulphate solutions. The values of R_{CRIT} were dependent of the chloride concentration in this case. $R_{CRIT}=1$ and $R_{CRIT}=2$ were found for 0.1M and 1M chloride solutions, respectively. Sulphate has been studied as an inhibitor of localized corrosion (pitting) of stainless steels. Sulphate cannot be reduced within an acidic solution and it is stable as SO_4^{2-} or HSO_4^-. The sulphates ions can act mainly as supporting electrolytes, slowing the accumulation of chloride. The above, involves the competitive migration and adsorption of sulphate and chloride on the alloy surface [16]. Sulphate accumulates at the alloys/solution interface by migration due to its higher charge in relation to ions chloride concentration, which prevents the increase of chloride concentration above a R_{CRIT} [13].

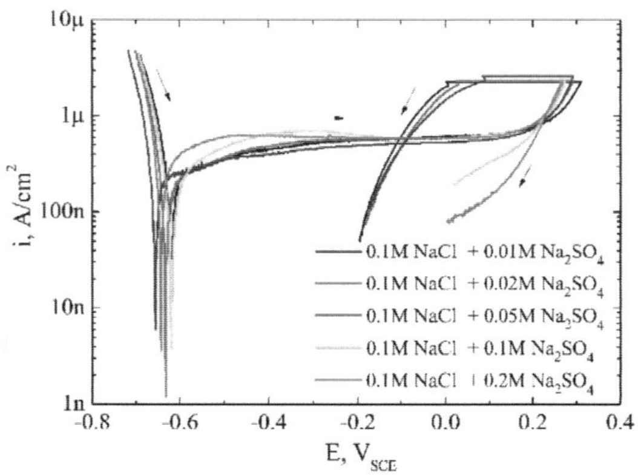

Figure 1. PD-GS-PD tests for Alloy 22 in 0.1M NaCl solutions with different Na_2SO_4 additions at 90°C.

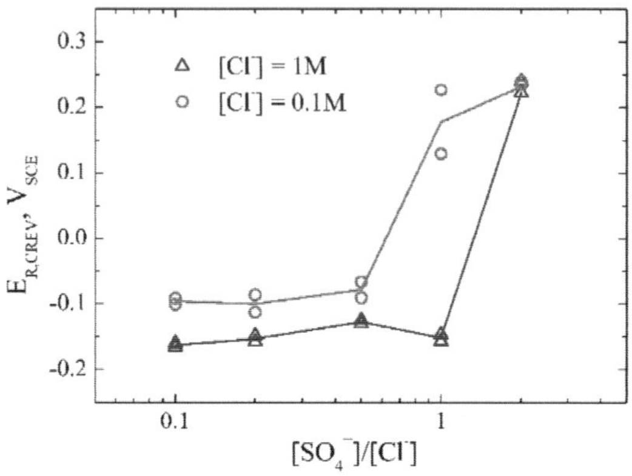

Figure 2. $E_{R,CREV}$ as a function of R for Alloy 22 in chloride plus sulphate solutions (pH 5.5 to 6.5) at 90°C.

Crevice corrosion was found in all solutions with ratios lowers to R_{CRIT}. The attack was always located below the crevice formers (see Figure 3(a)). This type of attack is named crystalline, since the different crystal planes corrode at different rates. Alloy grains can be seen in some areas (see Figure 4). For PD-GS-PD method the Extent of the localized corrosion (crevice) attack was limited by the current set in stage 2. Corrosion products appeared around to areas affected by localized corrosion. There was no crevice corrosion for R≥ R_{CRIT} (see Figure 3(b)).

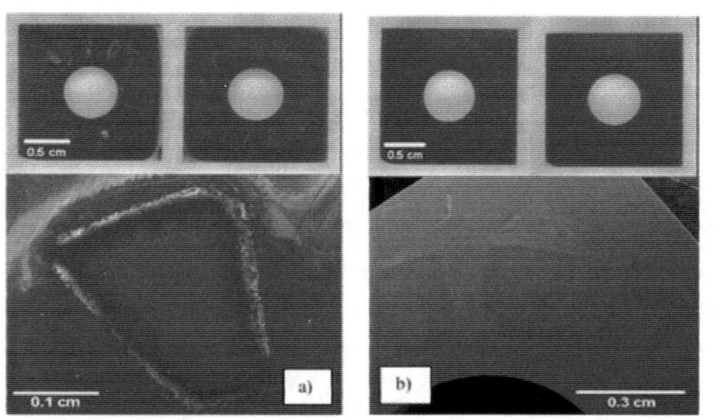

Figure 3. Photographs and SEM images of Alloy 22 after PD-GS-PD curves in different solutions: (a) 0.1M NaCl+0.05M Na_2SO_4 (b) 0.1M NaCl+0.1M Na_2SO_4.

Figure 4. SEM Images of Alloy PCA specimens after a PD-GS-PD curves in 0.1M NaCl+0.05M Na_2SO_4 solution.

Instantaneous corrosion rates were calculated using the parameter R_P, which was obtained by fitting simple equivalent circuits to the EIS data. Tests were performed after 1 hour and 24 hours immersion in naturally aerated solutions. There was a lower corrosion rate (˜0.1μm/year) for the 24-hour period when compared to the 1-hour period in all the analysed environments (see Figure 5).

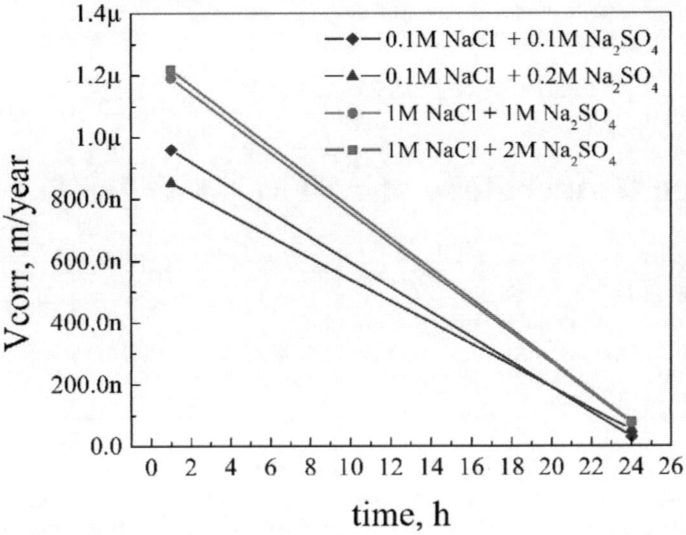

Figure 5. Corrosion rate as a function of immersion time for Alloy 22 in different aerated solutions at 90°C.

4. Conclusions

Crevice corrosion inhibition was observed in the Alloy 22 in solutions containing chloride and sulphates from a concentration ratios R=1 and R=2 for 0.1M and 1M chloride solutions, respectively.

With the use of PD-GS-PD technique, R_{CRIT} conservatives were obtained in comparison with other electrochemical techniques.

Low corrosion rates at 24 horas were determined, independent of the sulphate concentration

References

[1] R B Rebak 2000 *Corrosion and environmental degradation* vol 2, ed M Schutze (Weinheim: Wiley-VCH) pp 69-111
[2] R M Carranza 2008 *JOM* **60** 58
[3] G M Gordon 2002 *Corrosion* **58(10)** 811
[4] R M Carranza 2008 *Proceeding Corrosion/2008* (Houston: NACE Intl) 08580
[5] D S Dunn, Y M Pan, K T Chiang, L Yang, G A cragnolino, X He 2005 *JOM* **57** 49-55
[6] R B Rebak 2005 *Proceeding Corrosion/2005* (Houston: NACE International) 05610
[7] B A Kehler, G O Ilevbare, J C Scully 2001 *Corrosion* **57** 1042-1065
[8] G.O Ilevare 2006 *Corrosion* **62** 340-356
[9] M Miyagusuku, R M Carranza, R B Rebak 2015 *Corrosion* **71** 574-584
[10] D S Dunn, L Yang, C Wu, and G A Cragnolino 2004 *Scientific Basis For Nuclear Waste Management XXVIII* vol 824 ed J M Hanchar, S Stroes–Gascoyne and L Browning (Warrendale: Mater. Res. Soc. Proc.) pp 33-38.
[11] R B Rebak 2009 *Corrosion* **65(4)** 252-271
[12] G O Ilevare, K J King, S R Gordon, H A Elayat, G E Gdowsky, T S E Gdowsky 2005 *J Electrochem. Soc.* **152(12)** B547
[13] J R Galvele 1976 *J Electrochem Soc.* **123(4)** 464-474
[14] V A Mayer 2010 *Annual Book of ASTM standard* section 3 vol 2, ed S J Bailey, N C Emery, J Ermigiotti, D Gallagher, K Hanratty, E Moore, E A Olcese, K A Peters, J L Rosiak, D A Terruso, J Wright (West Conshohocken: ASTM International) pp 315-323
[15] A K Mishra, G S Frankel 2008 *Corrosion* **64(11)** 83
[16] C S Brossia, R G Kelly 1998 *Corrosion* **54** 145

Cyclic oxidation to high temperature of ASTM - A53 grade B

D Pérez[1], J L Marulanda[1] and A Santos[2]
[1] Universidad Tecnológica de Pereira, Pereira, Colombia
[2] Universidad Pontificia Bolivariana, Bucaramanga, Colombia

E-mail: dperez@utp.edu.co

Abstract. Samples of ASTM A53 steel grade B underwent 500 oxidation cycles at 600°C. Scanning Electron Microscopy (SEM), Energy Dispersive Spectroscopy (EDS) and X- Ray Diffraction (XRD) were performed to study the morphology, composition and phase structure of Fe_2O_3 y Fe_3O_4 layers on oxidation surface. Such layers were homogenous and poorly adhesive. Mass loss was significant due to absence of protective oxides and inevitable flaking, leaving the substrate exposed to corrosive environment.

1. Introduction

Industrial processes at high temperature, under corrosive and erosive conditions, not only affect life cycle of materials but also increase maintenance costs. Since carbon steel alloys are recurrently used, it is important to analyse their behaviour when integrity of parts is compromised to such an extent that a complete degradation might take place, thus the manufacturing process must cease leading to economic losses [1-3].

The usage of steel alloys (ASTM A53 grade B) requires heating and cooling cycles, as in the case with boilers and airplane turbines. During cooling, the most resistant alloys give rise to slow formation of oxide layers that retard the mechanical failure. The difference between thermal expansion coefficients of the metallic substrate and the oxides lead to several phenomena, namely: increase of compressive strength of substrate, as well as cracking, delamination and flaking of the oxide layers. Finally, substrate surface become exposed to the action of oxide gases and oxidation rate increases [4,5].

Cyclic Oxidation Tests (COT) are frequently used to assess materials employed under high-temperature conditions. COT usually consist in subjecting materials to heating and cooling cycles to create testing conditions shorter and more severe than actual working conditions. The most relevant indicator in COT is mass variation of the substrate. Samples are also exposed to thermal tensions that may produce some effects observable through SEM; such as cracking, blistering or flaking of oxide layer [2,6,7].

2. Experimental procedure

ASTM A53 grade B specimens (10cm×7cm×5cm) underwent cyclic oxidation in an automated convection oven. Each cycle had a heating stage at 600°C during 50 minutes followed by a cooling stage during 10 minutes, where samples were placed 30 cm away from the oven. Specimens mass was measured with an analytical balance (10^{-5}g accuracy) to monitor changes; measurements were taken before introducing samples in the oven and after cycles 5, 10, 25, 50, 100, 150, 200 and 500. Chemical composition of oxides formed was reported by means of EDS, additionally SEM was carried out to obtain surface and cross section images, topography, porosity and homogeneity of oxide layers.

3. Results and analyses

Figure 1 summarizes the mass variation of specimens in accordance with measurements of the analytical balance. The first 50 cycles exhibited the largest amount of oxides generated, and thus the highest oxidation rates. SEM topography explains such mass gain by reporting the emergence of oxide spinels until 5th cycle and the subsequent formation of the protective oxide layer after the 10th cycle. From the 50th cycle on, oxides generation and oxidation rates decrease; until the 150th cycle where samples begin to lose mass due to iron oxides flaking off the surface [8-10].

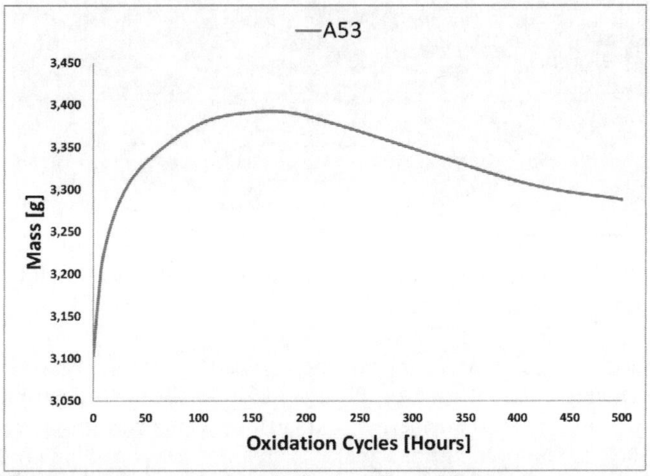

Figure 1. Mass variation during cyclic oxidation.

Cross section of a sample after the 10th cycle based on SEM topography (Figure 2) reveals three zones: the resin corresponding to the darkest zone, the substrate (point 1) whose chemical composition is 99.2% iron and 0.8% oxygen; and finally, a 100-μm-thick protective oxide layer (points 2, 3, 4) with good adherence to substrate, although exhibiting high porosity that will result in cracking. The respective EDS attests to the presence of protective compounds Fe_2O_3 y Fe_3O_4 in the oxide layer, given its results reported 80.4% iron and 19.6% oxygen.

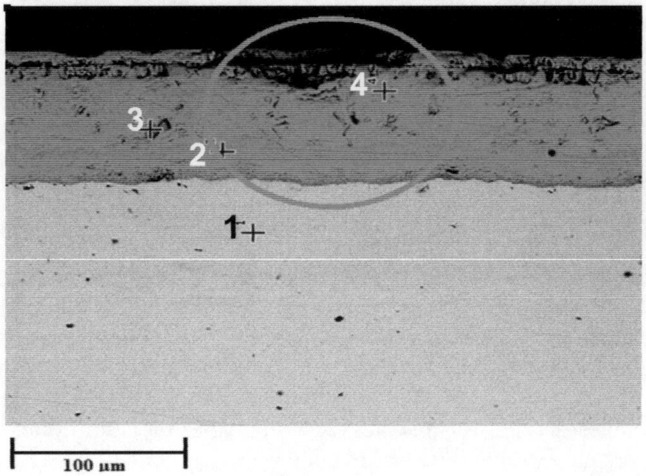

Figure 2. Cross section of sample after the 10th oxidation cycle.

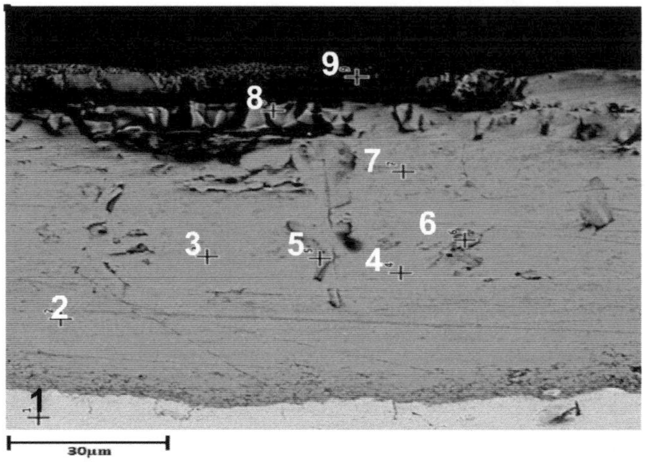

Figure 3. Magnified view on protective oxide layer after 10th oxidation cycle.

Magnified view illustrated in Figure 3 confirms the good adherence between substrate (point 1) and protective oxide layer (points 2 to 9). However, Figure 3 also details the little homogeneity on points 8 and 9 of the oxide layer due to constant thermal expansion, which in turn induces flaking on upper surface of the oxide layer. These conditions make easier the oxygen penetration and/or the inter-diffusion of the oxide layer with Fe. This phenomenon occurred due to iron has two energy levels, where oxide structure determines the physical-chemical properties of surface [11-14]. For this magnified zone, EDS yielded the following percentages 82.3% iron and 17.7% oxygen.

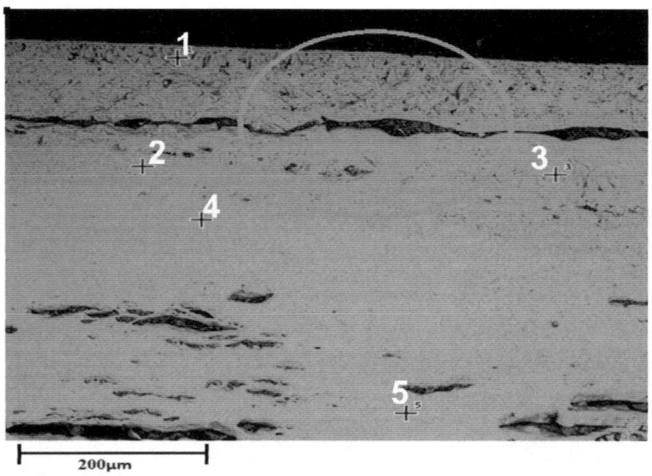

Figure 4. Cross section of sample after the 50th oxidation cycle.

The oxide layer reached the greatest thickness (450μm) after the 50th cycle, as seen in Figure 4. The preview is in accordance with the highest oxidation rates described in Figure 1. Nevertheless, a first horizontally oriented crack separating points 1 and 2 can be considered as the precursor of the subsequent flaking. Chemical composition at this point for the oxide layer presents 21.8% oxygen and 78.2% iron.

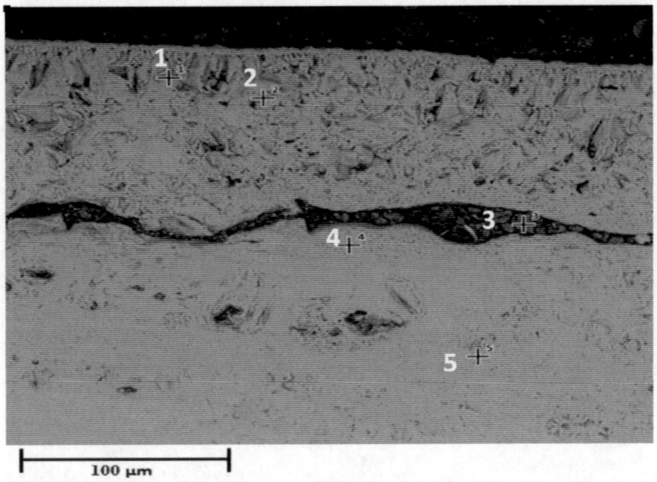

Figure 5. Magnified view on protective oxide layer after 50th oxidation cycle.

Figure 5 provides further details on the zone surrounding the first reported crack. As far as the porosity is concerned; whereas pores appearing after the 10th cycle had an average size of 7μm, those appearing after the 50th cycle were nearly 13μm in average. EDS found the following chemical compositions: 22.1% oxygen and 77.9% iron for points (1, 2, 4 and 5), as well as 75.5% oxygen, 14% calcium and 10.5% iron for the crack (point 3).

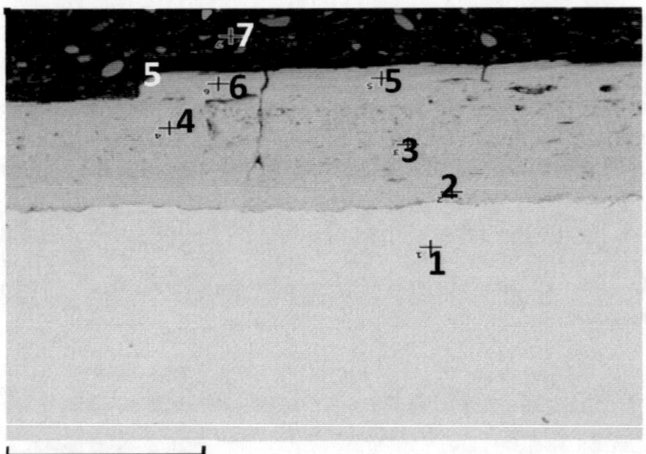

Figure 6. Cross section of sample after the 150th oxidation cycle.

Cross section of sample (Figure 6) points out that the oxide layer has become thinner (200μm), as result of flaking due to some severe horizontally oriented cracks. Having said that, cracking from this point on will develop vertical orientation leading to the inevitable exposure of the substrate to corrosive environment. There is not presence of pores nor cracks in the interphase between substrate and oxide layer according to Figure 6. The chemical compositions corresponding to this Figure are: for oxide layer (75.9% iron and 24.1% oxygen) and for the substrate (98.9% iron and 1.1% oxygen).

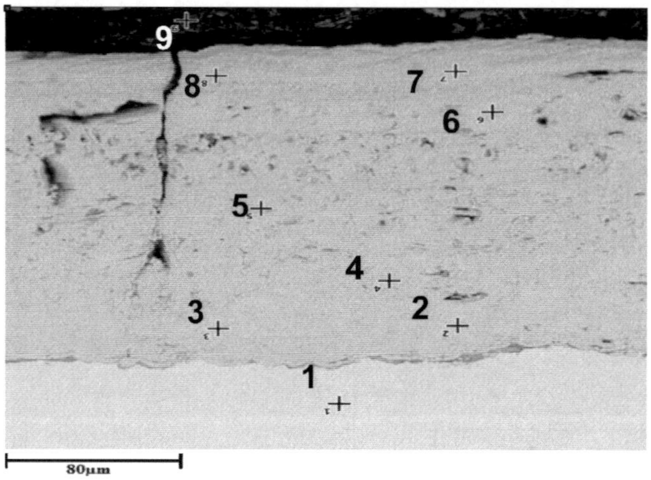

Figure 7. Magnified view on protective oxide layer after 150th oxidation cycle.

Magnified view illustrated in Figure 7 allows to see that the vertically oriented crack is about to come in contact with the substrate. The chemical composition corresponding to this magnified area revealed the highest percentage of oxygen (29.3%), while iron decreases to 70.7%. Experimental results are in strong agreement with literature stating that iron oxides increase with exposure time to corrosive environment at 600°C [10-17]. EDS analysis corresponding to the substrate after the 150th cycle still reports percentages of iron above 98.2% with low percentage of oxygen (1.8%).

4. Conclusions

ASTM A53 grade B steel was subjected to cyclic oxidation at 600°C to simulate shorter and more severe than actual working conditions. Oxides produced begin to lose their protective properties after 50 cycles, when the first significant cracking takes place. Eventually, the oxide layer flakes off the surface leaving the substrate exposed to the corrosive environment (after 150cycles).

Fe_2O_3 y Fe_3O_4 were the predominant oxides during the entire cyclic oxidation, their precursor spinnels started to form after the 5th cycles.

Findings in the experimental tests concur with similar studies on thermal expansion of materials concerning high porosity and cracking.

References

[1] X Cheng, L Fan, H Yin, L Liu, K Du, D Wang 2016 High-temperature oxidation behaviour of Ni-11Fe-10Cu alloy: Growth of a protective oxide scale *Corrosion Science* **112** 54-62

[2] Huabing Li, Bindin Zhang, Zhouhua Jiang, Shucai Zhang, Hao Feng, Peide Han 2016 A new insight into high-temperature oxidation mechanism of super-austenitic stainless steel S32654 *Journal of Alloys and Compounds* **686** 326-338

[3] Bradford S A 2001 *High temperature corrosion, Handbook A* vol 2, ed John E Bringas (Alberta: CASTI Publishing Inc) pp 62-76

[4] Marulanda Jose L, Castañeda Saúl I, Pérez Francisco J 2013 Study microstrucuture and compositions of oxidation in steam at 700 and 750°C of austenitic steels AISI 304, 316 and 317 *Revista Facultad de Ingeniería Universidad de Antioquia* **67** 98-111

[5] B Pujilaksono, T Jonsson, M Halvarsson, J Svensson, L Johansson 2010 Oxidation of iron at 400–600 C in dry and wet O_2 *Corrosion Science* **52** 1560–1569

[6] Norinsam K Othman, Jianquiang Zhang, David J Young 2010 Temperature and water vapour effect on the cyclic oxidation behavior of Fe-Cr alloy *Corrosion Science* **52(9)** 2827-2836

[7] E Dokumaci, I Özkan, B Önay 2013 Effect to boronizing on the cyclic oxidation of stainless steel *Surface & Coatings Technology* **232** 22-25

[8] Li Dong-Sheng, Dai Qi-xun, Cheng Xiao-nong, Wang Rong-rong, Huang Yan 2012 High- Temperature oxidation resistance of austenitic stainless steel Cr18Ni11Cu3Al3MnNb *Journal or iron and steel research, International* **19(5)** 74-78

[9] Balaji Raman, Derek M Hall, Stephen J Shulder, Michael F Caravaggio, Serguei N Lvov 2016 An experimental study of deposition of suspended magnetite in high temperature- high pressure boiler type environments *Colloids and surfaces A: Physicochemical and Engineering Aspects* **508** 48-56

[10] Xiao-jiang Liu, Yong-quan He, Guang-ming Cao, Tao Jia, Teng-zhiu Wu, Zhen-yu Liu 2015 Effect to Si content and temperature on oxidation resistance of Fe-Si alloys *Journal of Iron and Steel Research, International* **22(3)** 238-244

[11] Kamel Makhlouf, J W Jones 1993 Effects of temperature and frequency on fatigue crack growth in 18% Cr ferritic stainless steel *International Journal of Fatigue* **15(3)** 163-171

[12] D Pérez, J L Marulanda, J M Meza 2015 Mechanical characterization by CVD-FBR aluminum Coatings on austenitic stainless steel and the steam oxidation resistance *Revista Facultad de Ingeniería* **24(40)** 9-16

[13] N K Othman, J Zhang, D J Young 2010 Temperature and water vapour effects on the cyclic oxidation behaviour of Fe–Cr alloys *Corrosion Science* **52(9)** 2827-2836

[14] Shreir L S, Jarmon R A, Burstein G T 2000 *Corrosion, Metal/Enviroment Reactions* vol 1 (Oxford: British Library Cataloguing In Publicatin Data) pp 1:55-1:117

[15] Zurek J, Wessel E, Niewolak L, Schmitz F, Kern T, Singheiser L, Quadakkers W 2004 Anomalous temperature dependence of oxidation kinetics during steam oxidation of ferritic steels in the temperature range 550–650°C *Corrosion Science* **46(9)** pp 2301–2317

[16] Hirobumi Okabe 1981 High temperature oxidation of Fe-14Cr-4Al Alloy *Japanese Journal of Applied Physics* **20(3)** pp 165-241

[17] Michael Mueller, Klaus Hilper, Lorenz Singheiser 2005 High temperature corrosion of $MoSi_2$-HfO_2 composites in coal slag *Journal of Physics and Chemistry of Solid* **624** pp 270-278

IOP Conf. Series: Journal of Physics: Conf. Series **786** (2017) 012036 doi:10.1088/1742-6596/786/1/012036

Oxidation of 1020 steel in the abnormal glow discharge

J A García Zúñiga[1], A Sarmiento Santos[1] and E Y Soto Gómez[1]
[1] Universidad Pedagógica y Tecnológica de Colombia, Tunja, Colombia

E-mail: asarmiento.santos@uptc.edu.co

Abstract. 1020 steel is a material very used for surface treatment in the abnormal glow discharge. Because the composition of the gaseous atmosphere has an important influence on the results of plasma treatment, in this work the oxidation process of 1020 steel is verified on the abnormal glow discharge under different concentrations of air (20% to 100%) at temperatures of 600°C and 900°C. For each atmosphere used mass variation is measured during the process of surface oxidation, the structure and microstructure of the oxide film formed is observed and also its mechanical properties through its microhardness.

1. Introduction

Steel is one of the manufacturing and construction materials more versatile, more adaptable and of broad industrial use. At a relatively low price, steel combines the strength and the possibility of being worked, allowing various methods for manufacturing parts [1]. In addition, steel properties can be controlled according to specific needs by heat treatments, mechanical work or by varying their chemical composition [2]. The steel may be subjected to different thermochemical treatments in the plasma of an abnormal glow discharge (AGD), exploiting synergistic effects of photons, electrons, ions and free radicals presents in the AGD, which activate the plasma-surface interactions promoting its reactivity [3]. One of the processes that can be studied in the plasma of the AGD is the oxidation of steel which, under suitable conditions of temperature and oxygen partial pressure, as described in the phase diagram Fe-O [4], can form various iron oxides wustite (FeO), magnetite (Fe_3O_4) and hematite (Fe_2O_3). In this work the oxidation of 1020 steel [5] is made in the AGD in atmospheres containing varying concentrations of air at temperatures of 600°C and 900°C.

2. Methodology

1020 steel cylindrical samples of 10mm diameter and 4mm in height were sanded and then polished to obtain mirror shine following standard metallographic procedures for sample preparation [6]. Next, samples were placed for 40 minutes on the cathode of the AGD of pressure 2torr, 600°C, flows of the mixture of air and hydrogen of 2scc/s and of variable proportion of air from 20% to 100%. The confined anode geometry [7] was used as the electrode configuration and the discharge was initiated with a DC pulsed source. Mass variations (in percent) of these samples treated at 40 minutes, for each air content, were measured with a digital scale with a resolution of 0.0001g. Because changing the percentage of air of 60% for 40% a negative mass variation is observed, testing of these same concentrations of air at 900°C were performed for 24 minutes to verify this behaviour with changing temperature. Heating rates used were 100°C/min. Samples were analysed by XRD using cobalt K_α radiation with wavelength corresponding to 1,789Å. Samples oxidized at 900°C was conducted to microhardness analysis.

3. Results

Maintaining a constant temperature of 600°C, treatment time constant in 40 minutes and the flow of the air-H_2 mixture also constant in 2scc/s, mass variation (Δm in percent) versus air content (in percent) in the gaseous mixture as shown in Figure 1 was obtained.

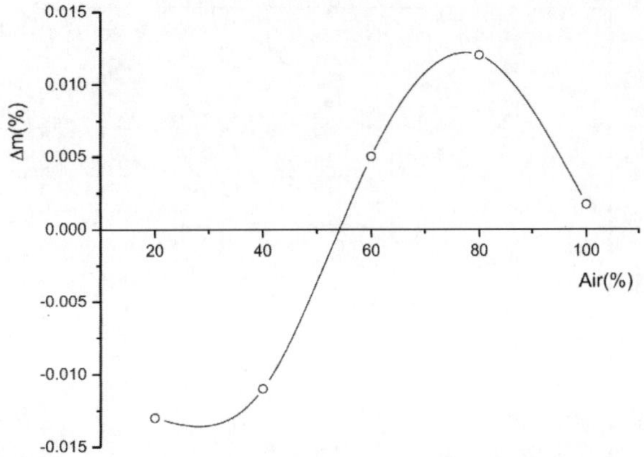

Figure 1. Sample mass variation as a function of air content (in percent) during the 1020 steel oxidation in AGD (Flow=2scc/s, T=600°C and t=40 minutes).

In Figure 1, mass loss in the samples is observed for low air contents. This is due to sputtering which is favoured by the large amount of hydrogen present in the atmosphere [8]. In an intermediate point between 40% and 60% of air content, a transition of the mass variation from loss to a gain occurs. This is due to the bigger iron oxides formation, in the sample surface, at bigger air content in the gaseous atmosphere. Subsequently, a maximum is observed in the mass gain of the sample corresponding to the least amount of hydrogen used and then it decreases in the absence of hydrogen in the gaseous mixture. This behaviour is due to a better efficiency of air discharges with small amounts of hydrogen as observed in other studies for nitrogen [9] and oxygen [10]. The behaviour of voltage and current, depending on the air content in the AGD during oxidation of 1020 steel under the conditions of Figure 1, is shown in Figure 2. In Figure 2, it can be seen that the voltage necessary to maintain the temperature of 600°C decreases with increasing air content in the discharge. This is due to increased heavier species that transferred their momentum to the cathode and to the presence of hydrogen that improves discharge efficiency by allowing the generation of more active species [11]. Conversely, the discharge current increases gradually with the air content to its maximum when the least amount of hydrogen to the gaseous atmosphere is supplied. In fact, as mentioned above by increasing the amount of air in the discharge heavy particles are increased and also the presence of hydrogen improves the efficiency of the same finding maximum efficiency for small amounts of hydrogen. In this case, one can observe the minimum current when the hydrogen of the gaseous mixture of the AGD is removed. Due to the change of sign of the mass variation from negative to positive for air concentrations of 40% and 60%, respectively (see Figure 1), the same tests are performed at a temperature of 900°C and t=24 minutes in order to verify the influence of temperature on this behaviour. Results of these tests are shown in Table 1. It can be seen (Table 1) the same behaviour present in Figure 1 where the variation of mass passes from loss for concentrations of 40% air to gain for air concentrations of 60%, due to the greater amount of iron oxides formed in the latter condition. Because it to increase the temperature of the sample was also necessary to increase the voltage of discharge (see Table 1), it is also obtained a greater mass loss due to sputtering with 40%

air in the atmosphere of the discharge with respect to seen in Figure 1, although the process be performed for a shorter time (24 minutes). In addition to the concentration of 60% air at the discharge, in Table 1 (at lower treatment time) a greater mass increase than that presented in Figure 1 is reported, which agrees with the endothermic nature of the reactions for the formation of iron oxides. [12].

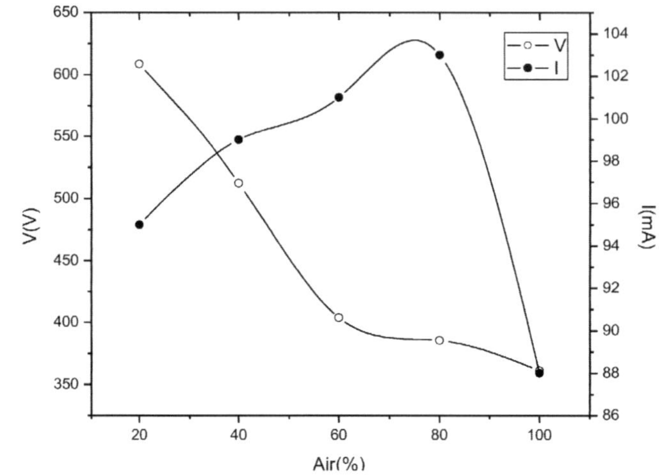

Figure 2. AGD voltage and current versus air content during 1020 steel oxidation (P=2torr, Flow=2scc/s, T=600°C and t=40 minutes).

Table 1. Mass variation depending of air content during 1020 steel oxidation in the AGD (P=2torr, Flow=2scc/s, T=900°C and t=24 minutes).

Air(%)	Δm(%)	V(V)	I(mA)
60	0.019	665	157
40	-0,021	725	149

Figure 3 shows micrographs at 1000X of 1020 steel samples oxide in AGD at 900°C for 24 minutes with air percentages of 40% and 60% in the Air + H_2 mixture of flow 2scc/s and pressure of 2torr. Micrographs of Figure 3 show a similar microstructure with larger deposits in the sample treated with 60% air in the gas atmosphere (see Figure 3(b)) due to the increased presence of oxygen interacting with the ferrous surface. In Figure 3(a) (40% air), however the sample having a mass loss (see Table 1) due to sputtering during the process, it has also been deposited a thin oxide film on the steel surface which is confirmed XRD as discussed below.

Diffractograms, using K_α cobalt line (λ=1,789Å), of 1020 steel before subjected to oxidation in AGD and after oxidation at 900°C for 24 minutes, 2torr pressure and flow rate of 2scc/s of Air+H_2 atmosphere with air percentages 40% and 60% are shown in Figure 4. In diffractograms of samples subjected to oxidation in the AGD with air concentrations of 40% and 60% (diffractograms of Figure 4(c) and 4(b), respectively), besides the peaks corresponding to 1020 steel (diffractogram of Figure 4(a)) present peaks associated with iron oxides Fe_3O_4 and Fe_2O_3 [13], with the absence of the corresponding FeO phase as reported for iron oxidation at 150°C in atmosphere of high purity oxygen [14]. In another study the formation of the three iron oxides (FeO, Fe_3O_4 and Fe_2O_3) at a temperature of 400°C and the formation of single Fe_3O_4 and Fe_2O_3 at a temperature of 600°C was reported [15]. Consistent with the results obtained for the oxidation of 1020 steel in AGD at 900°C, presented here, the formation of only iron oxides Fe_3O_4 and Fe_2O_3 was obtained, under the pressure conditions, gaseous atmosphere and time used.

(a) (b)

Figure 3. Micrographs at 1000X of 1020 steel samples subjected to the AGD at 900°C for 24 minutes with total flow of 2scc/s, 2torr of pressure of air $+H_2$ atmosphere with air contents of: (a) 40% (b) 60%.

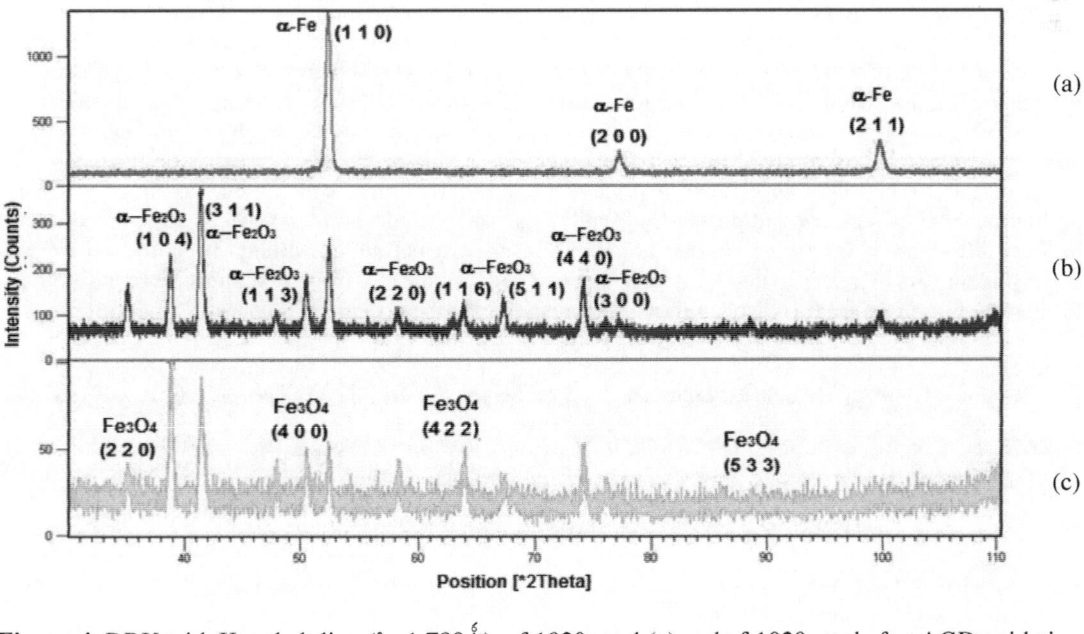

Figure 4. DRX with K_α cobalt line (λ=1.789Å), of 1020 steel (a) and of 1020 steel after AGD oxidation at 900°C during 24 minutes, at pressure 2torr and total flow of 2scc/s of the air $+H_2$ atmosphere with air contents of 60% (b) and 40% (c).

Table 2 shows microhardness measures (HV), made with an applied load of 25gf, depending on the percentage of air for the oxidation of 1020 steel in the AGD at temperature of 900°C (Flow=2scc/s and t=24 minutes). The measurements represent the average of 5 measurements made on the sample surface. In this table the value of microhardness for 1020 steel before subjected to oxidation treatment in the AGD is also included.

Table 2. Microhardness measurements (HV) with a load of 25 gf versus air content for 1020 steel oxidation in the AGD (P=2torr, Flow=2scc/s, T=900°C and t=24 minutes).

Air (%)	Microhardness (HV)	Deviation (HV)
1020 steel	130	4.0
60	415	120
40	580	17

Microhardness of 1020 steel is enhanced by the oxide formed on its surface as can be seen in Table 2. Microhardness of samples for AGD with air contents of 60% and 40% is in the range characteristic for magnetite phase [16,17]. Although also hematite (Fe_2O_3) is formed on the steel surface in both air concentrations (see Figure 4), the microhardness does not correspond to that of this phase [18] which could be due to non-formation of a consistent hematite layer due, to reduced treatment time of the samples in the AGD (24 minutes).

4. Conclusions

The 1020 steel surface oxidation has been made in the AGD with different air contents in the air + H_2 atmosphere at pressure of 2torr at temperatures of 600°C and 900°C. First, the change in mass of the samples increases with the content of air in the gaseous atmosphere finding a transition from mass loss to a mass gain in the range between 40% and 60% air, also it reaches a maximum when the least amount of hydrogen was used in the atmosphere and then decrease when the content of hydrogen in the gaseous atmosphere (100% air) is removed. Secondly, the potential to maintain the temperature of the cathode discharge decreases with air content in the atmosphere, namely, with the amount of heavy species in the gaseous atmosphere. On the other hand, the discharge current exhibits the same behaviour, as a function of air content, that the mass variation, which is increased with small amounts of hydrogen in the gaseous atmosphere. Finally, the surface of samples treated with air contents between 40% and 60%, have a similar structure and microstructure, constituted by both iron oxides corresponding to magnetite (Fe_3O_4) and hematite (Fe_2O_3) phases. The microhardness measured in these two types of samples is also similar and corresponds to the accepted value for magnetite.

References

[1] William D Callister, David G Rethwisch 2014 *Materials Science and Engeneering: An Introduction 9th edition* (United States of America: Jhon Wiley & Sons) Chapter 11
[2] James F Shackelford 2014 *Introduction to Materials Science for Engeneers 8th edition* (New Jersey: Pearson) Chapter 11
[3] M Venugopalan 1987 *Nuclear Instruments and Methods in Physics Research* **B23** 405
[4] Jean-Marc Steiler 1997 *Advanced physical chemistry for process metallurgy* (San Diego: Academic Press) p 311
[5] William F Smith 2006 *Fundamentos de la ciencia e ingeniería de materiales 4th edición* (México DF: McGraw-Hill Interamericana Editores)
[6] George F Vander Voort 2004 *ASM Handbook: Metallography and Microstructures* vol 9 (Ohio: ASM International)
[7] A Cárdenas, Y Pineda, A Sarmiento-Santos, E Vera 2016 *Journal of Physiscs: Conference Series* **687** 1

IOP Conf. Series: Journal of Physics: Conf. Series **786** (2017) 012036 doi:10.1088/1742-6596/786/1/012036

[8] Benjamin Bederson, Herbert Walther 2000 *Advances in Atomic, Molecular, and Optical Physics: Fundamentals of Plasma Chemistry* vol 43 (Unite Kingdom: Academic Press)
[9] S Bockel, G Baravian, A Ricard 1995 *Le Vide* **275** 32
[10] A Ricard, T Czerwiec, T Belmonte, S Bockel, H Michel 1999 *Thin Solid Films* **341** 1
[11] D E Rapakoulias and D E Gerassimou 1987 *J. Appl. Phys* **62** 2
[12] O Kubaschewski, C B Alcock, P J Spencer 1993 *Materials Thermochemistry 6th edition* (Oxford: Pergamon Press Ltd)
[13] Wei Wu, Zhaohui Wu, Taekyung Yu, Changzhong Jiang and Woo-Sik Kim 2015 *Science and Technology of Advanced Materials* **16** 1
[14] A P Grosvenor, B A Kobe, N S McIntyre 2004 *Surface Science* **565** 151
[15] Lu Yuan, Yiqian Wang, Rongsheng Cai, Qike Jiang, Jianbo Wang, Boquan Li, Anju Sharma, Guangwen Zhou 2012 *Materials Science and Engineering B* **177** 327
[16] L H Luong and T Heijkoop 1981 *Wear* **71** 93
[17] P A Munther and J G Lenard 1999 *Journal of Material Processing Technology* **88** 105
[18] J S Sheasby, W E Boggs, T E Turkdogan 1984 *Metal Science* **18** 3

IOP Conf. Series: Journal of Physics: Conf. Series **786** (2017) 012037 doi:10.1088/1742-6596/786/1/012037

Characterization of mechanical properties and electrochemical behaviour in a Hank´s solution of 316L/$Cr_{1-x}Al_xN$ system

D M Osorio[1], J C Caicedo[1], W Aperador[2], A M Benitez-Castro[3], A L Giraldo-Betancur[3], J Muñoz-Saldaña[3], J M Yañez-Limón[3], O Sanchez[4] and G Zambrano[1]
[1] Universidad del Valle, Cali, Colombia
[2] Universidad Militar Nueva Granada, Bogotá, Colombia
[3] Centro de Investigación y de Estudios Avanzados del Instituto Politécnico Nacional, Querétaro, México
[4] Servicio Nacional de Aprendizaje, Cali, Colombia

E-mail: osorio.diana@correounivalle.edu.co

Abstract. $Cr_{1-x}Al_xN$ hard coatings were successfully deposited by R.F. reactive magnetron co-sputtering in an Ar/N2 gas mixture using chromium and aluminium targets on 316L stainless steel substrates. Crystallographic orientations associated to the $Cr_{1-x}Al_xN$ FCC based in the conjugate complex of CrN and w-AlN phases, with ao=4.18Å lattice parameter for the ternary $Cr_{1-x}Al_xN$ compound were identified by X-Ray diffraction. The thickness and roughness of the deposited coatings are 1.00±0.05nm and 2.65±0.6nm, respectively. The mechanical properties were determined by nanoindentation leading to a hardness of 27.8±2.6GPa and elastic modulus of 346GPa. The corrosion resistance of the coated 316L/$Cr_{1-x}Al_xN$ system under simulated body fluid (SBF, Hank's solution) was determined via electrochemical impedance spectroscopy. A reduction in the corrosion rate of 99% in relation to uncoated 316L stainless steel substrate was found by Tafel. Thus, these coatings seem to be excellent candidates to be used in biomedical applications.

1. Introduction

CrAlN protective coatings have been studied for decades due to their excellent mechanical, tribological and electrochemical properties. Some of the first contributions were reported by Okimiya, et al. [1] and Makino et.al. [2]. It is well established that the aluminium addition to the CrN coatings contributes to a significant increase in hardness, wear and corrosion resistance [3–6], especially for high temperature applications. The properties of the CrAlN coatings are from five to six times better than the CrN or AlN coatings deposited at the same conditions [7].

For biomedical applications, the surface properties of a device play an important role. For instance, for surgically implanted biomaterials, the important reactions take place at the interface between the materials and tissues/body fluids determine their suitability[8,9]. The most frequently used methods to improve the surface properties of a component, are based in the use of biocompatible coatings, such as DLC, ZrO2, TiAlV, TiO2, HA, among others [10]. In general, the mechanical and electrochemical properties of hard coatings such as CrAlN improve the mechanical behaviour of steels exposed to wear conditions leading to an increase in their useful life [10–12]. However, there is a lack of information in the literature about the use of CrAlN as coating materials on materials for biomedical applications, such as of 316L stainless steel, considering that it has also lower cost compared to Ti and Ti alloys. The most determinant factor to be addressed is the electrochemical behaviour of the hard coatings when exposed to simulated body fluid (SBF) as a first biocompatibility test.

That is why the main goal of this work is to determinate the functionality of these films exposed to simulated body fluids such as Hank´s solution related to the mechanical and morphological properties of the top CrAlN protective layer.

2. Experimental Procedure

AISI 316L stainless steel coupons were polished and cleaned in an ultrasonic acetone bath prior to deposition. Additionally, silicon substrates with a (111) preferential orientation were used as control samples during deposition process. $Cr_{1-x}Al_xN$ films were deposited by R.F. (13,56MHz) magnetron sputtering on a rotating substrate holder at a constant velocity of 23rpm. Chromium and aluminium targets (99.99% of purity and 100mm of diameter) were used keeping deposition temperature constant at 250°C for 2h. Power applied to the Cr and Al targets was 200W and 350W, respectively. Other parameters were kept constant during deposition process as total pressure of 2.1×10^{-2}mbar, an Ar/N_2 (90/10) gas mixture with a -50V of applied bias voltage to the substrate. A schematic description of the experimental set up has been reported elsewhere [5]. To improve the bonding between the CrAlN ternary compounds to the substrate and reduce the misfit a system of two buffer layers was carried out by depositing Cr and then CrN on top (8 minutes deposition for each layer).

2.1. Films characterization

The $Cr_xAl_{1-x}N$ samples deposited on silicon substrate were characterized by using grazing incidence X-ray diffraction (GIXRD) at angles less than 1, with a RIGAKU (Dmax2100) diffractometer and using a Cu Kα radiation (λ=1.5418A, 30kV and 20mA), to determinate the main characteristics of crystalline structure and residual stresses of the films. Scanning electron microscopy (SEM) and energy dispersive spectroscopy (EDS) analysis, were done in a JEOL JSM9450LB at 13kV to determinate the thickness and elemental composition, respectively. Additionally, the roughness and grain size of the CrAlN top layer, an MFP-3D, Asylum Research atomic force microscope (AFM) was used. The nanohardness and reduced elastic modulus were obtained by nanoindentation, in an Ubi1-Hysitron Inc. device. The tests were carried out with a diamond Berkovich tip in a variable load test. The results were evaluated by the Oliver and Pharr method [13]. Particularly the elastic modulus of the film was obtained from the reduced elastic modulus considering the elastic constants (Young modulus, E and Poisson ratio, ν) of the diamond tip (E=1070GPa and ν=0.07)[14] and the CrAlN coatings (ν=0.25) [15]. Finally, electrochemical impedance spectroscopy (EIS) and Tafel polarization curves techniques were performed at room temperature, using a 300ml cell with a working electrode of CrAlN sample and an exposed area of $1cm^2$, an Ag/AgCl as a reference electrode, and a platinum wire counter-electrode immersed in a Hank´s solution as a simulated body fluid. From EIS measurements the Nyquist diagrams were obtained conducting frequency sweeps in the range 100kHz to 0.001Hz with a sinusoidal voltage perturbation and a signal amplitude of 10mV.

3. Results and discussion

The electrochemical behaviour of the prepared coatings as well as their mechanical behaviour strongly depends on the chemical composition, crystalline structure as well as roughness.

3.1. Chemical composition and crystalline structure

The concentrations of Al, Cr and N were measured by EDS on the $Cr_{1-x}Al_xN$ film and are listed in Table 1. In the film, impurities, such as carbon and oxygen, were less than 3%, and the Al content (defined as the metal ratio x=Al/(Al+Cr)) and the nitrogen-to-metal atomic ratio of nitrogen (N/(Cr+Al)), are 0.29 and 0.45, respectively. Figure 1 presents a typical cross section SEM micrograph of the $Cr_{1-x}Al_xN$ coating, where it was observed the homogeneity of the coating with a thickness around of 1μm.

Table 1. Atomic concentration of CrAlN top layer determined by EDS.

	Al	Cr	N
Wt. %	20±1	49±2	31±1

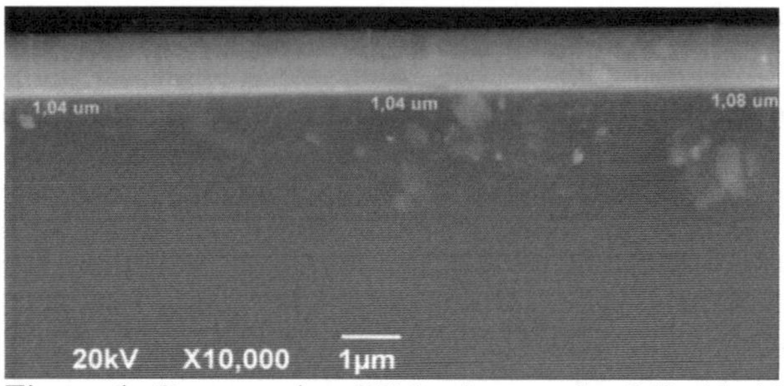

Figure 1. Cross section SEM micrograph of $Cr_xAl_{1-x}N$ coating.

Figure 2 shows the X-ray diffraction patterns of $Cr_{1-x}Al_xN$ sample at 1 grazing angle, where we can see that the films crystallized in the rocksalt structure (B1), with the preferential direction (111) perpendicular to the substrate, in accordance to JCPDF 00-003-1157 and 00-025-1495 card. Particularly, the diffraction peaks with the highest intensity observed at 38, 44, 63 and 76 correspond to both CrN and AlN of (111), (200), (220), (311) crystallographic planes [16]. An additional peak is observed at 33 that correspond to AlN (100).

The $Cr_{1-X}Al_XN$ ternary compound is formed when the AlN phase is incorporated into the CrN FCC structure, where the Al atoms are substituting Cr atoms leading to a variation in the lattice parameter, due to its atom size[6,17]. In this sense, it is clear that the (111), (200) and (220) diffraction planes have a shoulder on the left side, which indicates the FCC-(Cr, Al)N solid-solution phase formation [5,6,16].

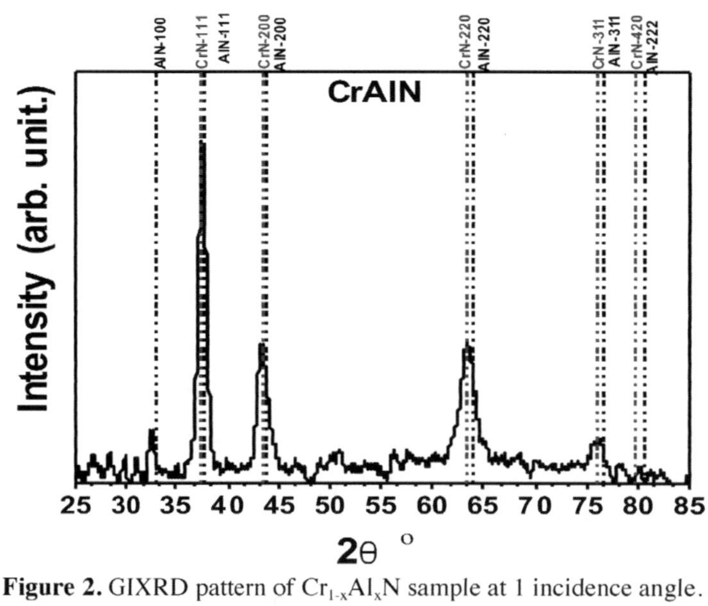

Figure 2. GIXRD pattern of $Cr_{1-x}Al_xN$ sample at 1 incidence angle.

3.2. Morphological analysis by AFM

Figure 3 shows the AFM image for CrAlN coating deposited on Si substrate with a scan area of 1μm^2. The surface roughness and grain size were obtained from AFM image with 2.7±0.6nm and 41±2nm,

IOP Conf. Series: Journal of Physics: Conf. Series **786** (2017) 012037 doi:10.1088/1742-6596/786/1/012037

respectively, indicating that the deposited coating exhibits a homogeneous surface. Moreover is possible to evaluate the electrochemical behaviour compared with the uncoated substrate since the electrical conduction in EIS depends on the conductivity of the electrolyte and microstructural characteristics of the $Cr_{1-x}Al_xN$ constituents including roughness, thickness, grain size, porosity, etc. Our focus is on the damage of the electrolyte to the coating depth.

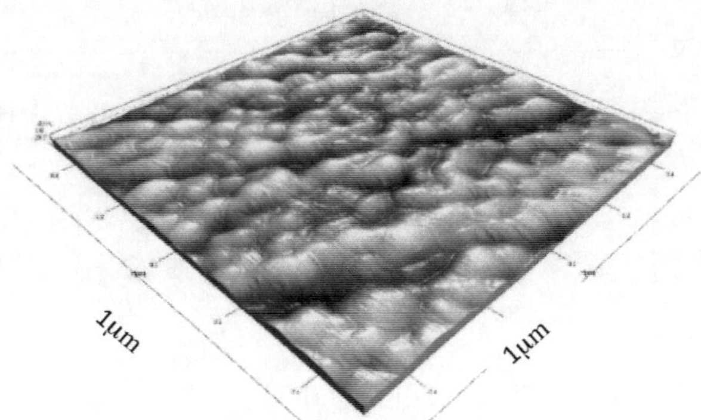

Figure 3. Image of $Cr_{1-x}Al_xN$ coating surface at 1µm×1µm window.

3.3. *Mechanical properties of $Cr_{1-X}Al_XN$ films*

The hardness and reduced elastic modulus measured by nanoindentation were evaluated with the Oliver and Pharr equation[13]. Figure 4 plots the applied load during the nanoindentation test as function of penetration depth. From it, can be seen that for all tests, the maximal penetration depth does not exceed 10% total coating thickness (~1µm). For this reason, the hardness was determined as the average of the experimental points (27.8±2.6GPa). Furthermore, the reduced elastic modulus was obtained following the principle of springs in series, where a linear fit in the graph of elastic modulus as function of contact radius/film thickness ratio, was carried out to obtain the value of reduced elastic modulus. The Young modulus of the film was obtained from the reduced elastic modulus considering the elastic constants of the diamond tip and the CrAlN coatings leading to a value of (346GPa). These hardness and elastic modulus values are in the range of those reported previously by O. M. Sanchéz et.al.[5] Under similar deposition conditions, depending on the concentration of aluminium in the $Cr_{1-x}Al_xN$ sample.

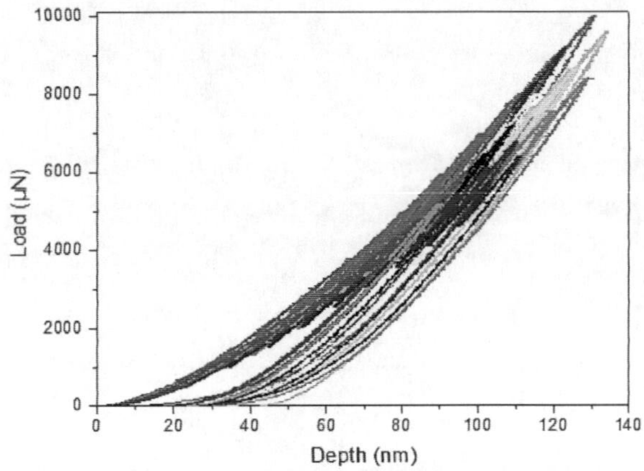

Figure 4. Summary of load-depth penetration curves obtained under variable load tests of samples coatings.

IOP Conf. Series: Journal of Physics: Conf. Series **786** (2017) 012037 doi:10.1088/1742-6596/786/1/012037

Nyquist plots (imaginary impedance Z_{Imag} vs real impedance Z_{real}) for uncoated 316L stainless steel and coated with $Cr_{1-x}Al_xN$ exposed to electrochemical evaluation in presence of a Hank´s solution as a simulated body fluid, are shown in Figure 5, where an equivalent circuit modelling of EIS data is used to extract physically meaningful properties of the electrochemical system by modelling the impedance data in terms of an electrical circuit composed of ideal resistors (R), capacitors (C), and inductors (L). To test a coating via EIS, it is necessary to fit a model data.

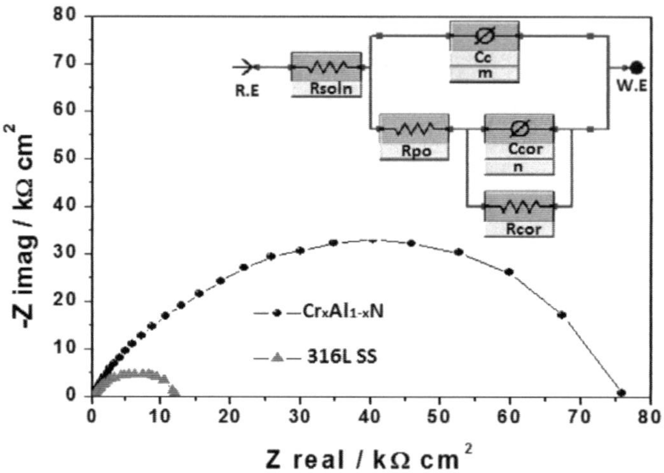

Figure 5. Nyquist plots for uncoated stainless steel 316L substrate and coated with $Cr_{1-X}Al_XN$.

The EIS data were interpreted based on proposed equivalent electrical circuits using a suitable fitting procedure elaborated with Echem Analyst™ software. Therefore, for the equivalent circuits, R_{soln}, R_{po} and R_{corr} are the ohmic resistance of the Hank´s solution, the polarization resistance of the coating/metal interface and the resistance between the solution and coating, respectively; and C_c, C_{corr} represent the capacitance of the intact coating and the double layer capacitance at electrolyte/coating interface.

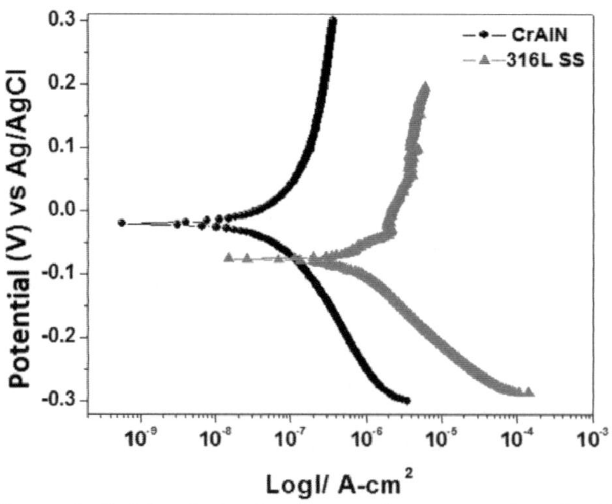

Figure 6. Tafel polarization curves for uncoated 316L SS and coated with $Cr_{1-X}Al_XN$.

IOP Conf. Series: Journal of Physics: Conf. Series **786** (2017) 012037 doi:10.1088/1742-6596/786/1/012037

On the other hand, Figure 6 shows Tafel polarization curves (corrosion potential as a function of the corrosion current density) for uncoated and coated with $Cr_{1-x}Al_xN$ 316L stainless steel. The corrosion resistance for protective materials in the polarization curve is determined by low current densities when the electric potential is increased. Therefore, the Tafel analysis is used to determine quantitatively the corrosion potential (E_{corr}), and the corrosion current density (i_{corr}). If the polarization or charge-transfer resistance is known, it is possible to calculate the electrochemical reaction rates. The corrosion current, i_{corr}, is related to the slope of the plot through the Stern–Geary Equation (1) [18]:

$$i_{corr} = \frac{B}{R_{po}} \tag{1}$$

Where: B and R_{po} are a constant and the polarization resistance, respectively. Thus, in addition to requiring a prior knowledge of B, the determination of i_{corr} requires the measurement of R_{po} *obtained from EIS measurements*. Moreover the B constant is defined by the equation (2):

$$B = \frac{\beta_a \beta_c}{2.3(\beta_a + \beta_c)} \tag{2}$$

Where β_a and β_b are the anodic and cathodic Tafel slopes, respectively, generated by the potentiodynamic polarization curves performed at the end of the experiment. Corrosion current can be related directly to corrosion rate through the equation (3):

$$\text{Corrosion rate (mpy)} = \frac{0.13 I_{corr}(E.W.)}{d} \tag{3}$$

Where, E.W., d and I_{corr} , are the equivalent weight of the corroding species (g), their densities (g/cm^2) and the corrosion current density ($\mu A/cm^2$), respectively. The electrochemical parameters evaluated from potentiodynamic polarization measurements for the 316L stainless steel substrate and the substrate with the system, are shown in Table 2, where the estimated i_{corr} values, from the R_{po} measurements using the Stern–Geary equation, are presented. As can see the values of corrosion rate are inversely proportional to the value of the polarization resistances (R_{po}) in the Nyquist diagrams, where 316L stainless steel substrate coated with the CrAlN is 99% more resistant to electrochemical attack than steel alone, increasing the polarization resistance of the system.

Table 2. Electrochemical parameters evaluated from potentiodynamic polarization measurements.

Sample	R_{po} (Ω cm^2)	Vc (mmy)
316L	3.31 x10^{-3}	1.97
316L/CrAlN	9.15x10^4	9.17x10^{-3}

4. Conclusions

$Cr_{1-X}Al_XN$ coatings were successfully deposited by R.F. reactive magnetron co-sputtering on 316L substrates and its mechanical properties were in the range of values reported previously under similar deposition conditions on other substrates. Therefore, is possible to observe the enhancement in the corrosion resistance for 316L stainless steel coated with $Cr_{1-X}Al_XN$ when it is exposed to simulate body fluids, particularly (Hank's solution), where the corrosion rate was reduced by 99% and the polarization resistance increased in six orders of magnitude compared with uncoated 316L stainless steel substrate ($9.15\times10^{4}\Omega cm^{2}$ coated to $3.31\times10^{-3}\Omega cm^{2}$ uncoated). Finally, it is important to notice that as a material to be employed in biomedical applications strongly requires exhaustive biocompatibility analysis using in-vitro tests such as MTT, genotoxicity, hemotoxicity, or even in-vivo testing. However, the current encouraging biocompatibility analysis with SBF allows us to consider $Cr_{1-X}Al_XN$ coatings as good candidates for this purpose.

Acknowledgments
This research was supported by funds for internal calls for projects at "Universidad del Valle", Cali, Colombia 2015 (CI 71023) and by the Center of Excellence for Novel Materials (CENM) and ASTIN SENA in Cali, Colombia.

References

[1] Okumiya M and Griepentrog M 1999 Mechanical properties and tribological behavior of TiN CrAlN and CrN – CrAlN multilayer coatings *Surf. Coatings Technol.* **112** 123–128
[2] Nogi Y M and K 1998 Synthesis of pseudobinary Cr-Al-N films with Bl structure by rf-assisted magnetron sputtering method *Surf. Coat. Technol* **98** 1008–1012
[3] Brizuela M, Garcia-Luis A, Braceras I, Oñate J I, Sánchez-López J C, Martínez-Martínez D, López-Cartes C and Fernández A 2005 Magnetron sputtering of Cr(Al)N coatings: Mechanical and tribological study *Surf. Coatings Technol.* **200** 192–197
[4] Beliardouh N E, Bouzid K, Nouveau C, Tlili B and Walock M J 2015 Tribological and electrochemical performances of Cr/CrN and Cr/CrN/CrAlN multilayer coatings deposited by RF magnetron sputtering *Tribol. Int.* **82** 443–452
[5] Sanchéz J E, Sanchéz O M, Ipaz L, Aperador W, Caicedo J C, Amaya C, Landaverde M H, Beltran F E, Muñoz-Saldaña J and Zambrano G 2010 Mechanical, tribological, and electrochemical behavior of Cr1-xAlxN coatings deposited by r.f. reactive magnetron co-sputtering method *Appl. Surf. Sci.* **256** 2380–2387
[6] Wang L, Zhang S, Chen Z, Li J and Li M 2012 Influence of deposition parameters on hard Cr–Al–N coatings deposited by multi-arc ion plating *Appl. Surf. Sci.* **258** 3629–3636
[7] Banakh O, Schmid P E, Sanjines R and Levy F 2003 High temperature oxidation resistance of $Cr_{1-x}Al_xN$ thin films deposited by reactive magnetron sputtering *Surf. Coatings Technol.* **164** 57–61
[8] Chu P K 2013 Surface engineering and modification of biomaterials *Thin Solid Films* **528** 93–105
[9] Ramsden J J, Allen D M, Stephenson D J, Alcock J R, Peggs G N, Fuller G and Goch G 2007 The Design and Manufacture of Biomedical Surfaces *CIRP Ann. Manuf. Technol.* **56** 687–711
[10] Montemor M F 2014 Functional and smart coatings for corrosion protection: A review of recent advances *Surf. Coat. Technol.* **258** 17–37
[11] Chang C T, Yang Y C, Lee J W and Lou B S 2014 The influence of deposition parameters on the structure and properties of aluminum nitride coatings deposited by high power impulse magnetron sputtering *Thin Solid Films* **572** 161–8
[12] Chen W, Zheng J, Meng X, Kwon S and Zhang S 2015 Investigation on microstructures and mechanical properties of AlCrN coatings deposited on the surface of plasma nitrocarburized cool-work tool steels *Vacuum* **121** 194–201
[13] Oliver W C and Pharr G M 1992 An improved technique for determining hardness and elastic modulus using displacement sensing indentation experiments *J. Mater. Res.* **7** 1564–83
[14] Fischer-Cripps and Anthony C 2011 *Mechanical Engineering Series* (New York: Springer-Verlag) chap 8 pp 150-152
[15] Lin J, Mishra B, Moore J J and Sproul W D 2008 A study of the oxidation behavior of CrN and CrAlN thin films in air using DSC and TGA analyses *Surf. Coatings Technol.* **202** 3272–3283
[16] Li T, Li M and Zhou Y 2007 Phase segregation and its effect on the adhesion of Cr-Al-N coatings on K38G alloy prepared by magnetron sputtering method *Surf. Coatings Technol.* **201** 7692–7698
[17] Cabrera G, Caicedo J C, Amaya C, Yate L, Muñoz Saldaña J and Prieto P 2011 Enhancement of mechanical and tribological properties in AISI D3 steel substrates by using a non-isostructural CrN/AlN multilayer coating *Mater. Chem. Phys.* **125** 576–586
[18] Stern M and Geary A L 1957 Electrochemical Polarization *J. Electrochem. Soc.* **104** 559

A study of corrosion inhibition of steel AISI-SAE 1020 in CO_2-brine using surfactant Tween 80

M L Cedeño[1], E Vera L[1] and Y Pineda T[1]
[1] Universidad Pedagógica y Tecnológica de Colombia (UPTC), Tunja, Colombia

E-mail: martha.cedeno@uptc.edu.co

Abstract. Surfactant inhibitors also called active surface agents are molecules composed of a polar hydrophilic group and a non-polar hydrophobic group, with characteristics of adsorption on metal surfaces, high efficiency of inhibiting, low price, low toxicity and easy production. In this work, the corrosion inhibition was study by CO_2 steel AISI-SAE 1020 with the addition of 0.01M Tween 80 surfactant to a brine solution (3% NaCl). Electrochemical Impedance Spectroscopy and potentiodynamic polarization testing investigated the phenomenon. The results revealed that the surfactant studied acts as an excellent corrosion inhibitor and inhibition efficiency (E%) increases with increasing fluid velocity. The morphology of the steel surface after exposure to the solution of 3% NaCl with and without surfactant indicates the inhibition phenomenon is due to the adsorption of the surfactant molecules, which insulate the surface of the corrosive medium and reduces the attack surficial.

1. Introduction

Naturally, the Carbon dioxide (CO_2) is finding in oil and gas wells. When it combined with water, a corrosive environment is generated due to the production of reducible species as hydrogen ions and carbonic acid, which cause electrochemical reactions with the metal. Mild steels, materials commonly used in the oil industry, are susceptible to CO_2 corrosion, which represents the biggest problem in oil components and structures. In order to counteract these disadvantages, they have been used modification methods aggressive means, such as injection of corrosion inhibitors. Surfactant inhibitors also called active surface agents are molecules composed of a polar hydrophilic group and a non-polar hydrophobic group, with characteristics of adsorption on metal surfaces, high efficiency of inhibiting, low price, low toxicity and easy production [1-3]. Some researchers [4-7] have reduced steel corrosion in acidic solutions by use of surfactants.

Nonionic Tween type surfactants are esters of polyoxyethylene sorbitan (anhydride sorbitol or hexitol), partially esterified with higher fatty acids, which due to its neutrality, it is not produce counter-ions, thus can be used as excellent corrosion inhibitors. Salah et al. [8] evaluated the inhibition of corrosion of magnesium in hydrochloric acid through the addition of the Tween 80 surfactant, they have found that it inhibits corrosion by physical adsorption on the surface of magnesium, and its ability to inhibit depends on the concentration of the surfactant. However, the efficiency of the inhibitor not only depends on its concentration but also of the structure of the molecule, as concluded in the work done by Abdallah et al. [9]. In this work it was determined that the corrosion inhibition efficiency increases with increasing length of the hydrocarbon chain and the presence of a double bond in the chemical structure. The effect of temperature on the efficiency of the inhibitor has also been studied by Guannan and Xianghong [10] who evaluated the inhibitory action of the Tween-20 nonionic surfactant in corrosion of cold rolled steel in sulfuric acid medium 0.5-7.0M; they have indicated that the corrosion decreases with increasing temperature. However, despite the widespread use on an industrial level and the excellent characteristics of tween surfactant as corrosion inhibitor and demulsifiers of water in oil [11], very little documentation is finding about the anticorrosive properties of this compound. Therefore, in this study investigated the effect of surfactant Tween 80 in inhibiting corrosion of steel AISI-SAE 1020, at different temperatures and speeds, using the rotating cylinder electrode.

Understanding the process of corrosion that occurs at the interface of a metal surface in contact with a medium containing dissolved CO_2, inhibitor and under various parameters such as temperature, speed, presence of corrosion products ($FeCO_3$) on the metal surface is great importance to ensure the integrity of the transportation pipelines.

2. Experimental development

2.1. Equip and preparation of solutions

To evaluate the corrosion of the AISI-SAE 1020 steel, electrochemical tests were performed using an EG & G PARC Model 636 rotary cylinder electrode (ECR), consisting of a rotary unit driven by a motor with rotation ranges between 0 and 9999rpm. The device was coupled to a glass cell with 5 openings, allowing the entry of CO_2, temperature gauge and three electrodes (Figure 1).

A platinum wire was select as counter electrode. A reference electrode of Ag/AgCl, this was connected to the solution by a salt bridge of KCl. Cylindrical samples were used as working electrodes, the material selected was AISI SAE 1020 steel, due to its similarity with the steels used in the oil industry The dimensions of working electrodes were 11.88mm and 11.13mm in diameter and high respectively. The steel composition is present in Table 1. Before performing each test, the test pieces were surface treated according to standard ASTM G1.

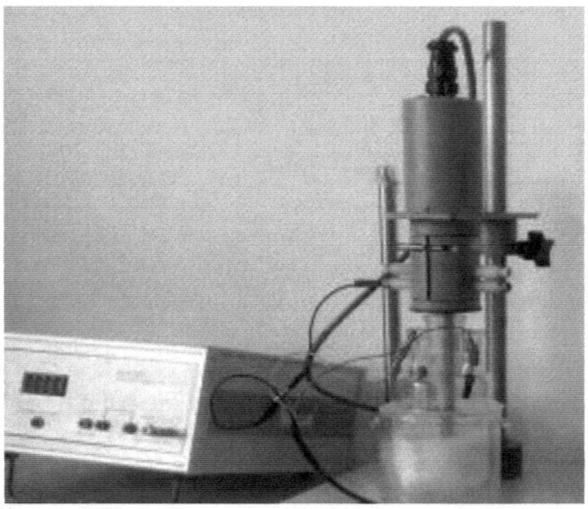

Figure 1. Experimental setup used.

Table 1. Chemical composition of tested steel (wt%).

C	0,024
Mn	0.448
P	0.0184
S	0.0121
Si	0.0017
Cr	0.0057
Mo	0.0039
Ni	0.0318
Nb	<0.0005
V	<0.0003
Ti	0.0009
Cu	0.0268
Fe	Balance

2.2. Equip and preparation of solutions

Impedance spectroscopy test and potentiodynamic polarization were performed. An AC voltage of 10mV was use in the first case, with initial and final frequencies of 100,000Hz and 0.1Hz respectively. On the potentiodynamic polarization test were used potentials between -250mV to 250mV with respect to corrosion potential, scan rate of 2mV/min in anodic direction. The Potentiodynamic polarization curves allowed determination of electrochemical characteristics of steel, such as the corrosion potential (Ecorr) and corrosion current density (Icorr).

2.3. Analysis of the layers formed

Once testing is complete, the specimens were rinse with distilled water and they dried with air jet for morphological analysis of the layers formed on the steel. The observations were make using a scanning electron microscope with secondary electron signal. The Semi-quantitative chemical analysis of energy dispersive X-ray (EDS) was perform with a Philips spectrophotometer.

3. Results and discussion

3.1. Electrochemical tests

Figures 2(a) and 2(b) show the Nyquist diagrams to speeds of 2500rpm and 5500rpm respectively. There is an almost semicircular aspect, indicating that the charge transfer process mainly controls the corrosion of the steel. Deviations from perfect circular shape are relating to the frequency dispersion of interfacial impedance, possibly by the inhomogeneity of the electrode surface or interfacial phenomena [12]. It is evidence that the addition of Tween 80 leads to larger semicircles; they are shift towards higher values on the real axis showing that the inhibitor significantly increases the resistance of the solution and to the polarization of the medium with CO_2. In mediums with inhibitor, the tendency to the formation of an inductive loop at low frequencies has been attribute to adsorption of surfactant molecules on the metal surface.

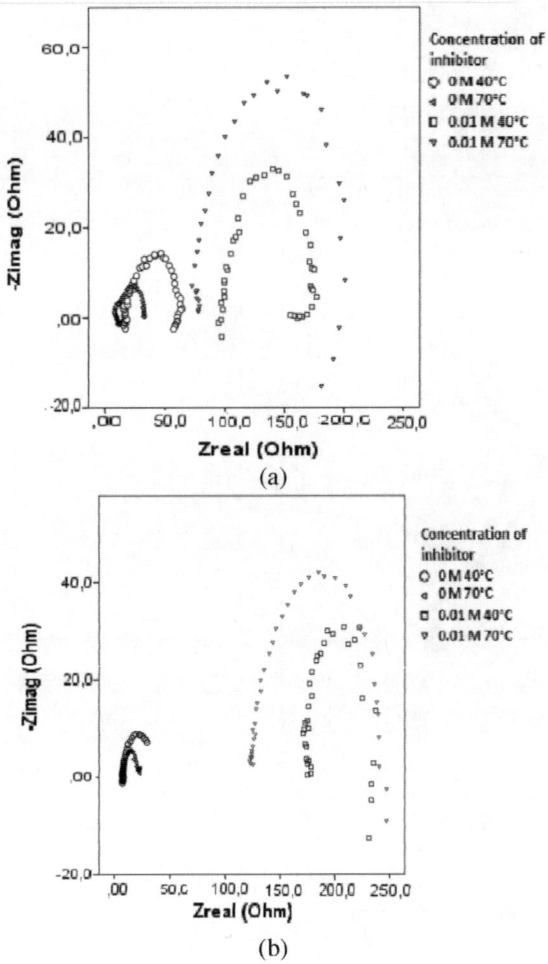

Figure 2. Nyquist diagrams for systems without inhibitor and with presence of 0.01M of this. (a) 2500rpm, (b) 5500rpm.

The potentiodynamic curves (Figures 3(a) and 3(b)) show that the presence of surfactant displaces the anodic and cathodic areas to densities lower currents, indicating that Tween 80 inhibitor interferes in reactions of oxidation-reduction, due that retards anodic dissolution and hydrogen generation, by adsorption of surfactant molecules on the steel surface.

On systems without inhibitor the change at 5500rpm 2500rpm, increases the supply of oxidizing agents in the Helmholtz double layer, resulting in an increase of the current density with the speed.

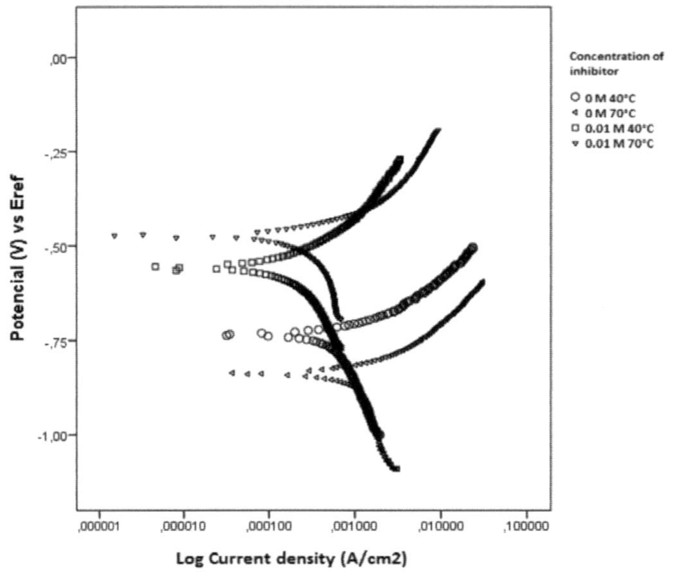

(a)

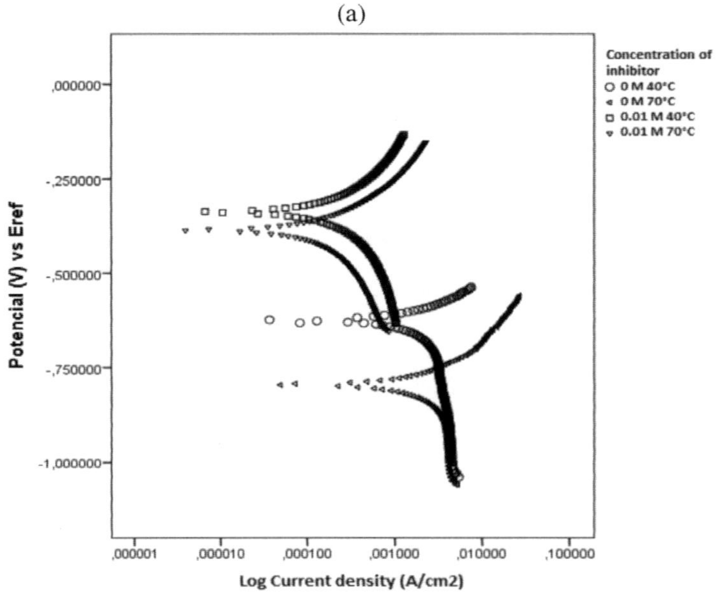

(b)

Figure 3. Polarization diagrams for systems without inhibitor and with presence of 0.01M of this. (a) 2500rpm, (b) 5500rpm.

3.2. Inhibition efficiency

Using equation 1 the inhibition efficiency was determined, which is shown in Figure 4. values of 73.13%, 89.67%, 96.63% and 92.55% for medium with 2500rpm and 40°C, 2500rpm and 70 were obtained, 5500rpm and 40°C, 5500 rpm and 70°C, showing that under the experimental conditions the addition of surfactant allows the corrosion rate (Vcorr) values are reduced considerably.

$$E(\%) = \frac{v_{corr}^{0} - v_{corr}^{i}}{v_{corr}^{0}} x\ 100 \quad (1)$$

The Figure 4 indicates that the inhibition efficiency increases with the increase in fluid velocity given the greater transfer of inhibitor molecules to the metal-solution interface. The lesser inhibitory capacity is acquired when the medium have 2500rpm and 40°C, this is attributed to the influence of temperature in the solubility of the surfactant in the brine, due that the agitation generates a better distribution of molecules in the medium, leading to a high homogeneity of the inhibitor film on the steel surface.

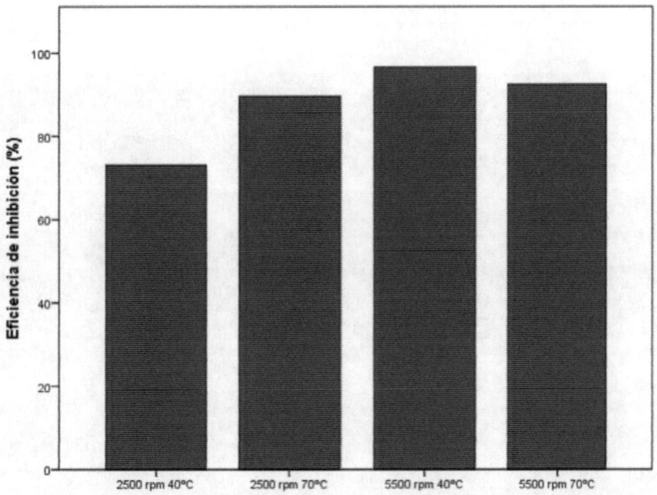

Figure 4. Polarization diagrams for systems without inhibitor and with presence of 0.01M of this. (a) 2500rpm, (b) 5500rpm.

3.3. Analysis of the layers formed

Images of Scanning Electron Microscopy (SEM), reveal the steel surface after being exposed to the solution in presence and absence of inhibitor, with temperature and speed of 70°C and 5500rpm respectively. The Figure 5(a) shows the micrograph of the specimen after immersion in the solution without inhibitor, it can be seen that the surface is strongly affected due to attack by species such as HCO_3^+ and H^+, leading to the formation of layers of iron carbonate not protective and presence of high roughness. Figure 5(b) shows that the metallic surface immersed in solutions with inhibitor is in better conditions, exhibiting smoother surfaces compared to the surface without inhibitor. Thus, it is concluded that the inhibitor tween 80 has a strong tendency to adsorb on the metal surface, protecting it of the corrosive medium.

IOP Conf. Series: Journal of Physics: Conf. Series **786** (2017) 012038 doi:10.1088/1742-6596/786/1/012038

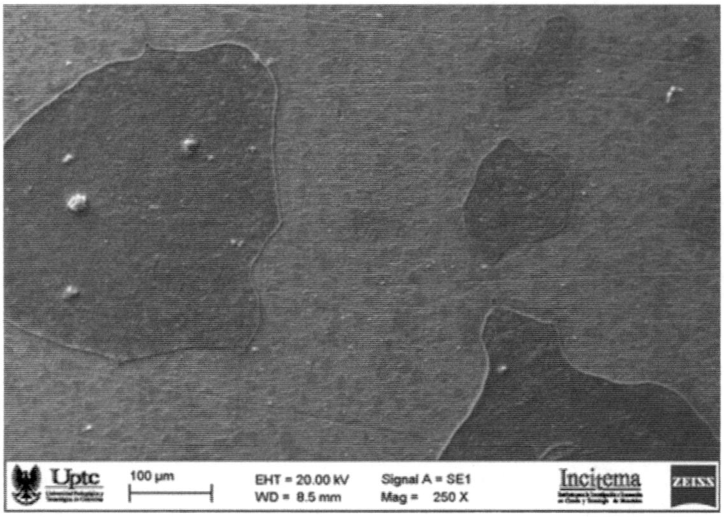

(a)

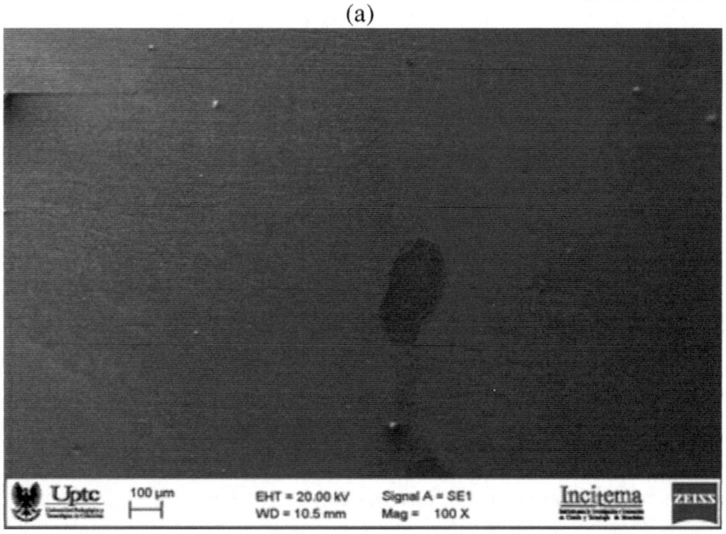

(b)

Figure 5. Steel surface after be exposed to the solution to 70°C and 5500rpm. (a) Without inhibitor, (b) with inhibitor.

3.4. Inhibition mechanism

According to the surface analysis and electrochemical experiments, it is suggest that the mechanism of corrosion inhibition in medium CO_2-brine occurs by adsorption of the functional groups of the surfactant molecules on the surface of 1020 steel, according to Malik et al. [13], this phenomenon can be express according to the following equation:

$$\text{Surfactant (sol.)} + nH_2O\text{ (ads.)} \rightarrow \text{Surfactant (ads.)} + nH_2O\text{ (sol.)} \quad (2)$$

Where n is the number of water molecules removed from the metal surface per molecule adsorbed surfactant.

It is consider that at a concentration of 0.01M, which is higher than the critical micelle concentration (CMC), the metal surface is cover with a monolayer of surfactant molecules and the additional molecules combine to form micelles or multiple layers, such as indicates [13].

4. Conclusions

Evaluation of inhibition of the steel 1020 corrosion by using surfactant Tween 80 shows that:

At a concentration of 0.01M Tween 80 to temperatures and speeds of 40°C, 70°C and 2500 and 5500rpm, was showed the rate of corrosion of steel is decreased significantly, due to the adsorption of the surfactant molecules on the metal surface. Increased solution resistance and polarization is obtain when the medium is in the presence of surfactant. High inhibition efficiencies are achieve above 70% and this increases with increasing speed. Suggesting that the Tween 80 is an excellent corrosion inhibitor in media with CO_2.

References

[1] Murira M C, Punckt C, Schniepp H C, Khusid B, Aksay I A 2008 Inhibition and Promotion of Copper Corrosion by CTAB in a Microreactor System *Langmuir* **24** 14269–14275 Doi: 10.1021/la8024759

[2] Kumar A 2008 Corrosion Inhibition of Mild Steel in Hydrochloric Acid by Sodium Lauryl Sulfate (SLS) *E-Journal of Chemistry* **5** 275–280 Doi: 10.1155/2008/574343

[3] El Azhar, M, Mernari B, Traisnel B, Bentiss F, Lagrenée M 2001 Corrosion inhibition of mild steel by the new class of inhibitors [2,5-bis(*n*-pyridyl)-1,3,4-thiadiazoles] in acidic media *Corrosion Science* **43** 2229–2238 Doi: 10.1016/S0010-938X(01)00034-8

[4] Migahed M A, Azzam E M S, Al-Sabagh A M 2004 Corrosion inhibition of mild steel in 1M sulfuric acid solution using anionic surfactant *Mat. Chem. Phys.* **85** 273–279 Doi: 10.1016/j.matchemphys.2003.12.027

[5] Popova A, Christov M, Raicheva S, Sokolova E 2004 Adsorption and inhibitive properties of benzimidazole derivatives in acid mild steel corrosion *Corros. Sci.* **46** 1333–1350 Doi: 10.1016/j.corsci.2003.09.025

[6] Hamid Z A, Soror T Y, El-Dahan H A, Omar A M A 1998 New cationic surfactant as corrosion inhibitor for mild steel in hydrochloric acid solutions *Anti-Corros. Meth. Mat.* **45(5)** 306–311 Doi: 10.1108/00035599810234605

[7] Osman MM, Omar A M A, Al-Sabagh, A M 1997 Corrosion inhibition of benzyl triethanol ammonium chloride and its ethoxylate on steel in sulphuric acid solution *Mater. Chem. Phys.* **50** 271-274 Doi: 10.1016/S0254-0584(97)01941-X

[8] Salah E, Walid M I 2015 Chemical and Theoretical Studies for Corrosion Inhibition of Magnesium in Hydrochloric Acid by Tween 80 Surfactant *Int. J. Electrochem. Sci.* **10** 8017–8027

[9] Abdallah M, El-Etre A Y 2003 Corrosion Inhibition of Nickel in Sulfuric Acid Using Tween Surfactants *Portugaliae Electrochimica* Acta **21** 315-326

[10] Guannan M, Xianghong L 2005 Inhibition of cold rolled steel corrosion by Tween-20 in sulfuric acid: Weight loss, electrochemical and AFM approaches *Journal of Colloid and Interface Science* **289** 184-192

[11] Roodbari N, Badiei A, Soleimani E 2011 Tweens demulsification effects on heavy crude oil/water emulsion *Arabian Journal of Chemistry* **9** S806-S811

[12] Song G, Bowles A L, Stjohn D H 2004 Corrosion resistance of aged die cast magnesium alloy AZ91D *Materials Science and Engineering* **366 (1)** 74–86

[13] Malik M A, Hashim M A, Nabi F, AL-Thabaiti, S A 2011 Anti-corrosion Ability of Surfactants: A Review *Int. J. Electrochem. Sci.* **6** 1927–1948

Effect of contents oil temperature and flow rate in the electrochemical corrosion of the AISI-SAE1020-steel

M L Cedeño[1], E Vera L[1] and Y Pineda T[1]
[1] Universidad Pedagógica y Tecnológica de Colombia (UPTC), Tunja, Colombia

E-mail: martha.cedeno@uptc.edu.co

Abstract. Primary causes of corrosion in components and equipment used in the petroleum industry are due to the density differences present in the multiphase system Water/Hydrocarbon/CO_2 as well as the presence of weak particles of carbonic acid. The present research is focus on the study of the corrosion rate of the steel AISI-SAE 1020 under a saturated CO_2 multiphase system. The effects of fluid speed, temperature and oil content on the steel corrosion were carried out in an electrode of rotator cylinder and also using electrochemical impedance spectroscopy, and potentiodynamic polarization measurements. The results show that the effect of oil content in the rate of steel corrosion is inversely proportional with the speed of the rotor. Our observations indicate that increasing the rotor speed in systems containing 60% oil or higher produce a simultaneous increase in the degradation rate of materials. Similarly, temperatures higher than 60°C generate layers of siderite that reduce the electrochemical effect.

1. Introduction

Since the beginning of the exploitation of oil, the oil industry has had to confront joint production of water and oil that causes formation of unwanted emulsions. Water and oil penetrate the bottom of the well as two separate phases, which due to the rise in the tube, pumping, push gas and expansion relief valves, are subjected to different stresses that result stirring and shearing, causing the dispersion of these phases. Actually, two-thirds of world oil production was obtained in the form of emulsion, which may present as oil-in-water (o/w) or water-in-oil (w/o).

It has been known that the formation of emulsions or multiphase flows are the main cause of corrosion in the oil industry, due to the density difference between the phases; in addition, the transportation stream increases the impact of corrosively [1]. Despite this, this phenomenon is not well defined. The electrochemical study and corrosion in the presence of different phases is relatively a new theme, which began to emerge less than a decade. Efforts to understand the electrochemical of emulsions still fail to clarify the phenomena and mechanisms of corrosion. Some researchers believe that the aqueous phase, which may include components that accelerate the corrosion process, such as dissolved salts, CO_2, H_2S, acetic acid and other acids, is the main cause of corrosion in these systems. Y. Gonzales et al. evaluated the behaviour of a carbon steel in a solution of oil and water to 3% NaCl, saturated with CO_2 [2], finding that the addition of oil results in better corrosion resistance that the system only with brine. K Efird [3] et al., they studied the effect of the percentage of salt water in the crude, demonstrating that the corrosion rate increases with the content of this. Additionally, some authors indicate that the proportion of hydrocarbon affects the development of corrosion in the metal, increasing the degradation rate with the amount of oil in the system, favours adsorptive processes [1] [4]. Echeverria suggests that the exposed part of metallic substrate in contact with systems water/oil seems to present small anodic and cathode areas, where the metal-aqueous medium interface is the anodic and cathodic that correspond to a metal- oleic medium interface, generating a behaviour less resistive [5].

To contribute to the understanding and clarification of the mechanisms and corrosion phenomena in multiphase systems with the presence of CO_2 that represents 28% of corrosion failures in the oil industry in this investigation, it was used a rotating cylinder electrode that shows the effect of the temperature, oil content and the fluid velocity in the corrosion of steel AISI-SAE 1020, in systems oil/water/CO_2. It is very important to understand the corrosion process that occurs at the interface of a metal surface in contact with a multiphase medium containing dissolved CO_2 and under various parameters such as temperature, relative water-hydrocarbon and presence of corrosion products ($FeCO_{3)}$ on the surface metal to ensure the integrity of the transportation pipelines.

2. Experimental development

Experiments to evaluate the electrochemical corrosion of steel AISI-SAE 1020 were performed using a rotating cylinder electrode (ECR) brand EG & G PARC Model 636, consisting of a rotating unit driven by a motor that is fit to the retainer of the specimen, with ranges of rotational speeds between 0 to 9999rpm. The device was attached to a glass cell, composed of 5 slots, allowing the entry of $CO_{2,}$ temperature gauge and three electrodes (Figure 1).

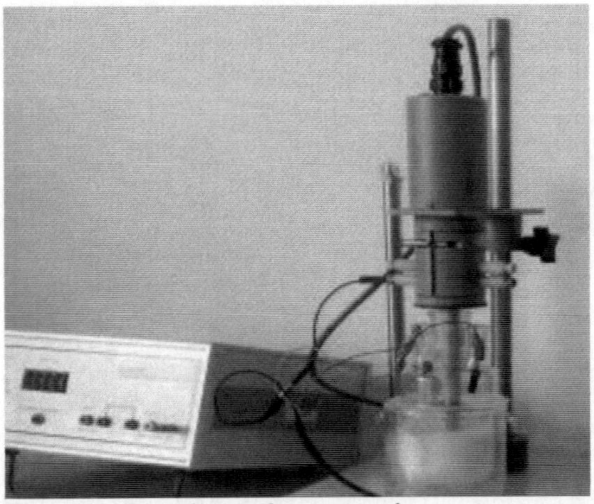

Figure 1. Experimental setup used.

Table 1. Chemical composition of tested steel (wt%).

C	0.024
Mn	0.448
P	0.0184
S	0.0121
Si	0.0017
Cr	0.0057
Mo	0.0039
Ni	0.0318
Nb	<0.0005
V	<0.0003
Ti	0.0009
Cu	0.0268
Fe	Balance

A platinum wire was select as counter electrode and reference electrode as an Ag/AgCl, which was connect to the solution by a salt bridge of KCl. The working electrode consisted of a specimen of AISI-SAE 1020 steel with dimensions of 11.88mm and 11.13mm in diameter and high respectively. The steel composition is present in Table 1. Before performing each test, the test pieces were surface treated according to standard ASTM G1.

The selected medium consisted of a mixture (o/w) deionized water to 3% NaCl, mineral oil ($0.86g/cm^3$) as a substitute for crude, in proportions of 0%, 20%, 40% and 60% by weight. In each experimental test, the mineral oil was add after the steel was immersed in the NaCl solution deaerated with N_2 and saturated with CO_2 for half an hour, to avoid any interference of the oil phase in contact with the surface of the working electrode. The working electrode was always located at the same distance from the interface area. The evaluation of the corrosion phenomenon was evaluate at temperatures of 20°C, 60°C and 75°C and speeds of 2500 and 5500rpm.

Resistance tests were performed with polarization potential with sweeps between -250mV and +250mV respect to the corrosion potential. Frequency sweeps used with electrochemical impedance technique were between 100,000 and 0.001Hz, with an applied potential of 10mV AC. For the analysis of the formed layers were obtain images of the steel surface by the technique of Scanning Electron Microscopy (SEM).

3. Results and discussion

3.1. Effect of oil content

Figures 2(a) and 2(b) show the polarization curves and Nyquist plots for 1020 steel in systems with oil content of 0%, 20%, 40% and 60% at 20°C and 2500-rpm speed. According to Figure 2 and the determined values of the Tafel slopes, it shows that the oil content has little influence on the anodic reaction. This concludes that the phenomenon significantly is dominated by reducing species (cathodic area). The displacement of the cathode area to the right indicates higher current densities corrosion with increasing the amount of oil, so that higher degradation rates of steel in the biphasic, mixture was obtain up to 40%, above which the corrosion rate decreases considerably. This behaviour is similar to the one found in [4] showed that the cathodic current density for emulsions with ratios of 20% to 40% oil tends to be slightly higher than the solution free of oil, while for emulsions content between 45 and 70% by weight, the electrochemical activity decreases.

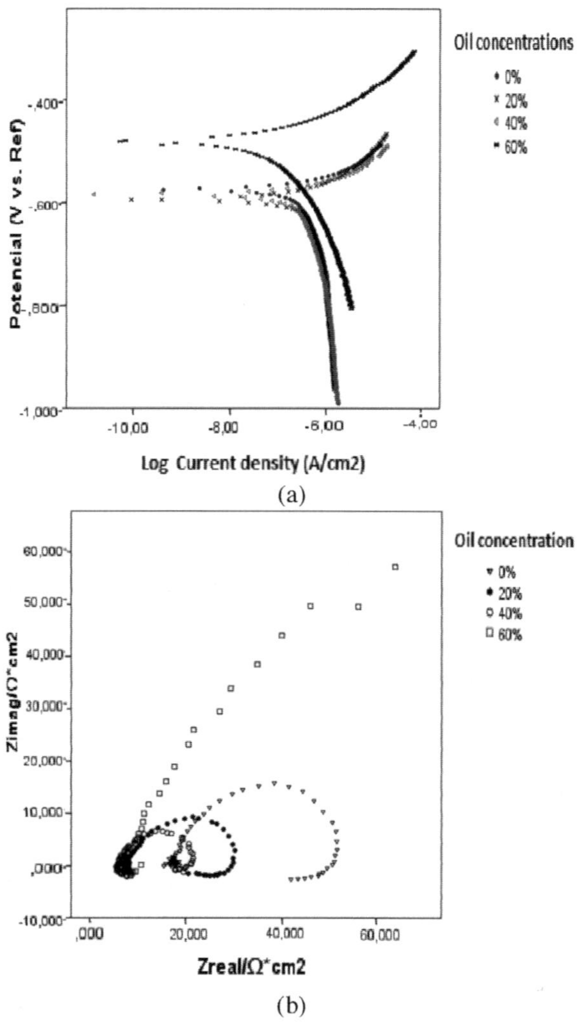

Figure 2. Curves of 1020 steel in emulsion with different concentrations of oil (a) Polarization. (b) Nyquist. 20°C, 2500rpm.

Nyquist plots showed the existence of semicircles with radii larger for systems with 0% and 60% oil, indicating values polarization resistance highest. In the presence of relations oil between 0 and 40%, a capacitive semicircle in the region of high and medium frequencies, characterizes the active state of the interface when exposed to a saturated solution of CO_2 [6] and inductive loop at low frequencies [7,8]. This is attributed to the surface roughness of the material, formation of corrosion products on the metal surface. In addition, for the ratio of 40/60 observes a diffusion phenomenon, represented by a poorly defined semicircle large diameter, which can be associate by the decreased concentration at the surface of an intermediate species adsorbed [9].

In Figure 3(a), the polarization curve of 1020 steel in emulsions with oil contents of 0%, 20%, 40% and 60% at 20°C and 5500rpm is present. It is note that for w/o 60/40 and 60/40 the curves shift to more positive potentials, indicating that tends to protect the metal. This behaviour is attributing to the formation of unstable emulsions, which produces more distribution of oil in the solution and increase deposition capacity of an organic film in the metal surface, corresponding to the oil phase. This film changes the wettability of steel and has a direct effect on corrosion [10].

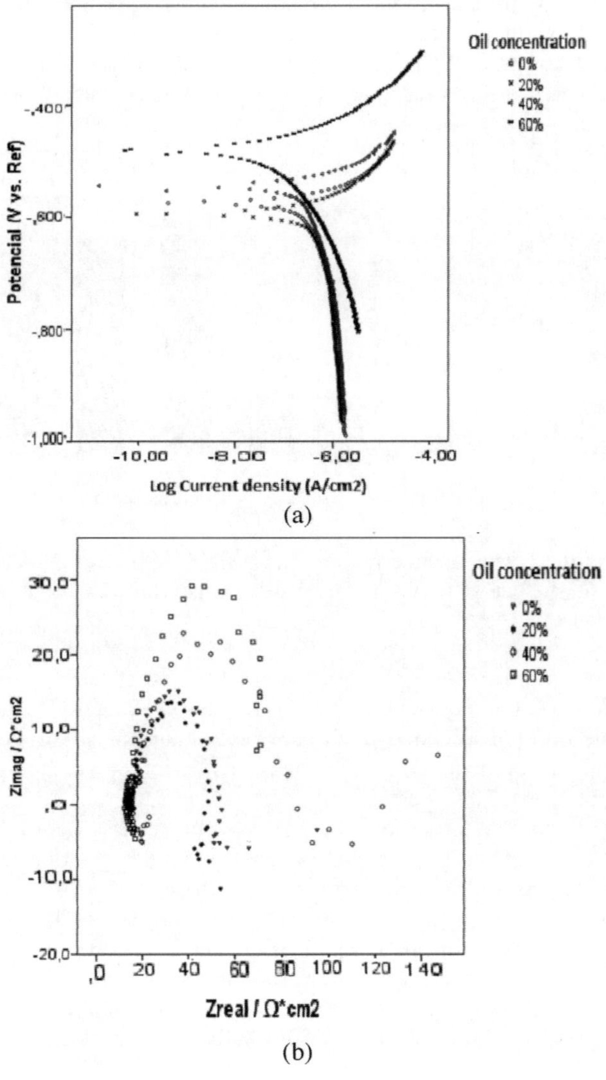

Figure 3. Curves of 1020 steel in emulsion with different concentrations of oil (a) Polarization. (b) Nyquist. 20°C, 5500rpm.

Nyquist diagrams for the system at 5500rpm (Figure 3(b)) show two time constants, a capacitive semicircle in the region of high and medium frequencies, which could be a result of charge transfer and ion migration. A second capacitive arc can be attributing to a dissolution process from the possible layer of oil on the steel surface.

3.2. Speed effect

Figure 4(a) shows the effect of the corrosion rate as a function of the rotation speed. Due to the increased in the supply of oxidizing agents in metal-electrolyte interface generated by the fluid velocity, occurs a simultaneous increase in the rate of degradation of the material with the speed, except for the system with 60% oil. The free oil mixture produces the greatest effect on the corrosion rate. This is attributed to the low conductivity of the oil phase due it difficult ion mobility, in the absence of this, the displacement of oxidizing species into the metal surface tends to develop faster. The decrease in corrosion rate of 40/60 emulsions is attributed to the formation of water in oil emulsion, where the oil phase is continuous and has higher resistivity.

Figure 4(a) shows too that at 5550rpm there is a proportional relationship between the oil content and the corrosion rate. Higher percentages of oil produce lower corrosion rates. An opposite behaviour is observe at low speeds and with systems up a maximum of 40% oil. The phenomenon is attribute to unstable emulsions formed to 5500rpm. They improving the distribution of oil in the solution and increase the likelihood of formation of organic films (corresponding to the oil phase) on steel, this film changes the wettability and has a direct effect on corrosion.

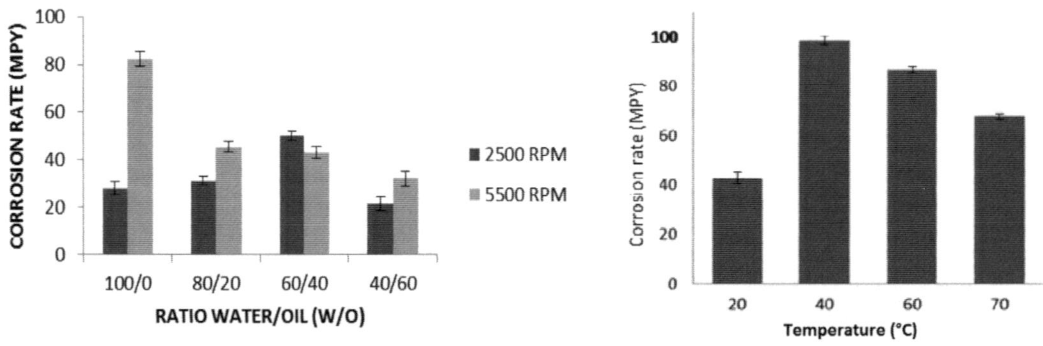

(a) (b)

Figure 4. (a) Effect of rotation speed on the corrosion rate for the AISI-SAE 1020 steel, NaCl 3% CO_2, 20°C, and different ratios of oil. (b) Effect of relationship of the temperature on the rate of corrosion.

3.3. Effect of temperature

The temperature affects the kinetics of corrosion, the concentration of CO_2 dissolved and diffusion of species to the metal surface. To evaluate this phenomenon in the corrosion rate of carbon steel, a system with rotational speed of 5500rpm, which generated emulsions, was select. A percentage of 40% oil and temperatures of 20°C, 40°C, 60°C and 70°C was use.

Figure 4(b) shows the values obtained of corrosion rate for this case. To temperatures below 60°C, the increase of this variable leads to decrease in the polarization resistance. This behaviour can be attribut to the temperature that increases the probability of occurrence of cathodic reactions what can take control the corrosion processes. However, decreasing the rate of degradation from 60°C must be for the formation of a film with protective features, in this case, siderite ($FeCO_3$). Under laboratory conditions, the minimum temperatures required for the formation and growth $FeCO_3$ layer (with protective characteristic) on the metal surface ranging between 50°C and 70°C. Schmitt G. [11] mentions that at low temperatures (<60°C) the solubility of the $FeCO_3$ is high and no protective layers are generated, while for high temperatures (≥60°C) solubility decreases and increases the rate of precipitation of the film of carbonate on the steel surface, thus reducing corrosion. A. B. Forero [12] showed that for the API5L X70 steel, immersed in a solution of 1% NaCl saturated with CO_2, the temperature causes an increase in the corrosion rate up to 60°C, above of this temperature begins to decrease. They attributed this behaviour to the formation of layers more stable.

3.4. Analysis of the layers formed
Figure 5 shows the films formed on the steel to temperatures studied. According to the results of SEM and EDS (Energy Dispersive Microanalysis) it corresponds to the layer formed siderite ($FeCO_3$) primary, which increases in size depending on the temperature.

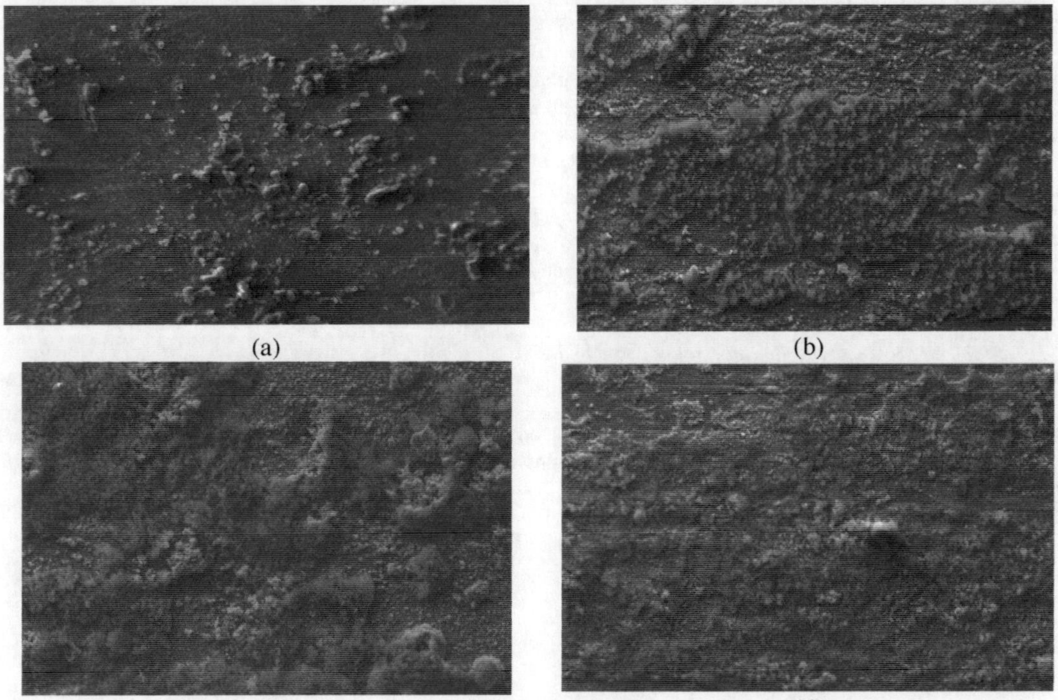

(a) (b) (c) (d)

Figure 5. Layers formed. (a) 20°C, (b) 40°C, (c) 60°C and (d) 70°C.

For the temperature of 20°C, it shows that the steel is practically devoid of $FeCO_3$ this temperature is not high enough to generate and build the growth of crystals, due the carbonates are not easily former or dissolved in the moving fluid. At 40°C there is more amount of iron carbonate, the corrosion rate increases significantly (Figure 4(b)), this is attribute to that the carbonate crystals are porous and act as anodic sites in the corrosion process. The images of 60°C and 70°C show greater homogeneity of the formed layers, which reduce the corrosion rate at these temperatures.

4. Conclusions
Evaluation of the effect of oil content in steel degradation demonstrates the existence of an inverse relationship between the corrosion rate and the amount of oily phase to 5500rpm speed. At low speeds, the opposite behaviour was observed.

Increasing the speed of rotation favours the process of reducing the corrosive species, increasing susceptibility of the material in the environment studied, except for medium containing 60% oil. Free oil mixture produces the greatest effect on the corrosion rate of steel with the speed variation.

At 60°C and 70°C was observed forming protective layers, which managed to reduce the corrosive phenomenon.

References

[1] Clavel Maqueda M, et al. 2011 Estudio del Proceso de Corrosión de Acero al Carbono en una Solución Tipo NACE ID196 en Presencia de Hidrocarburo *XXVI Congreso de la Sociedad Mexicana de Electroquímica 4th Meeting of the Mexican Section ECS* (México: Sociedad Mexicana de Electroquímica) pp 1-6

[2] González Y.; Pineda Y.; Vera E. 2009 Mediciones EIS para Monitoreo de la Corrosión Bajo Condiciones de Flujo Multifásico *Revista Latinoamericana de Metalurgia y Materiales* **S1(1)** 298-304

[3] Efird K D Y, Jasinski R J 1989 Effect of the Crude Oil on Corrosion of Steel in Crude Oil/Brine Production *Corrosion* **45(2)** 165-171

[4] Quiroga H; Retamoso C, Macdonald D 2000 The corrosion of carbon steel in oil-in-water emulsions under controlled hydrodynamic conditions *Corrosion Science* **42(3)** 561-575

[5] Echeverría Félix 1995 Determinación de velocidades de corrosión en sistemas emulsionados aceite en agua (Bucaramanga: Universidad Industrial de Santander)

[6] Farelas F; Galicia M B, Brown S 2010 Evolution of dissolution processes at the interface of carbon steel corroding in a CO_2 environment studied by EIS *Corrosion Science* **52(2)** 509-517

[7] Andre D, Meiler M K, Steiner Ch Wimmer 2011 Characterization of high-power lithium-ion batteries by electrochemical impedance spectroscopy *I Experimental investigation, J. Power Sources,* **196** 5334-5341

[8] Yadav M, Kumar S, Sinha R 2015 New pyrimidine derivatives as efficient organic inhibitors on mild steel corrosion in acidic medium: electrochemical, SEM, EDX, AFM and DFT studies *Journal of Molecular Liquids* **211(3)** 135–145

[9] Song G, Bowles A.L, Stjohn D H 2004 Corrosion resistance of aged die cast magnesium alloy AZ91D *Materials Science and Engineering A.* **366(1)** 74–86

[10] Nesic S, Brown B 2005 Corrosion under scale forming conditions *Corrosion NACE International* **05626** 1-29

[11] Schmitt G, Horstemeir M 2006 Fundamental aspects of CO_2 Metal Loss Corrosion-Part II: Influence of Different Parameters on CO_2 Corrosion Mechanisms *NACE International Corrosion Conference Series* San Diego

[12] Forero A B, Núñez M G, Bott I S 2014 Analysis of the Corrosion Scales Formed on API 5L X70 and X80 Steel Pipe in Presence of CO_2 *Materials Research* **17(2)** 461-471

Implementation of electrochemical elements for an alternative detection of ochratoxin A

D H Aristizabal[1,2], D A Giraldo[1,2], S Sanchez[1], G Taborda[2] and A Baeza[3]
[1] Servicio Nacional de Aprendizaje (SENA), Manizales, Colombia
[2] Universidad de Caldas, Manizales, Colombia
[3] Universidad Nacional Autónoma de México, Ciudad de México, México

E-mail: dhernab@sena.edu.co

Abstract. Ochratoxin A (OTA) is a nephrotoxic metabolite, hepatotoxic and carcinogenic produced mainly by Aspergillus and Penicillium fungi. Usually, the mycotoxin analysed through the technique of high performance liquid chromatography. This method is expensive and takes a lot of time. Therefore, expected to automate a device of low cost, minimal instrumentation micropolarographic (MIMP) for the analysis. For this purpose, the right programming environment and the manufacturing of the software for the graphical user interface selected automation of the MIMP through the design and simulation of the circuit, imprint and assembled in a container with connection ports. Lastly, validation and analysis of Ochratoxin A by cyclic voltammetry from analogous MIMP, automatic MIMP and a potentiostat, in order to corroborate the registered data. This way, voltammetry analysis of coumarin is obtained and the simulation of the MIMP electric circuit. Therefore, this are the key data during the investigation, because the molecule of OTA can be monitored through an acid hydrolysis of amides, composed by a phenylalanine and coumaric acid group, generating the significant decrease in the costs and time of analysis.

1. Introduction

Ochratoxin A is a secondary metabolite with some toxicity level, produced by fungi genus Aspergillus and Penicillium, which causes immune and hepatic diseases [1-5]. Therefore, the OTA considered worldwide as a contaminant in food products. Found in some cereals, spices, fruits, wine, beer and coffee [6]. The food analysis of the metabolite used in immune-affinity tests with chromatographic techniques, which represents raised costs [7]. Thus, the automation of a minimum micropolarographic instrumentation device in electrochemistry designed, to evaluate the consequences in food of this metabolite, using chemical analysis [8]. For this purpose, the use of electronic circuits programmed by means of Arduino microcontrollers is required and analyse the data captured through graphical user interfaces (GUI) [9,10]. In this regard, some of the principal development environment programs to design specific graphical user interfaces in the electroanalytical chemistry field used, according to the need. This could be associated with the data acquisition speed, analysis techniques or comfort for the design of the GUI. Matlab has developed some applications for these interfaces and has become one of the most used software for design GUI, applied in voltammetric monitors of geochemical level in the analysis of marine waters, to follow the levels of sulphur, gases, metals and dissolved gases. The data process and registration should be done as soon as possible by the constant change of the water column in the ocean, therefore the implementation of this type of software allow us to know in depth the natural phenomena 'in vivo', since there could not access a raw team [11]. Other programming environment as KDevolv used in the research work [12], which proposes to improve significantly the chronoamperometry technique used in electrochemical analysis using programming code. With the construction of the software, the objective is to pause the experiments or reverse the sign of the applied current, in order to minimize the undesirable effects of the burden of double layer capacitance and to offset uncompensated ohmic loss. Finally, the programming environment employed during the process of automation of the MIMP, due to it is free of use and pleasant to work. The proposed methodology is a contribution to process sample collection for specific laboratories. The instruments currently used for the analysis of samples are highly priced units and required trained personnel to carry out the analysis and to understand the data. These analytical systems are bulky to be used 'in-situ'. In many cases, the longtime delays involved in this process are unacceptable, and online processing is required. Most electrochemical sensor systems researches focus in analog systems. However, the development of this unit is get a graph of the Ochratoxin type A detection [13].

2. Materials and methods

2.1. Electronic Application

The MIMP system is composed of: electrical supply 12V Stern transformer; resistance of 5kΩ; rheostat of 10kΩ guarantees the resistance variability; voltmeter, allows the measurement of the imposed potential; three electrode electrochemical cell (also an auxiliary electrode, one for reference and one for indicator) and an ammeter, allows the measurement of the current flowing through the electrochemical cell. Two multimeters STERN used as a voltmeter and ammeter. The entire circuit assembled on a disposable plastic box [8]. With the description of the MIMP, it begins the automation process that implements two operational amplifiers with adder-subtracter configuration, thanks to an array of 1KΩ elements (see Figure 1). This is powered by two 9V batteries, controlled using a digital potentiometer X9C103P in a range of +/-2V using an Arduino nano board. Thanks to the arrangement of resistances, the system generates controlled potential sweeps to $100mVs^{-1}$, which processed by reading on Arduino pin configuration-controller and sent to the free processing software, which finally generates the voltammetry and presents the information recorded in both the wave of oxidation, and the reduction of the analytic. (see Figure 1).

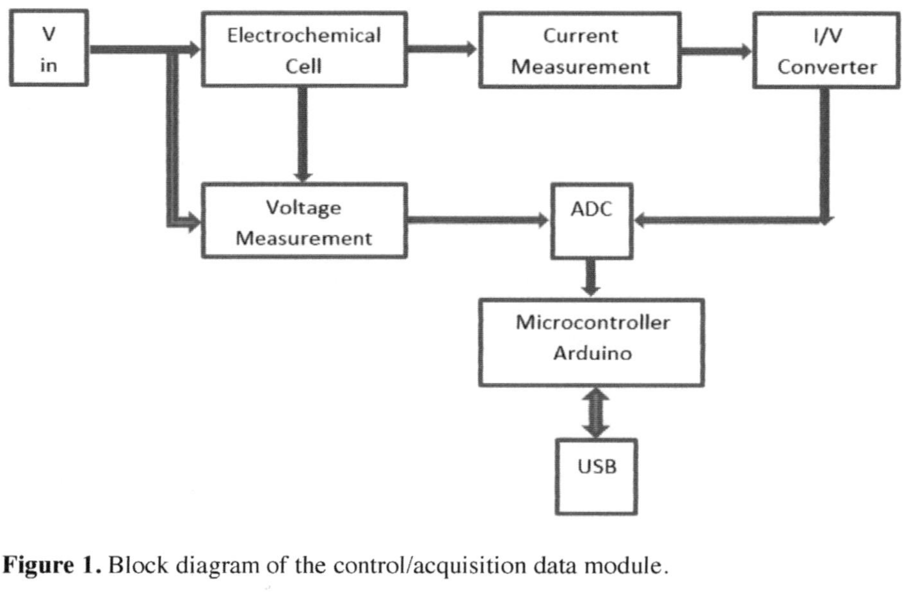

Figure 1. Block diagram of the control/acquisition data module.

IOP Conf. Series: Journal of Physics: Conf. Series **786** (2017) 012040 doi:10.1088/1742-6596/786/1/012040

2.2. *Chemical Analysis*

The voltammetry analysis with the MIMP performed in a glass cell of 4mL with a set of three electrodes: reference electrode $Ag°/(Ag^{+}Cl^{-})$, graphite electrode and vitreous carbon working electrode counter and starts with the electroactivity domain. For this purpose, adding 1mL of PBS 0.1mol L^{-1} buffer at pH 8.0, 0.5mL of electrolyte support 0.1mol L^{-1} of hexadecyltrimethylammonium bromide (CTAB) dissolved in methanol, 0.5mL of Acetonitrile-methanol mixture in proportion 60:40 reagent grade and nitrogen bubbling for 2 minutes. Then, new solutions are prepared for the analysis in triplicate in concentrations of 0,001molL^{-1}, 0,002molL^{-1} and 0,004molL^{-1}, coumarin dissolved in the same form in a mixture of Acetonitrile-Methanol in proportion 60:40 and a sweep rate used of 50mVs^{-1}. After each measurement, the work electrode polished with alumina, to guarantee that the surface area of interaction remains constant. Similarly, it comes with Gamry potentiostat interface of a 1000, with a slight change in the concentration of the solutions of analysis, diluting the concentration 1000 times to avoid overruns on the readings, because the equipment has greater sensitivity compared to the MIMP.

3. Results and discussion

The voltage generated by a dual source of 9V, was restricted to 2V and this voltage then measured between an electrode of reference and the voltage on the working electrode, using an operational amplifier. To do this, it was used a subtracter configuration which determines the difference between these two voltages and thus acquire the measurement required for the construction of voltammograms. In parallel form, it studies the participation of the current in the circuit, determined as an electro counter. Is in here that occurs a conversion from current to voltage through Ohm's law, a turning point during the investigation, because the Arduino microcontroller only receives voltage through its input pins. Additionally a new restriction appears. The values of the potential must be positive for this purpose. An offset implemented, which consisted in applying 2V both of the depleted potential and in the converted from the current. With these ensured that the Arduino board will only conduct readings between 0V and 4V, since the operational amplifier offset settings comes -2V the add 2V, reading will be 0V, otherwise when is received 2V in addition to 2V, at the end the result will be 4V, read by the microcontroller. These data in cyclic voltamperometric should treated with a simple programming Arduino code, since for the cycle there is also a need for negative values. Therefore, the data received by using Arduino analogue pins through the analog-digital conversion, multiplied by -2 to secure the cycle voltamperometry of +/-2V generated by the sweep of potential that gives the lines used [14].

Using the MIMP and applying the cyclic voltamperometry technique achieved to determine coumarins (see Figure 2). In this way, potential of reduction of the compound obtained in -1.5V as those reported in the work carried out by [15] (see Table 1). This results due to the transfer of two electrons and two protons in the carbonyl group to generate the hydroxyl group as a reduced species. Thus, the radical intermediate species may be more stable thanks to the alkaline medium. Regarding the corroboration of the preliminary analysis of the analyte with the 1000 Gamry interface reference, similar potential reductions with linear correlations from R^2=0.9998 are observed with regard to the increment of the cathodic peak current and increasing of the concentration. (see Figure 3 and Table 2).

Table 1. Cathodic peak current and potential half-wave to each concentration by the MIMP.

C (mol/L)	i_{pc} (μA)	$E_{1/2}$ (mV)
0.001	205	-1.573
0.002	230	-1.567
0.004	274	-1.570

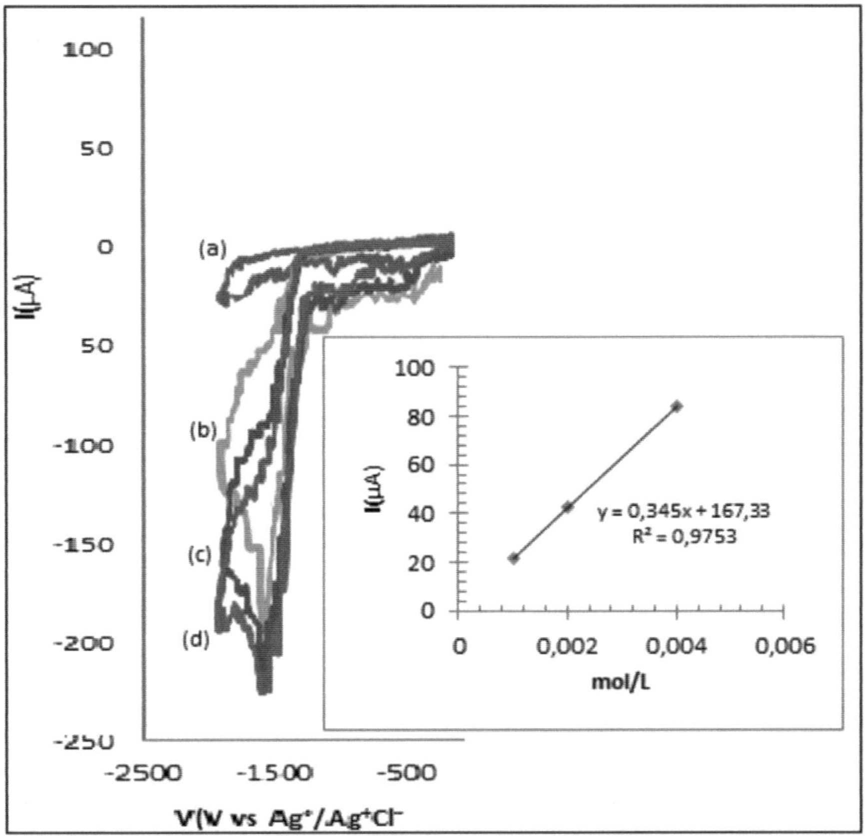

Figure 2. Cyclical voltammograms of (a) electroactivity domain, (b) 0.001mol L^{-1}, (c) 0.002mol L^{-1}, (d) 0.004mol L^{-1} of coumarin at a sweep rate of 50mVs^{-1} on a carbon electrode vitreous. Support electrolyte 0.1mol L^{-1} bromide hexadimethrine (CTAB), buffered solution of phosphates 0.5mol L^{-1} at pH: 8.0 and Acetonitrile-Methanol in proportion 60:40 reagent grade with the analog MIMP.

These findings are useful for the research because of the molecule of Ochratoxin A has a group of phenylalanine and a group coumaric, therefore recognize the potentials of reduction, which the metabolite found useful for the analysis of the compound.

Table 2. Cathodic peak current and potential of half wave to each concentration with the Gamry interface 1000.

C (mol/L)	i_{pc} (μA)	$E_{1/2}$ (mV)
0.0001	21.5	-1.515
0.0002	42.4	-1.496
0.0004	84.3	-1.505
0.0006	121.9	-1.510

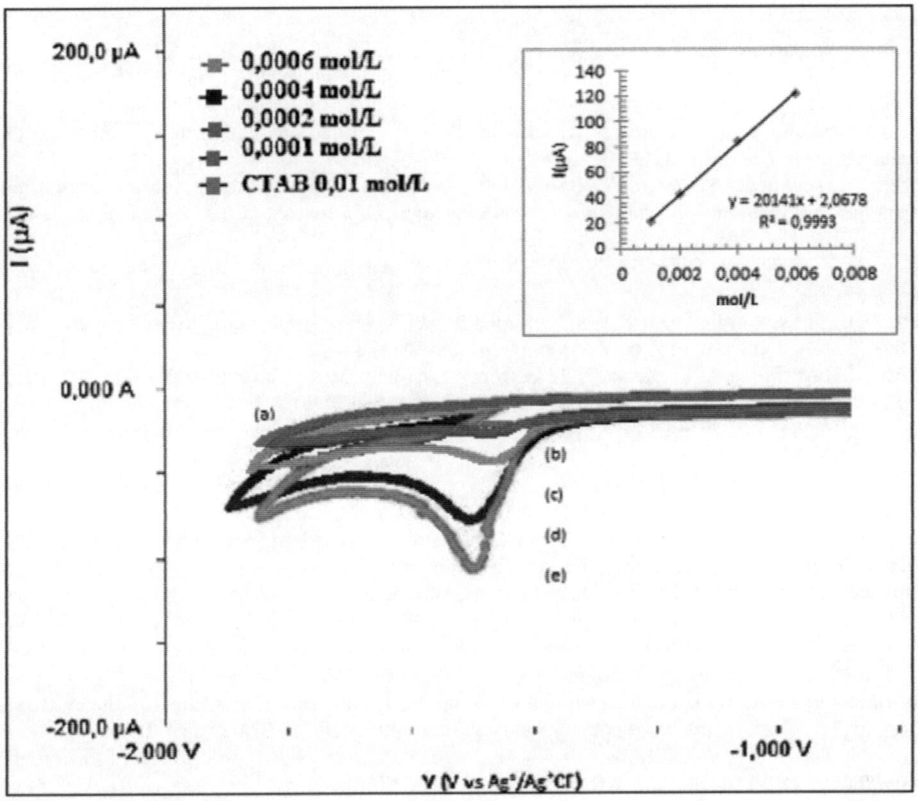

Figure 3. Cyclical voltammograms of (a) electroactivity domain, (b) 0.0001mol L^{-1}, (c) 0.0002mol L^{-1}, (d) 0.0004mol L^{-1}, (e) 0.0006mol L^{-1} of coumarin, at a sweep rate of 50mVs^{-1} on a carbon electrode vitreous. Support electrolyte 0.1mol L^{-1} hexadimethrine bromide (CTAB), buffered solution of phosphates 0.5mol L^{-1} at pH: 8.0 and Acetonitrile-Methanol in proportion 60:40 reagent grade with the potentiostat Gamry interface 1000.

4. Conclusions and expected results

Until now, the partial automatizing of MIMP is in progress, which turns out to be an important tool of chemical analysis for monitoring mycotoxins present in foods, in relation of traditional commercial equipment, both for chromatography analysis and electrochemestry. This way, appears new contributions and alternatives for electrochemical analysis to low-cost and less time, from analog electronics and programming applied to microcontrollers and software design of graphical interfaces free of use. In relation to the expected results, the data is plot from current and voltage through the free software application 'Processing' through code lines graph, obtaining curves from the analytical signal. In this way, the implementation of electroanalytical techniques will be fast and economic, concerning the conventional design until the time of the MIMP.

In appreciation to

The authors offer their gratitude to the National Learning Service (SENA), Tecnoparque Manizales, the University of Caldas in Colombia and the National Autonomous University of México, both for its academic knowledge and infrastructure collaboration.

References

[1] Visconti A, Pascale M, Centonze G 1999 Determination of ochratoxin A in wine by means of immunoaffinity column clean-up and high-performance liquid chromatography *J. Chromatogr. A* **864(1)** 89–101

[2] Leoni L, Furlani R, Valente Soares R, Oliveira P 2001 Ochratoxin A in Brazilian green coffee *Ciênc. Tecnol. Aliment. Campinas* **21(1)** 105–107

[3] Lobeau M, De Saeger S, Sibanda L, Barna-Vetró I, Van Peteghem C 2005 Development of a new clean-up tandem assay column for the detection of ochratoxin A in roasted coffee *Anal. Chim. Acta* **538(1–2)** 57–61

[4] Tafuri A, Ferracane R, Ritieni A 2004 Ochratoxin A in Italian marketed cocoa products *Food Chem.* **88(4)** 487–494

[5] Vega F E, Posada F, Peterson S W, Gianfagna T J, Chaves F 2006 Penicillium species endophytic in coffee plants and ochratoxin A production *Mycologia* **98(1)** 31–42

[6] Ferraz M B M, Farah A, Iamanaka B T, Perrone D, Copetti M V, Marques V X, et al. 2010 Kinetics of ochratoxin A destruction during coffee roasting *Food Control* **21(6)** 872–877

[7] Daško L, Belajová E, Rauová D, Kováč M 2005 Determination of ochratoxin A in beer *Czech. J. Food Sci.* **23(2)** 69–73

[8] Vilasó J E 2014 Propuesta de instrumentación voltamperométrica de bajo costo para uso docente en la carrera de Química *Natura* **1(1)** 14–18

[9] Stevic Z, Rajcic M, Nikolovski D, Bugarovic S, Antic D 2011 Hardware and Software of a System for Electro-Chemical and Bio-Electro-Chemical Investigations *Acta Tech. corviniensis - Bull Eng.* **1** 53–55.

[10] Baltuano O, Hernández Y, Montoya E 2012 Desarrollo de un potenciostato analítico de bajo costo y altas prestaciones *Informe Científico Tecnológico* **12** 165–169

[11] Bristow G, Taillefert M 2008 Voltint: A Matlab-based program for semi-automated processing of geochemical data acquired by voltammetry *Comput. Geosci.* **34(2)** 153–162

[12] Martiínez-Ortiz F 2004 On the use of a real time application interface under Linux in the electrochemistry laboratory. Application to chronopotentiometry *J. Electroanal Chem.* **574(2)** 239–250

[13] Amin M, Isa M M, Sidek R M, Yusof N A 2015 An embedded processing of differential pulse voltammetry (DPV) data using ARM processor (LPC1768) *Proceeding IEEE Int Circuits Syst Symp ICSyS* **2016** 80–84

[14] Maffrand C, Zarate D, Zón M A, Fernandez H, Romero M R 2015 Arbitrary Waveform Generator Using FPGA for Applications in Ultrafast Scan Voltammetry *Front Comput Sci Technol (FCST) Ninth Int. Conf.* **2015** 161–165

[15] Miyano D M, Lima T, La-scalea M A, Oliveira H P M, Codognoto L 2014 Electrochemical Study of Simple Coumarin and its Determination in Aqueous Infusion of *J. Braz. Chem. Soc.* **25(3)** 602–9

IOP Conf. Series: Journal of Physics: Conf. Series **786** (2017) 012041 doi:10.1088/1742-6596/786/1/012041

Implementation of coulometric titration system at constant current for developing of certified materials as primary standards

H Suarez[1], R Cristancho[1], F Peralta[1] and H Torres[1]
[1] Instituto Nacional de Metrología, Bogotá D.C., Colombia

E-mails: hasuarez@inm.gov.co

Abstract. Coulometry is a primary method for measuring high purity substances. This is used to certify the primary reference materials required in a chemical analysis process. This paper describes the coulometric titration system developed by the National Metrology Institute (NMI) of Colombia for the certification of hydrochloric acid 0.1mol/kg reference materials. In addition, it also shows preliminary studies for future development of potassium chloride (KCl) and ethylenediaminetetraacetic acid disodium salt (EDTA-Na_2) certification.

1. Introduction

Coulometry is a high-level electroanalytical technique for measuring the amount of substance in compounds and solutions. This technique has been used to determine the purity of primary reference materials required by industry and in chemical testing laboratories [1]. Coulometry is one of the most widely used analytical techniques in research laboratories and National Metrology Institutes (NMI) worldwide due to its high precision and accuracy for measuring the trace level or purity of different materials [2]. Coulometric analysis uses electrolysis to generate a titrant that reacts stoichiometrically with an analytic, thus making possible a determination of the amount of substance. For this reason, The Consultative Committee for Amount of Substance (CCQM) of the International Bureau of Weights and Measures-(BIPM) has considered coulometry to be a primary method that can be used to determine the purity of primary reference materials.

Electroanalytical methods for analyzing compounds are supported by current, potential and charge measurements of a system. When an analytic is reduced or oxidized, these electrical signals are directly related to the amount of analysed substance present in the system, which has a direct relation to the mole unit of the International system-SI. With coulometry, this relation consists of a determination of the amount of substance with electric charge added to the electrochemical system described by Faraday's law:

$$n = \frac{Q}{zF} = \frac{\int Idt}{zF} \quad (1)$$

Where n is the amount of substance, Q is electric charge, I is electric current, t is electrolysis time, z is the charge number of reaction and F is Faraday constant.

In accordance with Faraday's law, there are two forms of coulometry: (i) controlled-potential, in which electric current decreases exponentially over time; and (ii) controlled-current, in which electric current is constant. The latest form is used for determining analytic content in the matrix, such as salts or metals in solution. Coulometry provides high selectivity and good sensitivity for the identification and quantification of trace-level, which is hard to determine by other electroanalytical methods [3,4]. Additionally, it is able to provide good results with low uncertainty and has been implemented in a larger number of NMIs [5].

The goal of this paper is to provide constant current coulometry studies as developed for the Electrochemical Analysis Working Group at the National Metrology Institute (NMI) of Colombia. First, characterization studies of aqueous hydrochloric acid solutions 0.1mol/kg will be presented. Then, preliminary studies to determine the purity of KCl and EDTA-Na_2 will be presented.

2. Methods

2.1. Coulometry measurement system
Coulometry system development for the NMI of Colombia is similar to the system used by Ural Scientific and Research Institute for Metrology of Russia [6]. This coulometry system consists of: (i) a titration vessel with two independent compartments (these compartments are separated by agar gel) (ii) A Metrohm Autolab potentiostat/galvanostat PGSTAT128N (Metrohm AG Switzerland, Inc.) is used as a source of current, as a potentiometric endpoint indicator and for time control; (iii) argon UHP grade; (iv) a microbalance (Mettler Toledo XP56, resolution 0.001mg); (v) a magnetic stirrer; (vi) a pH unitrode electrode, Ag/AgCl reference electrode, silver (Ag-), platinum (Pt-) and copper (Cu-) electrodes; (vii) and Metrohm Autolab Nova 2.0 software. Water used has electrolytic conductivity ≈0.08μs/cm.

2.2. Coulometric titration of HCl
An aqueous solution of HCl (0.1000mol/kg) was prepared for diluting high-purity hydrochloric acid solution (Merck Titripur-1.0N) in ultrapure water. Approximately 5g of HCl dissolution, weighed in microbalance, was transferred to a titration vessel and KCl (2.0mol/kg) as electrolyte. The Pt-electrode (AlfaAesar-99.997% purity and 5mm×0.5mm of diameter) was used as cathode and Ag-electrode (AlfaAesar-99.99% purity and 80cm×1.0mm of diameter) in spiral shape was used as anode. The electric current value is kept for 1800 seconds at 30mA. The endpoint was determined by pH-measure. Systems have been bubbling using argon (99.999% purity UHP).

2.3. Coulometric titration of KCl [1]
An aqueous solution of KCl (0.1005mol/kg) was prepared for diluting high-purity KCl (CRM 999c from NIST) in ultrapure water. Approximately 5g of KCl dissolution, weighed in microbalance, was transferred to a titration vessel and perchloric acid (3.5mol/kg) as electrolyte. The Ag-electrode was used as cathode and a high-purity silver rod electrode (99.999% purity and 8cm^2 of area) was used as anode. The endpoint was determined by amperometric detection using the Ag/AgCl reference electrode.

2.4. Coulometric titration of EDTA [7]
An aqueous solution of EDTA (0.0958mol/kg) was prepared for diluting EDTA-Na_2 (Merck-99% purity) in ultrapure water. Approximately 1g of this dissolution, weighed in microbalance, was titrated with Cu^{2+} electrochemically generated at 30mA from a Cu-electrode. Since Cu^{2+} is captured as soon as it is formed, it was not necessary to separate electrode compartments. An electrochemical reaction was performed with phthalate buffer solution of 4.00pH-units (20°C), where the EDTA predominate species is $[H_2EDTA]^{2-}$. The endpoint was determined by measuring the increase in absorbance of the $[CuEDTA]^{2-}$ product using a Cary 4000 UV-visible spectrophotometer at 320 and 730nm.

3. Results

3.1. HCl dissolution

The hydronium ions (H_3O^+) derived from HCl dissociation reacted with hydroxide ions (OH^-) to produce water. The experimental procedure developed in this step consists of the electrochemical generation of hydroxide ions for determining hydrochloric acid concentration; thus, calibration curves were taken from 7 to 10pH units to determine the charge quantity required (1.5 Coulombs) and to evaluate the negative presence of species that interfered with the titration process (Figure 1). Due to this, resistivity increases with an increment in potential difference at low-concentration which could cause redox effects in water.

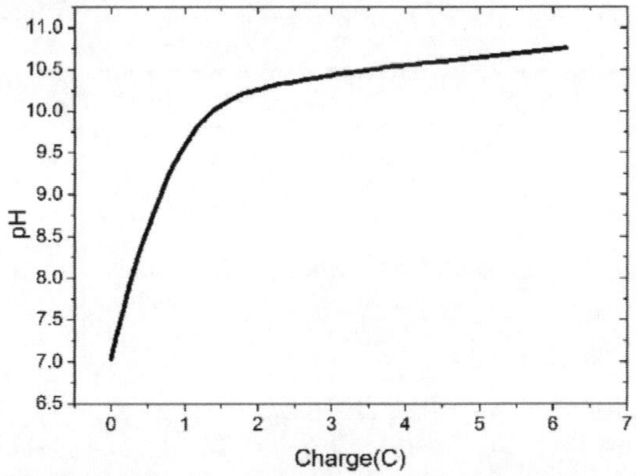

Figure 1. Coulometric titration for a typical blank solution.

To determine hydrochloride acid concentration (0.1mol/kg), coulometric titration was brought into contact with argon (Figure 2). The degasification process was efficient because measurements do not exceed expected value (0.1000mol/kg). The presence of CO_2 gas affects this measure [8]. The average value of HCl concentration was 0.099639mol/kg with 0.36% of error and 0.11% of RSD (percent relative standard deviation). The value found was comparable with the literature [6,9], but this measurement was not sufficiently accurate to consider for standard use. Due to diffusion problems that generate a decrease in accuracy when acid diffusion is too high between anodic and cathodic compartments, it is not possible to titrate the H_3O^+ ions located in the anodic compartment. Thus, it is necessary to reduce diffusion rates with an extreme control of temperature [10].

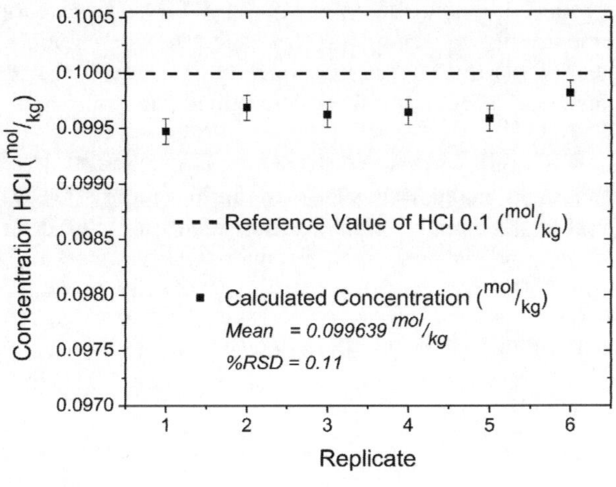

Figure 2. HCl concentration determination by coulometric titration at constant current.

3.2. KCl dissolution

A KCl concentration determination is made by means of a precipitation reaction between Cl^- and Ag^+ ions. The cations were produced electrochemically. Once generated, chlorine anion reacts quickly, forming a white precipitate. Due to electrical current, potential and time measurements, the exact Ag^+ concentration was known. By stoichiometry it was possible to calculate KCl concentration (Figure 3). To assure results it was necessary to analyse several blanks.

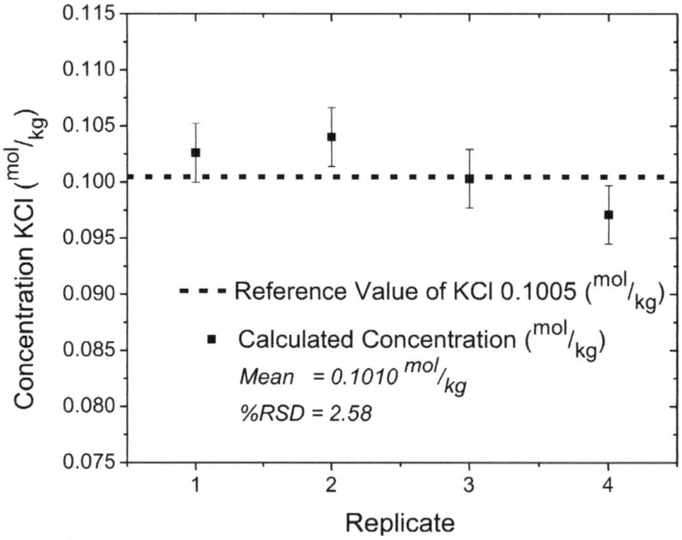

Figure 3. KCl concentration determination by coulometric titration at constant current.

The average value of KCl concentration was 0.1005mol/kg with 2.5% of RSD. This value was not comparable with the literature which had RSD of 0.02% [11,12]. The cause of this was information transfer problems between the electronic interfaces used during titration; in addition, it is recommended that a rinsing-degasification system be developed to avoid diffusion problems [13].

3.3. EDTA dissolution

The reaction mechanism of EDTA-Na_2 titration was similar to that of previous studies [7], with the difference being that copper cations were produced by ionic complex $[Cu(EDTA)]^{2-}$. The reaction was followed by spectrophotometric measurements and the endpoint was detected when ionic specie $[Cu(H_2O)_6]^{2+}$ appeared after that ionic specie $[H_2(EDTA)]^{2-}$ was completely consumed. The relation between absorbance and charge data obtained at 320 and 730nm determined the point of species consumption and generation (Figure 4).

In EDTA-Na_2 titration, absorbance increases sharply as the $[Cu(EDTA)]^{2-}$ ion is formed. However, at the intercept point between lines, this increment changes because there are not more $[H_2(EDTA)]^{2-}$ and Cu^{2+} generated electrochemically. $[Cu(H_2O)_6]^{2+}$ complex starts to form with water. To determine the exact value of charge, it was necessary to find intercept between lines. One is taken at 320nm (L=1, L=2) and the other at 730 nm (L=3, L=4). The value of this intercept was 9.10 Coulombs (This value is obtained through the mathematical interpolation between lines: L1=0.0205x+0.0152; L2=0.0087x+0.12289; L3=0.0042x+0.001; L4=0.002x+0.0209) and it corresponds to a concentration of 0.0943mol/kg (Error=1.56%). This obtained concentration was close to real value (0.0958mol/kg). Therefore, it was necessary to conduct more tests to obtain clear results.

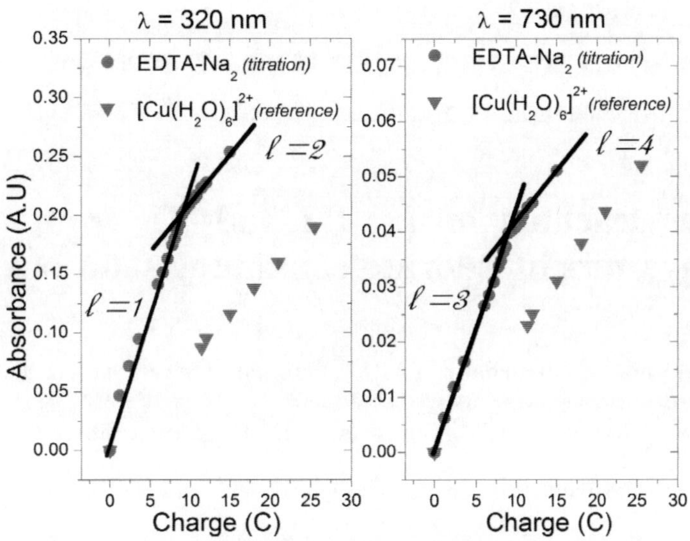

Figure 4. Comparison between EDTA-Na_2 curve titration with Cu^{2+} and $Cu(H_2O)_6^{2+}$ electrochemical formation at 320 and 730nm.

4. Conclusion

The coulometric titration system developed at NMI of Colombia will allow for certification of Reference Materials of HCl (0.1mol/kg) because of the homogeneous and reproducible measurements obtained. On the other hand, a comparison with KCl solution (0.1mol/kg) prepared from the NIST standard showed an error of 2.58% RSD. Although this difference is high, the experiment has allowed us to identify possible sources of error for consideration in future tests. Finally, the preliminary studies for certification of KCl and EDTA-Na_2 were successful. Although the coulometry system is in an early stage of implementation, the results obtained indicate that the measurement system has been properly implemented.

Acknowledgments

Special thanks to the NMI Inorganic Analysis Metrology Group for their technical support in spectrophotometric analysis of samples obtained from coulometric titration of EDTA-Na_2.

References

[1] Borges P P and da Silva W B 2014 *J. Appl. Electrochem.* **44** 1411–20
[2] De Bièvre P, et al 2011 *Pure Appl. Chem.* **83** 1873–935
[3] Harrar J E 1987 *TrAC Trends Anal. Chem.* **6** 152–7
[4] Felber H, et al 2002 *Metrologia* **39** 08002–08002
[5] Medvedevskikh S V, et al 2015 *Meas. Tech.* **57** 1295–8
[6] Terentiev G I, et al 2014 *Am. J. Anal. Chem.* **5** 559–65
[7] Williams K R, et al 2011 *J. Chem. Educ.* **88** 315–6
[8] Pratt K W 1994 *Anal. Chim. Acta* **289** 125–34
[9] Máriássy M, et al 2009 *Metrologia* **46** 199–213
[10] Claesson J and Lindberg J 1972 *J. Electroanal. Chem. Interfacial Electrochem.* **40** 255–63
[11] Ponciano I de M, et al 2016 *Rev. Bras. Eng. Agrícola e Ambient.* **20** 343–9
[12] Vanchikova E V, et al 2015 *Eurasian Soil Sci.* **48** 169–76
[13] Máriássy M, et al 2013 *Metrologia* **50** 08012–08012

Simultaneous leaching of Pt, Pd and Rh from automotive catalytic converters in chloride-containing solutions

M Hasani[1], A Khodadadi[1], S M J Koleini[1], A H Saeedi[1] and A M Meléndez[2]
[1] Tarbiat Modares University, Tehran, Iran
[2] Universidad Industrial de Santander, CMN, Piedecuesta, Santander, Colombia

E-mail: akdarban@modares.ac.ir , amelende@uis.edu.co

Abstract. Dissolution of platinum group metals (PGM; herein Pt, Pd and Rh) in different chloride-based leaching systems from spent auto catalysts was performed. Response surface methodology and a five-level-five-factor central composite design were used to evaluate the effects of 1) temperature, 2) liquid-to-solid ratio, 3) stirring speed, 4) acid concentration and 5) particle size on extraction yield of PGM by aqua regia. Analysis of variance was used to determine the optimum conditions and most significant factors affecting the overall metal extraction. In the optimum conditions, leaching of Pt, Pd and Rh was 91.58%, 93.49% and 60.15%, respectively. The effect of different oxidizing agents on the PGM dissolution in chloride medium was studied comparatively in the following leaching systems: a) aqua regia/sulfuric acid mixture, b) hydrogen peroxide in sulfuric acid (piranha solution), c) sodium hypochlorite and d) copper(II). Dissolution of Rh is increased in both aqua regia and hydrogen peroxide/hydrochloric acid solutions by adding sulfuric acid.

1. Introduction

The primary use of PGM is in the automobile industry. In the first half of 2016, the amount of PGM used in automobile catalysts as a percentage of total world demand for Pt, Pd and Rh was 40.3%, 78.7%, and 81.5%, respectively [1]. In a catalytic converter, PGM nanoparticles are evenly coated onto a ceramic honeycomb substrate [2,3]. Sintering, thermal deactivation and vehicle vibration can lead to deterioration of catalyst [2], because PGM particles gradually grow in size mainly by Ostwald ripening [4] and can be oxidized for instance in the case of Rh to form Rh_2O_3 [5]. Hence, replacement of catalytic converter is needed. As a result, automotive industries produce annually considerable amounts of spent catalysts containing PGM. Owing to their extensive use in catalysis, electronic devices and jewellery, among others, PGM demand is increasing. In addition to high production costs from naturally occurring supplies, low production of these metals due to their low abundance and complexity of PGM mineralogy in the Earth's crust [6], has made precious metals extraction from spent catalysts a viable and cost effective operation [7].

Aqua regia has been frequently used to dissolve PGM mainly due to rapid leach rates and high metal recoveries [8]. However, process needs further improvement due to consumption of both acid and complexing agent in side reactions, such as dissolution of alumina and other oxides. A lack of knowledge still exists in deeply understand the oxidative dissolution of Pt, Pd and Rh in solutions containing chloride ions [9]. The difficulty of using a phenomenological model may be surmounted by the use of a statistical model, to predict the best conditions for the leaching system within the typical range of representative process variables. Therefore, the aim of experimental design here used was to maximize the recovery of the PGM all at once, taking into account the ranges of five variables and interactions among these. Optimum results obtained in the leaching by aqua regia are compared to the results achieved by using HCl and other oxidizing agents.

2. Experimental

2.1. Preparation of sample

About 800g of PGM-containing honeycomb type support of one spent automobile catalysts was used. The ceramic support was crushed, dried and sieved. Chemical analysis of particles sample was determined by inductively coupled plasma atomic emission spectroscopy (Varian Liberty RL ICP-AES). X-ray fluorescence spectrometer (PW2404, Philips) was used for the analysis of support.

2.2. Leaching experiments

Solutions were prepared in ultra-pure water (resistivity 18.2MΩ). Analytical grade chemicals were used without further purification. Aqua regia was freshly prepared by mixing concentrated hydrochloric acid (37%) and concentrated nitric acid (65%) at a volume ratio of 3 HC l:1 HNO_3. Hydrogen peroxide, solution 30% w/w (110vol), reagent grade was used. The batch reactor used was a 0.5L glass container equipped with a thermometer, a mechanical turbine-type stirrer and a flux condenser. The condenser was used to recover and send back to reactor the vapour. Any escaped gases (such as Cl_2, NOCl or NO_x) at the exit of the condenser were scrubbed into a cold water container. Temperature was kept constant using a water bath. All analyses of PGM were done using an atomic absorption spectrophotometer (Varian-AA240).

2.3. Experimental design

A five-level-five-factor central composite rotatable design was used, requiring 32 experiments in the factorial portion, 10 in the axial points, and 3 canter experiments for 45 experiments. Response surface methodology (RSM) was used for statistical optimization of leaching process for the extraction of PGM, which is influenced by number of operating parameters with a minimum number of experiments as well as to analyse the interaction between the parameters. Statistical design-expert software DX7 was used to analyse the significance of the experimental results. The experimental variables and their corresponding levels are listed in Table 1. Five different parameters were varied: HCl concentration from 6M to 13M, temperature from 25°C to 90°C, sample size from -53 to -106μm, liquid-to-solid ratio from 2 to 10mLg^{-1} and stirring speed from 100 to 400rpm.

Table 1. Experimental range and levels of the independent variables.

	Low axial (-α=-2)	Low factorial (-1)	Centre (0)	High factorial (+1)	High axial (+α=+2)
HCl concentration (M)	6	7.4	9.5	11.5	13
Temperature (^{0}C)	25	38.6	57	76.3	90
Liquid/solid (mLg^{-1})	2	3.6	6	8.3	10
Particle size (μm)	53	64.1	79.5	94.8	106
Stirring speed (rpm)	100	163	250	337	400

3. Results and discussion

3.1. Chemical composition of spent auto catalysts

The data in Table 2 reflect the typical composition of a honeycomb-type automobile catalysts [3], which is made of cordierite ($2MgO·2Al_2O_3·5SiO_2$), coated with alumina (Al_2O_3) and a combination of PGM (Pt, Pd, Rh), rare earth oxides (La_2O_3, CeO_2) and other oxides.

Table 2. Major and trace elements concentrations in spent catalytic converter.

Oxide[a]	MgO	Al_2O_3	SiO_2	CaO	Cr_2O_3	Fe_2O_3	NiO	ZnO	ZrO_2	BaO	La_2O_3	CeO_2
wt%	9.80	38.15	34.28	1.16	0.69	4.56	3.90	1.09	1.85	0.73	0.84	2.73
Element[b]	Pt	Pd	Rh									
ppm	914	876	114									

[a] Results obtained by X-ray fluorescence (XRF) analysis.
[b] Results for Pt, Pd and Rh after digestion were obtained by ICP-AES.

3.2. Optimization of aqua regia leaching process

Leaching experiments of sample were performed with aqua regia during 4h. The variables investigated were: acid concentration, reaction temperature, stirring speed, liquid-to-solid ratio, and particle size of sample. Table 3 shows the values of the five independent variables used in batch experiments to determine the response variations in the extraction percentage of Pt, Pd and Rh.

In order to identify the significance of the main and interactive effects of the variables combination, as it affects the extraction yield, data obtained from leaching experiments were statistically analysed using Fisher's ANOVA methodology. The results of regression analysis on reduced cubic model are given in Table 4. Values of "Prob>F" less than 0.0500 indicates high significant regression at 95% confidence level. According to the F- and p-values, concentration of HCl, temperature, and liquid-to-solid ratio were found more effective on the extraction of Pt, Pd, and Rh in the aqua regia leaching process. Figure 1 displays the correspondence between experimental and predicted values of PGM recovery yields. Fits showing good linear agreement between both values confirming the robustness of the model. In addition, the reliability of the model was checked by the coefficient of determination R^2. Predicted R^2 (Pt:0.9818, Pd:0.9860, Rh:0.9903) and adjusted R^2 (Pt:0.9111, Pd:0.9314, Rh:0.9645) values are close to one, which indicates that variables are highly correlated. In addition, experimental and predicted values are reasonably close to each other, standard deviation for Pt, Pd and Rh was 6.75, 4.52 and 3.91, respectively, indicating that the model fits the experimental data well.

According to the parameter values estimated from 45 batch experiments and statistical analysis (Tables 3 and 4), an empirical relationship between response and tested variable in coded units was obtained for each PGM (Equations (1)-(3)) by response surface methodology (RSM), as follows:

$$Pt(\%)=37.96+12.69A+9.69B+16.42E+3.95D2 \quad (1)$$

$$Pd(\%)=63.65+12.93A+12.37B+11.30E-3.86A2-3.80B2-6.82AB2 \quad (2)$$

$$Rh(\%)=26.21+16.07A+9.19B+11.08E+3.47AE \quad (3)$$

where Pt(%), Pd(%) and Rh(%) is the extraction yield percentage of PGM; A, B, D and E are the coded values of variables; concentration of hydrochloric acid in $molL^{-1}$ (A), temperature in °C (B), particle size in μm (D), and liquid-to-solid ratio in mLg^{-1} (E). Particle size does not have an important effect on the extraction of Pd and Rh (Equations (2) and (3)), while it has only a quadratic effect for Pt (Equation 1). All three empirical models show that interactions between variables have no significant effects. Only in Equation (3) interaction with small positive effect between concentration of hydrochloric acid (A) and liquid-to-solid ratio (E) for Rh recovery is observed.

Table 3. Experimental design matrix for Pt, Pd and Rh extraction. A=HCl, B=temperature, C=stirring speed, D=particle size and E=liquid-to-solid ratio.

Trial	Run	A (M)	B (°C)	C (rpm)	D (μm)	E (mLg^{-1})	Pt (%)	Pd (%)	Rh (%)
1	6	7.50	39.0	164	64	3.60	63.26	71.83	57.44
2	19	11.50	39.0	164	64	3.60	29.97	56.51	22.76
3	41	7.50	76.0	164	64	3.60	46.64	72.78	33.39

4	4	11.50	76.0	164	64	3.60	62.74	76.39	41.88
5	10	7.50	39.0	337	64	3.60	44.69	67.18	29.19
6	5	11.50	39.0	337	64	3.60	38.71	46.49	26.71
7	36	7.50	76.0	337	64	3.60	71.47	65.98	44.19
8	15	11.50	76.0	337	64	3.60	59.66	70.57	30.61
9	7	7.50	39.0	164	90	3.60	56.71	65.15	44.52
10	3	11.5	39.0	164	90	3.60	60.41	70.6	29.21
11	29	7.50	76.0	164	95	3.60	46.72	69.19	32.91
12	14	11.5	76.0	164	95	3.60	90.09	84.63	76.97
13	18	7.50	39.0	337	95	3.60	59.08	74.98	29.29
14	33	11.5	39.0	337	95	3.60	88.39	89.27	62.12
15	11	7.50	76.0	337	95	3.60	25.32	40.96	21.69
16	24	11.5	76.0	337	95	3.60	57.55	65.09	28.09
17	44	7.50	39.0	164	64	8.30	42.13	51.89	11.48
18	32	11.5	39.0	164	64	8.30	10.14	37.84	3.77
19	31	7.50	76.0	164	64	8.30	58.1	60.62	44.41
20	38	11.5	76.0	164	64	8.30	38.91	56.39	10.29
21	16	7.50	39.0	337	64	8.30	91.56	94.61	63.19
22	22	11.5	39.0	337	64	8.30	44.66	67.25	29.22
23	25	7.50	76.0	337	64	8.30	87.33	79.22	60.93
24	30	11.5	76.0	337	64	8.30	55.11	54.9	43.32
25	35	7.50	39.0	164	95	8.30	9.65	30.54	4.98
26	13	11.5	39.0	164	95	8.30	44.69	67.18	29.19
27	8	7.50	76.0	164	95	8.30	36.39	36.6	25.54
28	37	11.5	76.0	164	95	8.30	43.61	61.83	12.56
29	2	7.50	39.0	337	95	8.30	40.68	59.21	26.2
30	23	12.00	39.0	337	95	8.30	52.24	73.2	36.67
31	20	7.50	76.0	337	95	8.30	27.05	51.11	21.69
32	42	11.5	76.0	337	95	8.30	28.46	46.53	21.69
33	1	6.00	57.0	250	75	6.00	40.73	56.48	11.48
34	17	13.00	57.0	250	75	6.00	11.74	37.84	4.98
35	45	9.50	25.0	250	75	6.00	19.5	27.24	2.03
36	43	9.50	90.0	250	75	6.00	39.65	42.01	26.62
37	39	9.50	57.0	100	75	6.00	7.18	22.39	2.75
38	28	9.50	57.0	400	75	6.00	68.49	60.59	43.09
39	26	9.50	57.0	250	53	6.00	8.48	32.43	3.89
40	21	9.50	57.0	250	106	6.00	7.11	37.41	3.68
41	27	9.50	57.0	250	75	2.00	73.19	76.24	45.39
42	9	9.50	57.0	250	75	10.00	41.54	41.95	27.77
43	34	9.50	57.0	250	75	6.00	42.74	60.97	23.81
44	12	9.50	57.0	250	75	6.00	62.39	81.02	30.35
45	40	9.50	57.0	250	75	6.00	70.56	70.59	44.29

Table 4. Analysis of variance (ANOVA) for the reduced cubic model.

Source	Sum of Squares			df[a]			Mean Square			F Value			p-value Prob> F		
	Pt	Pd	Rh	Pt	Pd	Rh	Pt	Pd	Rh	Pt	Pd	Rh	Pt	Pd	Rh
Model term	22156.3	12904.2	14098.9	35	35	35	633.0	368.7	402.8	13.9	18.1	26.4	0.0001	< 0.0001	< 0.0001
A	957.5	994.1	1535.1	1	1	1	957.5	994.1	1535.1	21.0	48.7	100.5	0.0013	< 0.0001	< 0.0001
B	907.0	909.9	502.1	1	1	1	907.0	909.9	502.1	19.9	44.6	32.9	0.0016	< 0.0001	0.0003
C	17.8	92.1	25.8	1	1	1	17.8	92.1	25.8	0.4	4.5	1.7	0.548	0.0626	0.2256
D	7.9	33.8	41.4	1	1	1	7.9	33.8	41.4	0.2	1.7	2.7	0.6866	0.2302	0.1341
E	1603.5	759.7	729.6	1	1	1	1603.5	759.7	729.6	35.2	37.3	47.8	0.0002	0.0002	< 0.0001
AE	0.0	3.2	386.0	1	1	1	0.0	3.2	386.0	0.0	0.2	25.3	0.9744	0.6998	0.0007
A^2	140.9	263.8	65.8	1	1	1	140.9	263.8	65.8	3.1	12.9	4.3	0.1126	0.0058	0.0678
B^2	8.3	167.4	20.2	1	1	1	8.3	167.4	20.2	0.2	8.2	1.3	0.6788	0.0186	0.2802
C^2	227.8	49.5	67.2	1	1	1	227.8	49.5	67.2	5.0	2.4	4.4	0.0522	0.1536	0.0654
D^2	275.8	33.0	38.0	1	1	1	275.8	33.0	38.0	6.0	1.6	2.5	0.0362	0.2354	0.1491
E^2	11.9	33.9	3.0	1	1	1	11.9	33.9	3.0	0.3	1.7	0.2	0.622	0.2292	0.6693
AB^2	22.5	233.2	24.3	1	1	1	22.5	233.2	24.3	0.5	11.4	1.6	0.5	0.0081	0.239
Residual[b]	410.3	183.6	137.5	9	9	9	45.6	20.4	15.3	---	---	---	---	---	---

[a] Degrees of freedom.
[b] Obtained from experimental error estimated by the 3 replicates at central point.

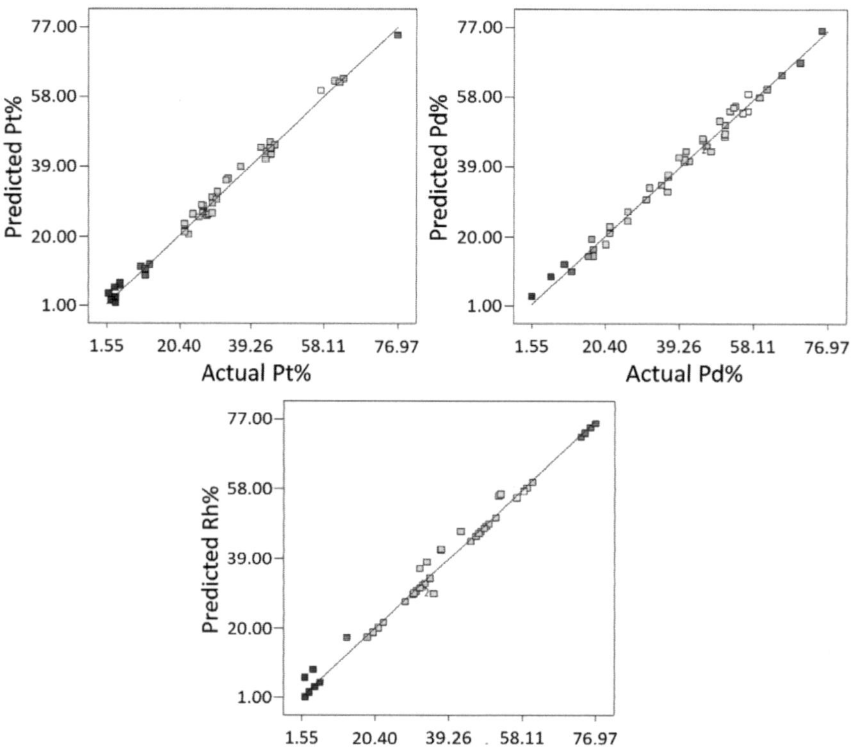

Figure 1. Predicted vs experimental values of PGM extraction.

Metal dissolution depends more on the interaction between hydrochloric acid concentration and temperature rather than on other parameters. In this way, main effects of interactive relationship between independent variables on the PGM extraction yields are displayed in Figure 2. It can be observed that, in the range from 9.5M to 11.5M HCl, an increase in the temperature from 38.65°C to 76.35°C improves the extraction of Pt, Pd and Rh above about 30%. Consequently, at a certain temperature, an increase in the concentration of HCl enhances the extraction of metals. In the case of Pt extraction, increasing the concentration of HCl and liquid-to-solid ratio above 9.5M and 6.5mLg^{-1}, respectively, has a positive effect on the leaching efficiency (data not shown); nevertheless, to maximize extraction of all of the metals simultaneously, the optimized values of the whole process could be different from that of a single process. Thus, optimum conditions have been found to be: 11.3M HCl, temperature of 70°C, liquid-to-solid ratio of 9.8mLg^{-1}, stirring speed of 339rpm and particle size less than 94 μm. Under these conditions the rate of simultaneous PGM metals leaching reached to 93.9%, 95.01% and 64.5% for Pt, Pd, and Rh, respectively. Thereby, Pt and Pd were more easily leached than Rh from spent catalyst.

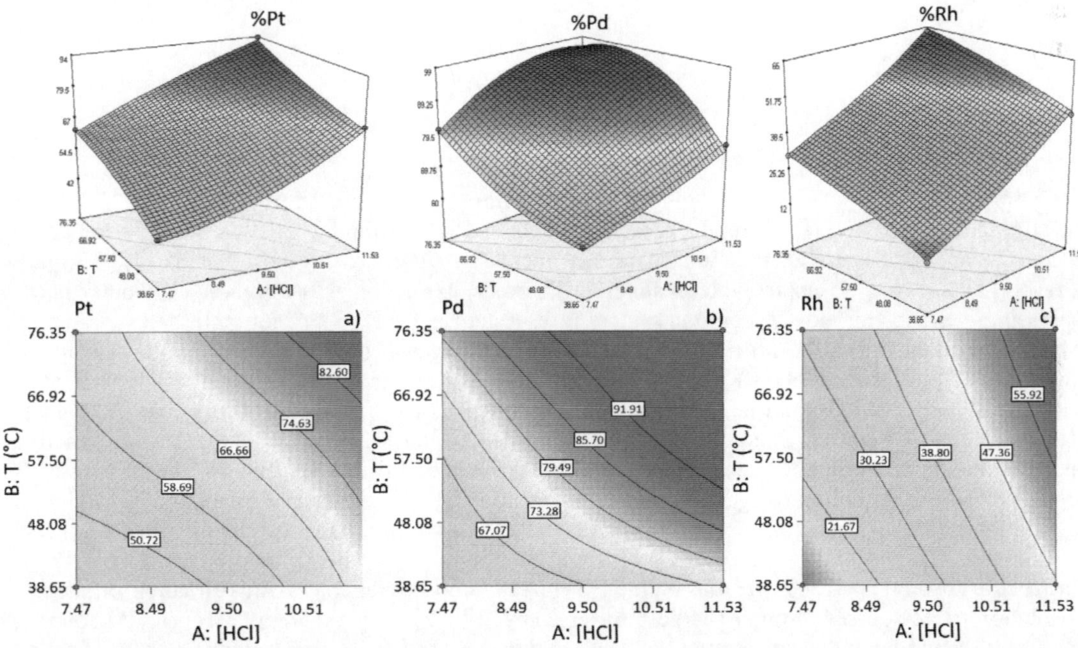

Figure 2. Response surface and contour plots for the effect of temperature (°C) and concentration of HCl (molL^{-1}) on the (a) platinum, (b) palladium and (c) rhodium extraction.

3.3. Comparison of different oxidizing agents in chloride media for PGM recovery

Dissolution of Pt, Pd and Rh is facilitated by the presence of chloride ions in acid media owing to formation of strongly complexed PGM ions [9,10]. Thus, in order to ensure a fair comparison between actions of different oxidizing agents, and evaluate the aqua regia efficiency, a series of leaching tests were performed under similar experimental conditions (Table 5). In all these experiments, the optimum parameters have been reported in literature [3,11]. Because for complexes the extent of formation is only depending on ligand concentration and the medium acidity, and considering that leaching systems contains high levels of dissolved chloride ions (11.5M Cl^- in all leach solutions) in strong acid media, it is expected that coordinatively saturated metal-ligand complex chlorides $PtCl_6^{2-}$, $PdCl_6^{2-}$ and $RhCl_6^{3-}$ are formed [10,12,13].

Table 5. Compositions for each leach test.

Lixiviant system	Reagent concentration (M)	
	HNO_3	H_2SO_4
A	3.85	0
		12.0
		18.0
	H_2O_2 (vol)	H_2SO_4
B	110	0
		12.0
		18.0
	NaClO	
C	3.0	
	5.0	
	8.0	
	Cu(II)	
D	0.3	
	0.5	

Comparison of the effectiveness of different leaching systems are shown in Figure 3. Adding sulfuric acid to aqua regia (optimum conditions obtained above) and HCl/H_2O_2 mixture (leaching systems A and B, Table 5), ruthenium extraction increased significantly, by more than half (Figure 3(a),(b)). In contrast, no appreciable changes in Pt or Pd extraction was observed. On other hand, increasing the oxidizing power from the system B to system A leads to a slight increase of Pt and Rh extraction, and the opposite is true for Pd. Thereby, in leaching system A (18M H_2SO_4) the dissolution of Pt, Pd and Rh reached 94.08%, 94.53% and 89.95% (Figure 3(a)), while system B (18M H_2SO_4) achieved 89.22%, 95.71% and 83.02%, respectively (Figure 3(b)). It is noteworthy that the highest extraction yields were obtained for leaching systems A and B. Indeed, oxidants and reactive intermediate species generated in leaching systems A (nitrate, nitrosyl ion and chlorine gas) and B (hydrogen peroxide, atomic oxygen and chlorine gas) are strong oxidants. In this way, PGM can be passivated by these chemical species, however, chloride ions are strongly chemisorbed on PGM surface.

Specific adsorption of chloride ions on the PGM can inhibit irreversible oxide formation [9], hence, dissolution process arises through charge transfer coupled with complexation of Cl^- at PGM/solution interface. Increasing the concentration of sodium hypochlorite, a strong oxidizing agent, has no noticeable effect on the recovery of any metal (Figure 3(c)). In addition, the lowest recoveries were obtained with this lixiviant system, below 65%. This is explained because sodium hypochlorite exhibits chemical instability in acidic solutions, and although sodium hypochlorite generates chlorine gas, another strong oxidizing agent, their solubility decreased with the increase of temperature (leach tests were carried out at a temperature of 80°C) [14]. Copper(II) in system D is a weak oxidizing agent compared to aforementioned oxidants; however, extractions are larger than that obtained in system C with sodium hypochlorite/chlorine gas. The effect of increasing Cu(II) concentration on PGM extraction is marginal. Certainly, the lowest copper concentration here used was found as the optimum value to obtain the highest PGM recovery in a recent study [11].

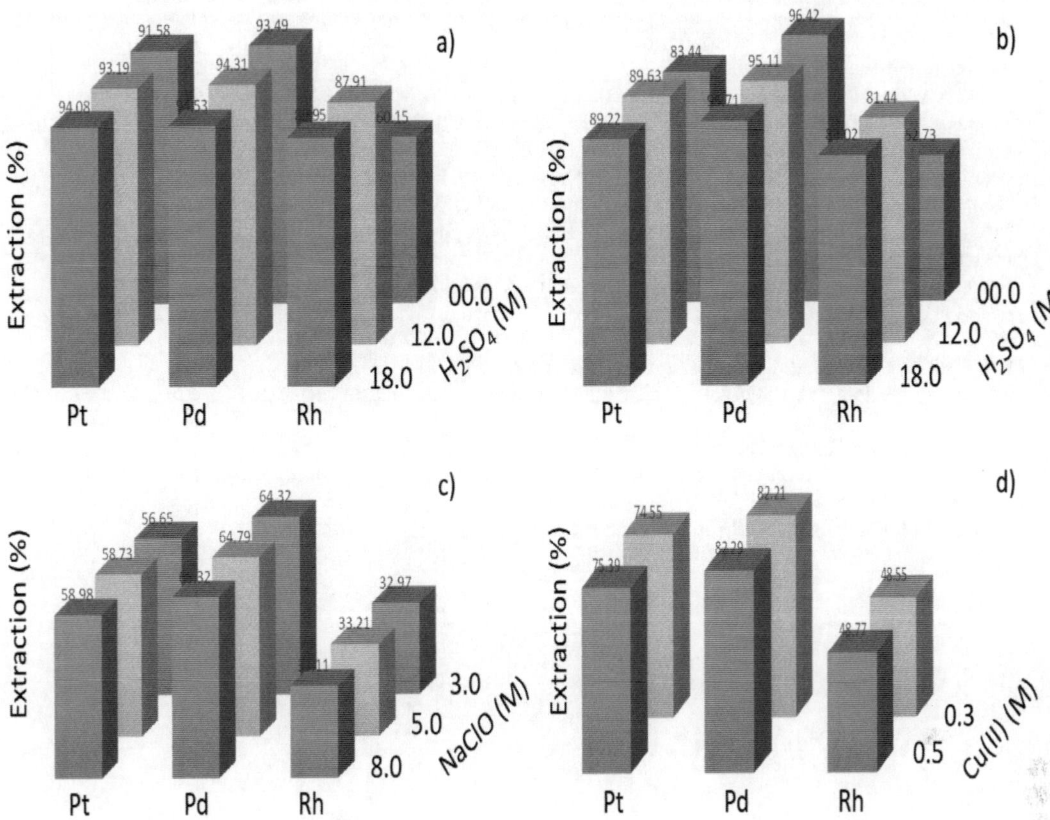

Figure 3. PGM extraction yields as a function of leach solution composition (Table 5): (a) A, (b) B, (c) C and (d) D. Conditions used in all leaching tests: 11.5M HCl, stirring 400rpm, temperature 80°C, particle size -63 μm, solution/solid ratio 8.5mLg^{-1} and time 240min.

4. Conclusions

In order to obtain optimum simultaneous extraction of Pt, Pd and Rh from catalytic converters of spent exhaust systems, a series of leaching tests were carried out to investigate the key variables in evaluating leaching behaviour by response surface methodology. Mineral acid concentrations and temperature were the most influential factors in the PGM leaching. At an optimum composition of the leach solution, aqua regia/sulfuric acid mixture was the best medium combination for PGM dissolution at almost 95% of Pt and Pd extraction, and yield of about 90% for Rh. The optimum leaching conditions for 4h were: leaching temperature 70°C, ratio of liquid to solid 9.8, stirring speed 337rpm, particle size less than 94μm, acids concentration 11.3M HCl and 3.4M HNO_3 (3 HCl:1 HNO_3 volume ratio). Strong acidification of either aqua regia or piranha in chloride medium is beneficial to enhance the PGM extraction performance. Soluble chloride complexes of PGM prevent the formation of insoluble products that could passivate the metal surface.

Acknowledgements

Authors are grateful to Tarbiat Modares and Tehran Universities for supporting this research. The Universidad Industrial de Santander (VIE-UIS project 1336) also supported this project.

References

[1] Johnson Matthey 2016 *Summary of platinum, palladium, and rhodium: supply & demand in 2016* PGM market report http://www.platinum.matthey.com/documents/new-item/pgm%20market%20reports/pgm-market-report-may-2016.pdf/

[2] Lloyd S M, Lave L B and Matthews H S 2005 *Environ. Sci. Technol.* **39** 1384

[3] Jimenez de Aberasturi D, Pinedo R, Ruiz de Larramendi I, Ruiz de Larramendi J I and Rojo T 2011 *Minerals Engineering* **24** 505

[4] Hansen T W, DeLaRiva A T, Challa S R and Datye A K 2013 *Acc. Chem. Res.* **46** 1720

[5] Newton M A, Fiddy S G, Guilera G, Jyoti B and Evans J 2005 *Chem. Commun.* 118

[6] Mpinga C N, Eksteen J J, Aldrich C and Dyer L 2015 *Minerals Engineering* **78** 93

[7] Fornalczyk A 2012 *JAMME* **55** 864

[8] Jha,M K, Lee J, Kim M, Jeong J, Kim B and Kumar V 2013 *Hydrometallurgy* **133** 23

[9] Lopes P P, Strmcnik D, Tripkovic D, Connell J G, Stamenkovic V R and Markovic N M 2016 *ACS Catal.* **6** 2536

[10] Nikoloski A N and Ang K-L 2014 *Min. Proc. Ext. Met. Rev.* **35**, 369

[11] Nogueira C A, Paiva A P, Oliveira P C, Costa M C and Rosa da Costad A M 2014 *J. Hazard. Mater.* **278** 82

[12] Mahmoud M H H 2003 *JOM* **55** 37

[13] Harjanto S, Cao Y, Shibayama A, Naitoh I, Nanami T, Kasahara K, Okumura Y, Liu K and Fujita T, 2006 *Mater. Trans.* **47** 129

[14] Massucci M, Clegg S L and Brimblecombe P 1999 *J. Phys. Chem. A* **103** 4209

Platinum leaching from automotive catalytic converters with aqua regia

M Hasani[1], A Khodadadi[1], S M J Koleini[1], A H Saeedi[1], Y Pérez-Pacheco[2] and A M Meléndez[2]
[1] Tarbiat Modares University, Tehran, Iran
[2] Universidad Industrial de Santander, CMN, Piedecuesta, Santander, Colombia

E-mail: akdarban@modares.ac.ir

Abstract. Herein, kinetics extraction of platinum from spent auto catalysts, using nitric acid as an oxidant in hydrochloric acid solution, was investigated. The parameters such as temperature, hydrochloric and nitric acid concentrations, stirring speed, particle size and liquid/solid ratio, were analysed. The kinetic data were analysed using the shrinking core model. A variant of this model fits the kinetic data more appropriately. At a temperature of 90°C, the values of R^2 in surface chemical reactions and diffusion were 0.819 and 0.937, respectively. With the alternative model, however, 0.991 was obtained. The activation energy for the dissolution was 35.75kJ/mol.

1. Introduction

Several investigations have been reported concerning the extraction of platinum from spent catalysts through pyro or hydrometallurgical processes [1]. Pyrometallurgical processes require high energy consumption, making the process less attractive to the industry. Instead, hydrometallurgy has emerged as the most suitable processing method for the recovery of a variety of metals. In the case of platinum, dissolution process has been frequently studied in the presence of chloride ions, Cl^- [2], and has been claimed that chloride is the only cost-effective medium to platinum dissolution [3-6]. In the commonly applied hydrometallurgical processes, platinum from used auto catalytic converters is leached in aqua regia, a simple 1:3v/v mixture of concentrated hydrochloric and nitric acids.

Aqua regia produces strong oxidizing agents: nitrosyl chloride, NOCl, and chlorine gas, Cl_2 [7]. The nitrosyl chloride further hydrolyses to nitrous acid, HNO_2, and chloride. Nitrous acid undergoes decomposition by a disproportionation reaction, producing nitrogen dioxide, NO_2, and nitric oxide, ·NO. The latter is an unstable radical, which can disproportionate into nitrogen dioxide and nitrous oxide. The simultaneous presence of several oxidizing agents makes the aqua regia very reactive, at the same time making complicated the mechanism of platinum dissolution. Thus, for instance, it has been suggested that HNO_2 in HNO_3 activates passive metals [8-10]. Therefore, in an overall form, Pt dissolution can be described by the following reaction (Equation 1).

$$3Pt+18HCl+4HNO_3 \rightarrow 3PtCl_6^{2-}+6H^++4NO+8H_2O \qquad (1)$$

In the present study, kinetics of auto catalyst dissolution by aqua regia has been studied. The influence of stirring speed, particle size, sulfuric and nitric acid concentrations, temperature and liquid/solid ratio was investigated.

2. Experimental

A ceramic-honeycomb-type automobile catalytic converter was crushed, dried and sieved to four size fractions. Both, chemical analysis of sample and a complete description of the leaching experiments can be consulted in another paper in this same volume. The leaching was performed for four hours. At specified time intervals, 10mL of the solution was taken from the leach solution and diluted using ultra-pure water (resistivity 18.2MΩ). All analyses of platinum were done using an atomic absorption spectrophotometer. The experimental conditions are shown in Table 1. Keeping the other parameters constant, the effect of each parameter on the dissolution rate was evaluated.

Table 1. Parameter values used for the leaching of platinum.

Parameter	Value
Temperature (°C)	40, 60, 80, 90, 100
[HCl] (M)	6.0, 8.5, 10.0, 12.0, 13.0
[HNO_3] (M)	5.0, 7.5, 8.5, 9.5, 11.0
Particle size (μm)	-53, -75+53, -106+75
Liquid to solid ratio (mL/g)	10, 20, 40
Stirring speed (rpm)	200, 400, 600

3. Results and discussion

3.1. Factor affecting Pt extraction with aqua regia

The effect of the solid/liquid ratio on platinum dissolution was studied for (10, 20 and 40) mL/g at 80°C in a solution containing 12.0M HCl and 9.5M HNO_3. The stirring speed and the particle size were kept constant at 400rpm and -53+45μm, respectively. Platinum extraction increased with a decrease in the amount of solid (data not shown); this was caused by the increase of acid available per unit of solid. No significant difference in Pt extraction ~91% was obtained between the ratio of 10 and 20mL/g. Hence, to study the other leaching parameters, a 20mL/g solid/liquid ratio was chosen.

To investigate the influence of particle size on the dissolution rate, leaching experiments were carried out for different particle sizes (-106+75, -75+53, -53) μm while other parameters were kept constant. Dissolution rate increased significantly with decreasing particle size, due to smaller particles providing larger contact surface area between solid and the leaching reagent (data not shown), thereby other leaching parameters were studied using the fraction of particle size less than 53mμ.

The influence of stirring speed on auto catalyst leaching for platinum extraction was studied at (200, 400 and 600) rpm and 80°C in solution containing 12.0M HCl and 9.5M HNO_3. The platinum recovered at 200rpm was almost 78%, whereas 91.1% and 93.0% when a stirring speed was maintained at 400rpm and 600rpm, respectively.

To determine the influence of temperature, leaching experiments were performed at five different temperatures: (60, 70, 80, 90 and 100) °C. The fixed laboratory parameters at this stage were: stirring speed 400rpm, 12.0M hydrochloric acid, 9.5M nitric acid, particle size -53μm and phase ratio 20ml/g. With an increase in temperature, the platinum dissolution rate increases. As can be seen in Figure 1, an increase of temperature increases the platinum extraction.

The effect of HCl concentration on platinum dissolution was studied for (6.0, 8.5, 10.0, 12.0, and 13.0) M HCl at 90°C in solution containing 9.5M HNO_3. The results are shown in Figure 2. Platinum dissolution increased from 23% to 90.7% after 200min as the acid concentration was increased from 6.0 to 12.0M. With an increase in acid concentration to 13.0M, the dissolution rate increased to 92%.

The effect of HNO_3 concentration on platinum extraction is shown in Figure 3. Experiments were carried out at five different concentrations: (5.0, 7.5, 8.5, 9.5, and 11.0) M HNO_3. The results show that an increase in the oxidant concentrate accelerates platinum dissolution. At low concentrations of nitric acid, the platinum dissolution rate is very weak, confirming confirm the role of oxidant in oxidative dissolution of platinum.

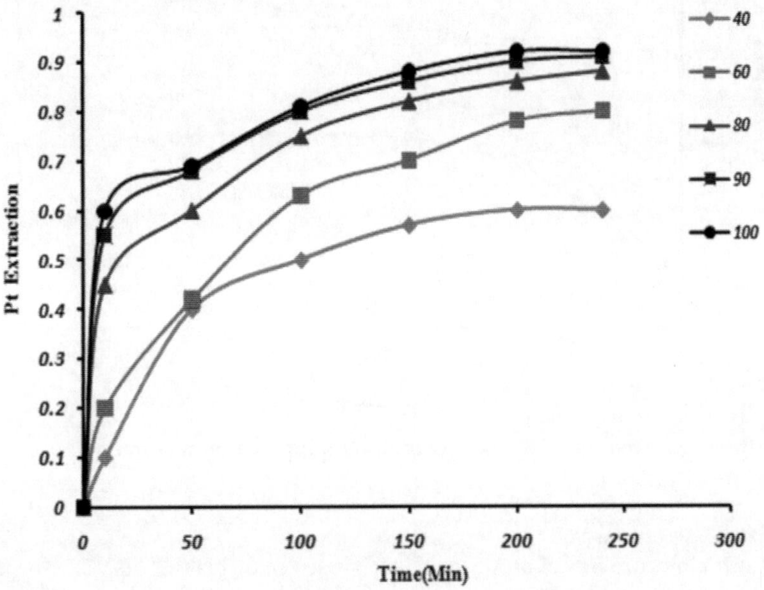

Figure 1. Effect of temperature on the platinum extraction.

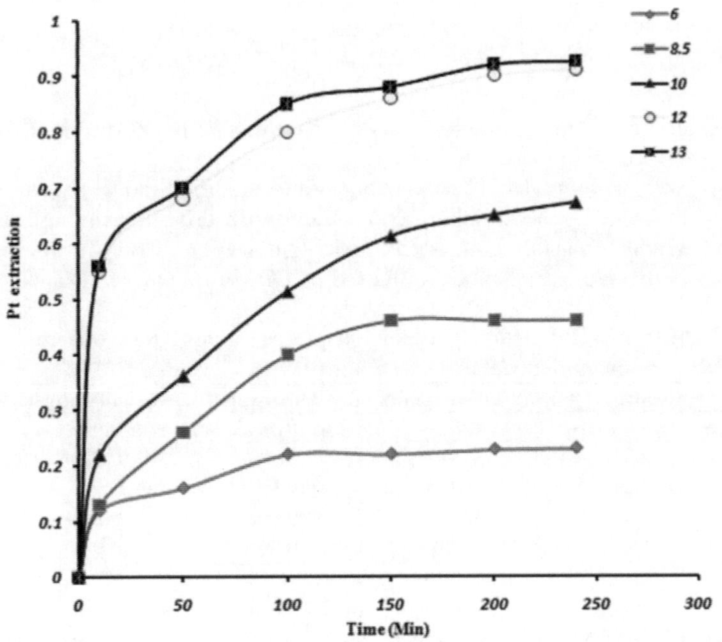

Figure 2. Effect of HCl concentration on the platinum extraction.

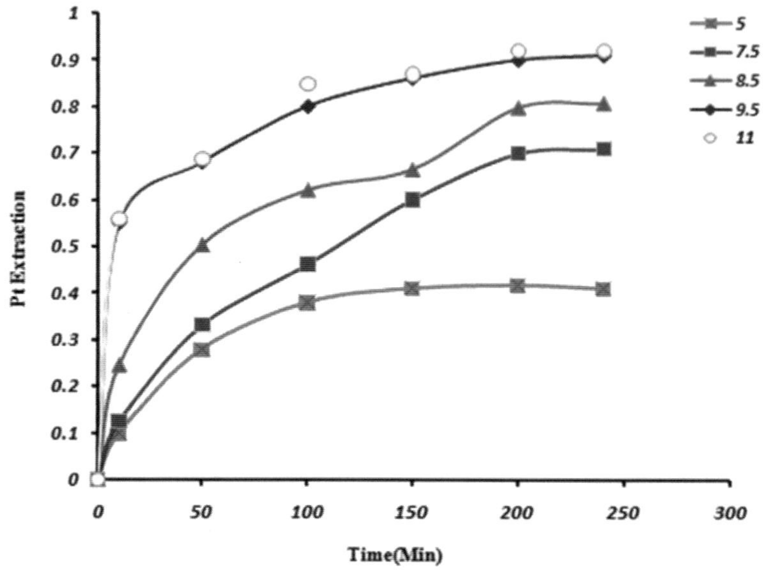

Figure 3. Effect of HNO_3 concentration on the platinum extraction.

3.2. Kinetics of leaching process

A variant of the shrinking core model proposed by Dickinson and Heal [11] is a more complete type of diffusion-control model, where both solid and acid concentrations have a role in controlling the reaction rate, and diffusion occurs in two directions. An equation of this model is given as follows.

$$\frac{1}{3}\ln(1-X)+\left[(1-X)^{\frac{-1}{3}}-1\right]=K_m t \tag{2}$$

where K_m is a kinetic constant. In Figure 4 was plotted versus time at different temperatures. The slope of this line is the specified kinetic constant.

Conventional shrinking core model [12] was compared to modified model (Equation 2). According to the results given in Table 2, the shrinking core model with diffusion through the product layer (compared to the chemical reaction-controlled model) showed a better fit to the data, but the alternative model of the shrinking core model proposed by Dickinson provided an even better fit.

Table 2. Correlation coefficients R^2 values for different temperature and models.

Temperature (°C)	Surface Reaction	Diffusion	Variant of SCM
40	0.824	0.940	0.967
60	0.819	0.937	0.991
80	0.938	0.931	0.995
90	0.862	0.988	0.985
100	0.839	0.962	0.946

The activation energy of platinum leaching in hydrochloric acid medium determined from Arrhenius equation [12] was 35.75kJ/mol, which can be considered as the mean value, in the above classifications between diffusion (usually of the order of 5kcal or less) and surface chemical activation energy amounts [5,12]. Hence, considering the new model used in this research, according to which, the rate of leaching is both diffusion and chemical controlled, the amount of activation energy here obtained could be considered as a reflection of the intermediate model.

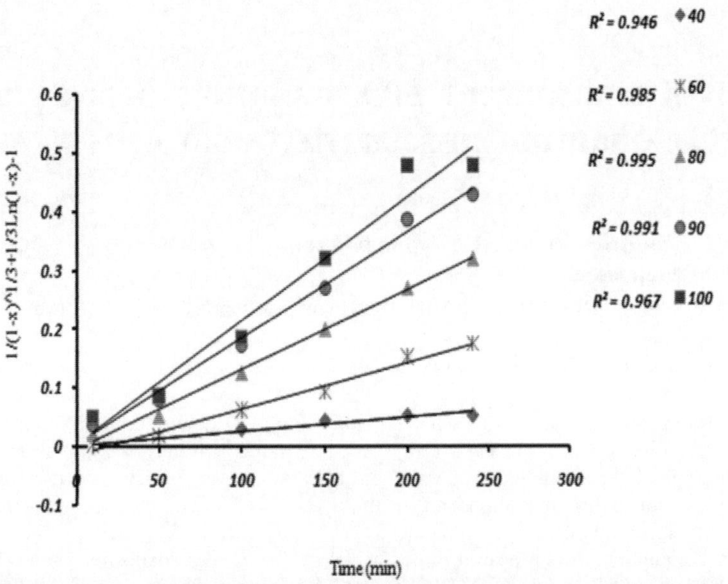

Figure 4. The variation in $\frac{1}{3}\ln(1-x)+\left[(1-x)^{\frac{-1}{3}}-1\right]=K_m t$ with time at various temperatures.

4. Conclusions

In the present study, the kinetics of platinum leaching from auto-catalyst in aqua regia was studied. The effect of experimental parameters such as temperature, hydrochloric and nitric acid concentrations, stirring speed, particle size and liquid/solid ratio were analysed. Acid concentration and temperature were the most important variables affecting the leaching process. The activation energy was calculated to be 35.75kJ/mol.

Acknowledgements

Authors are grateful to Tarbiat Modares and Tehran Universities for supporting this research. The Universidad Industrial de Santander (VIE-UIS project 1336) also supported this project.

References

[1] Fornalczyk A 2012 *JAMME* **55** 864
[2] Jha M K, Lee J, Kim M, Jeong J, Kim B and Kumar V 2013 *Hydrometallurgy* **133** 23
[3] Mahmoud M H H 2003 *JOM* **55** 37
[4] Harjanto S, Cao Y, Shibayama A, Naitoh I, Nanami T, Kasahara K, Okumura Y, Liu K and Fujita T, 2006 *Mater. Trans.* **47** 129
[5] Lillkung K, Aromaa J and Forsen O 2013 *Physicochem. Probl. Miner. Process.* **49** 463
[6] Nikoloski A N and Ang K-L 2014 *Min. Proc. Ext. Met. Rev.* **35** 369
[7] Massucci M, Clegg S L and Brimblecombe P 1999 *J. Phys. Chem. A* **103** 4209
[8] Banerji B C and Dhar N R 1922 *Z. Anorg. Allgem. Chem.* **122** 73
[9] Joss E J 1926 *J. Phys. Chem.* **30** 1222
[10] Hedges E S 1930 *J. Chem. Soc.* 561
[11] Dickinson C F and Heal G R 1999 *Thermochim. Acta* **340** 89
[12] Sohn H Y, Wadsworth M E 1979 *Rate Processes of Extractive Metallurgy* (New York: Plenum Press)

IOP Conf. Series: Journal of Physics: Conf. Series **786** (2017) 012044 doi:10.1088/1742-6596/786/1/012044

Fabrication of transparent TiO_2 nanotube-based photoanodes for CdS/CdTe quantum co-sensitized solar cells

A F Gualdrón-Reyes[1], A Cárdenas-Arenas[1], C A Martínez[1], V V Kouznetsov[1] and A M Meléndez[1]

[1] Universidad Industrial de Santander, CMN, Piedecuesta, Santander, Colombia

E-mail: amelende@uis.edu.co

Abstract. In order to fabricate a solar cell, ordered TiO_2 nanotube (TNT) arrays were prepared by double anodization. TNT arrays with variable lengths were obtained by changing the duration of the anodizing process of up to 3h. TNT membranes were transferred to indium tin oxide substrates and attached with a B-TiO_2 sol. TNT photoanode with the best photoelectrochemical performance was sensitized with CdS by SILAR method. On other hand, CdTe quantum dots prepared via colloidal synthesis were deposited on TNT photoanodes for 2h, 4h and 6h. In addition, TNT/CdS was loaded with CdTe quantum dots for 4 h. Morphology and chemical modification of TiO_2 were characterized by FESEM and XPS, while their photoelectrochemical performance was measured by open-circuit photopotential and photovoltammetry under visible light. TiO_2 nanotubes grown during 2.5h showed the highest photocurrent due to presence of Ti^{3+} donor states by N and F co-doping, increasing the number of photogenerated electrons transported to back collector. TNT/CdS/CdTe photoanode reach the highest conversion efficiency under AM 1.5G simulated solar illumination.

1. Introduction

In recent years, sunlight harvesting has been considered the most attractive green alternative to decrease the use of non-renewable energy sources and diminish the environmental contamination. In order to collect high amounts of photons from solar energy, fourth-generation photovoltaic devices such as quantum dots sensitized solar cells (QDSSC) have been used. Cadmium chalcogenides (CdS, CdSe and CdTe) based quantum dots (QDs) have been studied due to their narrow band gaps, high extinction coefficient and multiple electron-hole pair generation [1], which extends the light harvesting of the device to visible or IR regions and improves its energy conversion efficiency. To fabricate QDSSC, QDs are deposited on wide-band gap semiconductors as TiO_2, promoting the injection of photogenerated electrons from QDs to TiO_2 and their transport through the solar cell [1-3]. Transport of electrons is hindered by the presence of grain boundaries in photoanodes prepared for photovoltaic applications from TiO_2 nanoparticulated films [4]. To overcome this disadvantage, oriented TiO_2 nanotube photoanodes are used to address the electron mobility through their tubular structure. TiO_2 nanotube membranes can be easily transferred to conductive glass substrate allowing the front-side illumination of solar cell to enhance its photovoltaic performance [4]. Nevertheless, the preparation of TiO_2 nanotubes with adequate characteristics is still a huge challenge in order to deposit a sufficient QDs content to improve the light harvesting, the electron transport through the photoanode and the energy conversion efficiency of solar cell.

2. Experimental

2.1. Preparation of transparent TiO_2 nanotubes photoanodes
Ti foils (99.5% of purity, 2.5×2.0cm^2) were polished with different SiC emery papers. These foils were sonicated in ethanol, acetone, and rinsed with deionized (DI) water each one for 5min. Samples were anodized in an electrolyte solution containing 0.45wt% NH_4F and 2wt% DI water in ethylene glycol. A bias potential of 60V was supplied in a two-electrode cell, using a cathode of Cu. Anodization time was variated at 2h, 2.5h and 3h. Anodized samples were again rinsed with DI water, dried at 100°C, and calcined at 400°C for 2h with a heating rate of 3°Cmin^{-1}. In order to obtain transparent TiO_2 nanotubes (TNT) in a membrane form, a second anodization for 15min was carried out to the samples under the same experimental conditions to the first anodization. Materials were immersed in a 30wt% H_2O_2 solution during 2min. TNT membranes were removed from Ti and deposited on ITO substrate to prepare photoanodes using a boron doped TiO_2 sol as adhesive, whose preparation has been described elsewhere [5]. Photoanodes were again dried at 100°C and calcined at 400°C at a heating rate of 3°Cmin^{-1}.

2.2. Physicochemical and photoelectrochemical characterization of TNT photoanodes
The morphology of TNT was analyzed with a JOEL Quanta 650 FEG, and XPS spectra of materials were acquired by a SPECS PHOIBOS 100 spectrometer with a hemispheric analyzer operating at X-ray energy of 1486.6eV from Al Kα radiation. The peak fitting of high-resolution XPS spectra of C 1s, O 1s, N 1s, F 1s and Ti 2p was carried out with the CASA-XPS software. C 1s (284.6eV) signal was used as reference energy. The photoelectrochemical performance of TNT photoanodes was measured in a conventional three-electrode cell using an Autolab PGSTAT 302N potentiostat. A graphite rod (AGKSP grade) and Ag/AgCl(3.0M KCl) electrode were used as counter and reference electrode, respectively. The electrolyte solution was 1.0M Na_2S, which was deaerated for 20min by N_2 bubbling. A metal halide lamp (MHN-TD Phillips, 60mWcm^{-2}) was used as illumination source. Photopotential measurements were carried out at open circuit. Photovoltammetry measurements were performed at 10mVs^{-1} in positive direction from open circuit potential to 1.6V. The exposed geometric area of photoanodes was 1cm^2.

2.3. Preparation of TNT/CdS/CdTe photoanodes
CdTe QDs were synthesized by mixing 4mmol of $CdCl_2$, 3.2mmol of tioglycolic acid and 200mL of DI water in a 250mL three-neck flask under vigorous stirring and inert atmosphere. A 2.0 N NaOH solution was used to adjust the pH of the mixture to 10. The chalcogenide precursor was prepared mixing 0.6mmol $NaBH_4$ and 0.2mmol Te in 5mL of DI water under vigorous stirring and Ar inert atmosphere. The freshly prepared NaHTe solution was quickly injected into the Cd precursor solution and the final mixture was refluxed for 4h. The obtained QDs were purified by centrifugation at 5000rpm for 15min. CdTe QDs showed an average particle size of 4.2nm obtained by DLS. TNT photoanodes were immersed in 50mM dispersion of CdTe QDs during 2h, 4h and 6h. TNT photoanodes were also sensitized with CdS quantum dots by SILAR method. Thereby, photoanodes were dipped into methanol solutions of 0.1M $3CdSO_4{\cdot}8H_2O$ and 0.1M Na_2S each one during 1min. Eight SILAR cycles were performed according to literature [6]. Finally, photoanodes were immersed in 0.1M $Zn(CH_3COO)_2{\cdot}2H_2O$ and 0.1M Na_2S aqueous solutions (two SILAR cycles) to deposit a ZnS layer.

2.4. Photovoltaic measurements of assembled solar cells
In order to investigate the photovoltaic performance of TNT/CdS/CdTe photoanodes, solar cells were elaborated assembling the photoanodes with CoS based cathodes, which were prepared by mixing methanol solutions 0.5M $Co(NO_3)_2.6H_2O$ and 0.5M Na_2S, forming a black paste. The CoS paste was deposited on ITO and dried at 100°C during 15min. Solar cells were filled with polysulfide solution consisting of 1.0M Na_2S and 1.0M S. Typical current-voltage curves were acquired by using an

Autolab PGSTAT 204 potentiostat under solar illumination (AM 1.5G, $75mWcm^{-2}$) simulated with a Suntest CPS+ solar simulator (ATLAS, GmBH). The exposed geometric area of photoanodes was $0.15cm^2$.

3. Results and discussion

3.1. Morphology and chemical environment of TNT photoanodes

Figure 1 shows the typical top view and the cross section FESEM images of transparent TNT membranes obtained varying the anodization time. TNT average pore diameter was 85.8nm, 95.6nm and 116.0nm and an average length of 16.7μm, 20.1μm and 35.9μm was obtained at 2h, 2.5h and 3h of anodization time, respectively. This indicates to increase the anodization time keeping constant the NH_4F concentration and the applied potential, the compact oxide can be dissolved more deeply by fluoride ions, thereby enlarging the pores and the length of TNT. In addition, a well-oriented tubular structure is observed in the prepared materials.

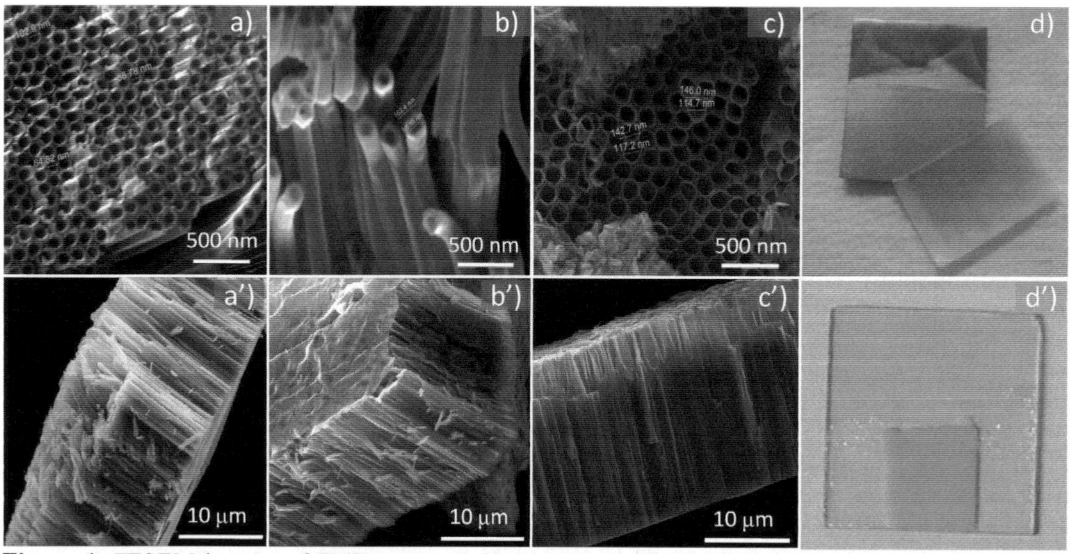

Figure 1. FESEM images of TNT grown at 2h (a), (a'), 2.5h (b), (b') and 3h (c), (c'), respectively. Photograph of a TNT membrane (d) before and after detachment from Ti (d) and (d') after deposition on ITO.

The chemical environment of typical TNT grown at 2.5h was determined by XPS analysis. Ti, O, C, N and F elements were identified in the general spectrum (Figure 2(a)). High resolution XPS N 1s spectrum (Figure 2(b)) exhibits two signals at binding energies of 399.4eV and 401.1eV, associated to the presence of N-O-Ti and N-Ti-O type bond into TiO_2, which indicates nitrogen was incorporated in interstitial and substitutional positions into TiO_2 lattice [7]. XPS peak at 684.2eV (Figure 2(c)) is ascribed to incorporation of fluorine in substitutional positions into TiO_2. It is well known that fluorine atoms are introduced into TiO_2 as fluoride ions, which induces the formation of Ti^{3+} donor states in semiconductor allowing its photoactivation under visible light [8]. This is confirmed by Ti $2p_{3/2}$ and Ti $2p_{1/2}$ photoelectron signals at 457.0eV and 461.1eV (Figure 2(d)), respectively, besides to Ti^{4+} signals from TiO_2 lattice (458.5eV) and the N-O-Ti bond (459.9eV). The coexistence of N and F species into semiconductor indicates the TiO_2 modification by co-doping, which would contribute to its photoactivity in the visible region.

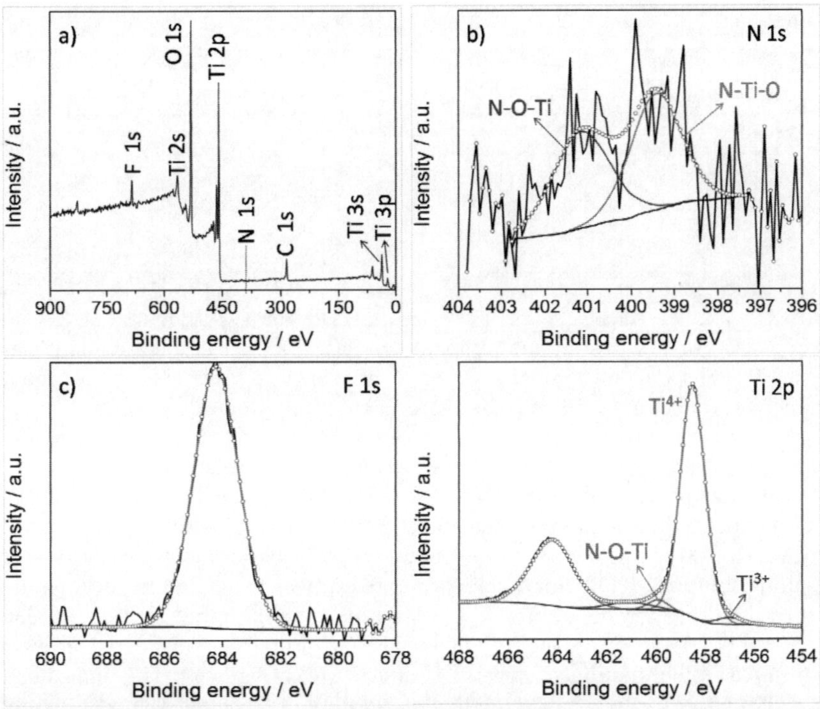

Figure 2. XPS general (a) and high-resolution XPS spectra of (b) N 1s, (c) F 1s, and (f) Ti 2p spectra of TNT photoanode grown at 2.5h.

3.2. Photopotential and photoelectrochemical measurements

Open-circuit potential measurements for the TNT based photoanodes are showed in Figure 3(a). To illuminate the photoanodes, a displacement of photopotential to more negative potentials is observed, indicating, first, the accumulation of photogenerated electrons in each material and second, a n-type semiconductivity. Increasing anodizing time leads to an increase in the photopotential, due to TNT enlargement, forming more lattice defects [9]. According to high resolution XPS Ti 2p spectrum, presence of fluoride ions induces the formation of Ti^{3+} donor states which are located below the bottom of conduction band of TiO_2 [8], allowing the photoactivation of semiconductor in visible region. Thus, to grow TNT to 35.9μm, more fluoride ions could be incorporated into TiO_2 lattice after anodization to form a high Ti^{3+} species content, promoting the accumulation of more electrons photogenerated under illumination.

In order to choose the electrode material with best photoactivity, TNT photoanodes were characterized by photovoltammetry (Figure 3(b)). To increase TNT length from 16.7μm to 20.1μm, an increment in photocurrent was observed. It is ascribed to formation of more Ti^{3+} donor species by fluoride ion modification. This lead to an increase of photogenerated electrons transported through the photoanode to reach the back collector, which enhances the conducting properties of material. For TNT of 35.9μm, the photoresponse decreased implying the formation of an excessive Ti^{3+} species content, which can act as recombination centers, hindering the electron mobility into the photoanode and thereby its photoelectrochemical performance. This is confirmed by the slow return to rest potential (in the dark) when light is turned off (Figure 3(a-iii)), indicating an increase of charge carrier trapping and recombination which adversely affect the electron transfer rate [10].

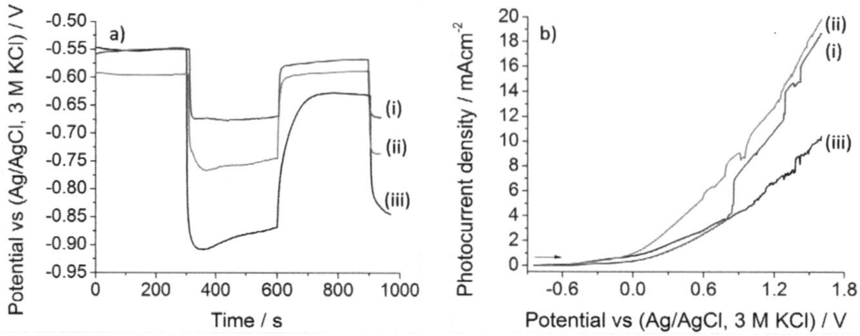

Figure 3. (a) Open-circuit photopotential and (b) photovoltammetry measurements performed in 1.0M Na_2S at $10 mVs^{-1}$ for TNT photoanodes obtained for (i) 2h, (ii) 2.5h and (iii) 3h, respectively.

3.3. Photovoltaic performance of assembled CdTe sensitized solar cells

In order to prepare the photoanode for solar cell, CdTe QDs were deposited on TNT membrane supported on ITO with the best photoelectrochemical performance (Figure 3(b)). To improve the solar cell performance a ZnS layer was coated on the photoanodes to avoid the charge carrier recombination with the polysulfide solution [1,11]. Solar cells performance was evaluated for different load of CdTe QDs deposited on TNT membrane. Figure 4 shows the behaviour of the short-circuit current density (J_{sc}), the open circuit voltage (V_{oc}) and fill factor (FF). To increase the deposition time from 2h (2-CdTe) to 4h (4-CdTe), an increment of both J_{sc} and V_{oc} was observed. This indicates that at 4h, a sufficient CdTe QDs content can enhance the light harvesting of TNT photoanode, thereby increasing the number of electrons transported to back collector. Nevertheless, to raise the deposition time at 6h (6-CdTe), a decrease of J_{sc} and V_{oc} was evidenced, which could be associated to the presence of an excessive CdTe QDs content acting as recombination sites and/or an obstruction for the polysulfide solution to reach the TNT. Thus, although the FF was increased to increment the CdTe QDs load as response of improvement of light harvesting, electron transport through the device is inhibited, obtaining low J_{sc} and V_{oc} values.

The abovementioned can restrain the charge carrier separation and diminish the number of electrons injected to TNT. With the purpose of photogenerate more electrons under artificial solar illumination, CdS QDs were deposited on TNT photoanodes prior to 4-CdTe QDs sensitization. An increase of photocurrent was obtained due to light harvesting from CdS besides to transport the photogenerated electrons injected from CdTe QDs. A high J_{sc} is directly related to an increase of energy conversion efficiency of cell (η) which reached a value of 0.16% for TNT/CdS/CdTe photoanode. It is noteworthy that conversion efficiency for TNT/CdS is lower than TNT/CdS/CdTe, hence sensitization with both CdS and CdTe shows a synergistic effect. The photovoltaic parameters of solar cells are given in Table 1. In spite of the V_{oc} shift to ~-0.30V, a low short circuit-current was registered; this may possibly be attributed to a deficient contact between TNT membrane and ITO substrate or to some internal resistance of the cell.

Table 1. Solar cell parameters obtained from J-V curves for different photoanodes here prepared.

Photoanode	$J_{sc}/mAcm^{-2}$	V_{oc}/V	FF	η/%
TNT	0.33	-0.18	0.13	0.01
TNT/2-CdTe	0.21	-0.18	0.13	0.01
TNT/4-CdTe	1.04	-0.24	0.24	0.08
TNT/6-CdTe	0.31	-0.22	0.33	0.03
TNT/CdS	1.15	-0.27	0.29	0.12
TNT/CdS/4-CdTe	1.39	-0.29	0.30	0.16

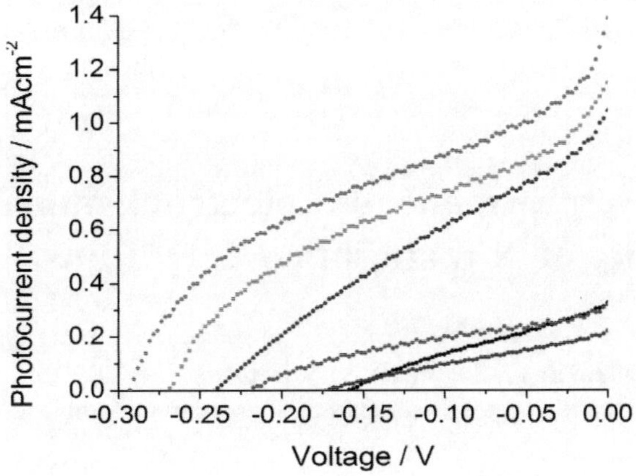

Figure 4. J-V characteristics of solar cells based on (a) TNT, (b) TNT/2-CdTe, (c) TNT/4-CdTe, (d) TNT/6-CdTe, (e) TNT/CdS and (f) TNT/CdS/4-CdTe photoanodes.

4. Conclusions

Highly oriented transparent TiO_2 nanotube photoanodes with different average pore diameter and length were prepared by double anodization by varying the TNT growth time. The materials were photoactive to visible region due to the presence of N and F into the TiO_2, causing the co-doping of semiconductor. The TNT photoanode obtained at 2.5h showed the best photoelectrochemical performance, due to the enhancement of electron flow to back collector by the presence of an adequate Ti^{3+} species content formed by N, F co-doping. A solar cell was fabricated modifying TNT with both CdS and CdTe QDs, achieving a maximum of conversion efficiency of 0.16% under ¾ sun illumination for 4h of CdTe deposition and using a CoS cathode. Synergistic effect was obtained by TNT sensitization with CdS/CdTe QDs assembly improving the light harvesting of solar cell. Current efforts are conducted to enhance the synergy of cadmium chalcogenides by band-gap engineering and to obtain a best cathode material.

Acknowledgements

This research was supported by Universidad Industrial de Santander and funded by Colciencias, project 8836-1828 and VIE-UIS project 1336. AFG-R acknowledges support from Colciencias through a doctoral research scholarship to pursue graduate studies. We also gratefully acknowledge to professor Darío Yesid Peña Ballesteros and GIC for the equipment support.

References

[1] Yang Z, Chen C, Roy P and Chang H 2011 *Chem Commun* **47** 9561
[2] Pawar S, Patil D, Jung H, Park J, Malic S, Hong C, Shin J, Patil P and Kim J 2016 *Electrochim Acta* **203** 74
[3] Jin M, Chen X, Gao Z, Ling T and Du X 2012 *Nanotechnology* **23** 485401
[4] Li Z, Yu L, Liu Y and Sun S 2014 *Electrochim Acta* **129** 379
[5] Gualdrón-Reyes A F, Meléndez A M, Niño-Gómez M E, Rodríguez-González V and Carreño-Lizcano M I 2015 *Rev Acad Colomb Cienc Ex Fis Nat* **39** 77
[6] Li Z, Yu L, Liu Y and Sun S 2014 *J Mater Sci* **49** 6392
[7] Zhou J, Ren F, Zhang S Wu W, Xiao X, Liua Y and Jiang C 2013 *J Mater Chem* A **1** 13128
[8] Czoska A, Livraghi S, Chiesa M, Giamello E, Agnoli S, Granozzi G, Finazzi E, Di Valentin C and Pacchioni G 2008 *J Phys Chem C* **112** 8951
[9] Lei B, Liao J, Zhang R, Wang J, Su C and Kuang D 2010 *J Phys Chem* C **114** 15228
[10] Ramírez-Ortega D, Meléndez A M, Acevedo-Peña P, González I and Arroyo R 2014 *Electrochim Acta* **140** 541
[11] Zhao K, Pan Z and Zhong X 2016 *J Phys Chem* C **7** 406

IOP Conf. Series: Journal of Physics: Conf. Series **786** (2017) 012045 doi:10.1088/1742-6596/786/1/012045

Effect of substrate surface treatment on electrochemically assisted photocatalytic activity of N-S co-doped TiO_2 films

N J Parada-Gamboa[1], J A Pedraza-Avella[1] and A M Meléndez[1]
[1] Universidad Industrial de Santander, CMN, Piedecuesta, Santander, Colombia

E-mail: amelende@uis.edu.co , angelemet@gmail.com

Abstract. To investigate whether different metal surface treatments, performed on meshes of stainless steel 304 and titanium, affect the photocatalytic activity (PCA) of supported modified anodic TiO_2 films, metallic substrates were coated with titanium isopropoxide sol-gel precursor modified with thiourea. Substrates were pretreated by some of the following techniques: a) sandblasting, b) pickling, c) hydroxylation and d) passivation. The as-prepared electrode materials were characterized by X-ray photoelectron spectroscopy (XPS), field emission scanning electron microscopy (FESEM), and voltammetry in the dark and under light UVA irradiation. PCA of modified N-S-TiO_2 electrodes was evaluated by electrochemically assisted photocatalytic degradation of methyl orange. The results of XPS revealed that N and S were incorporated into the lattice of TiO_2. FESEM showed that surface roughness and thickness of films varies depending on surface treatment. Voltammetric and XPS characterization of N-S co-doped TiO_2 films supported on stainless steel revealed that their surface contains alpha-Fe_2O_3/FeOOH. Accordingly, iron contamination of the films coming from stainless steel was detrimental to the degradation of methyl orange. Prior to sol-gel coating process, sandblasting followed by nitric acid passivation for stainless steel or hydrofluoric acid pickling process in the case of titanium improved the PCA of N-S co-doped TiO_2 films.

1. Introduction

In continuation of our previous studies on photoelectrocatalytic (PEC) performance of modified anodic TiO_2 films with non-metals elements for water decontamination [1], herein is determined the effect of substrate surface treatment on PCA of sol-gel modified TiO_2 films deposited on titanium and stainless steel (AISI 304) substrates. On one hand, it is known that titanium is covered by a native oxide film which is readily attacked by hydrofluoric acid [2], leading to a rough fresh surface after pickling [3]. Thus, this could be exploited to increase the surface area of sol-gel film and enhance the PCA of the photoanodes. On other hand, it is well known that TiO_2 thin films prepared on stainless steel by sol-gel are contaminated with significant amounts of iron and other metals, thereby decreasing its PCA [4-6]. In the field of photocatalysis, this problem has been solved by thickening of the TiO_2 film, avoiding the diffusion of metals from stainless steel to the TiO_2 film surface [6,7]. However, in photoelectrocatalysis, PEC activity diminishes with increase of film thickness [8,9], due to long and tortuous path of the electrons across the TiO_2 nanocrystal network where interface states can act as recombination centres, resulting in a decrease in the photocurrent and photocatalytic efficiency [9]. In the field of photoelectrocatalysis, very few efforts have been dedicated to enhance the PEC performance of above-mentioned type of photoanodes [1]. In an attempt to improve the PCA of surface-modified anodic TiO_2 films, herein are evaluated various surface treatments on metallic substrates (stainless steel and titanium) before the sol-gel application.

2. Experimental

2.1. Preparation of S-TiO_2 coatings

AISI-SAE 304 stainless steel (mesh stem: 0.9mm; mesh open size: 6.6mmx3.7mm–length of the diagonal diamond) and grade 2 ASTM B265 titanium (mesh stem: 0.9mm; mesh open size: 6mmx3mm) plates were used as substrates. Mesh plates were cut into geometrical dimensions of 50×45mm^2. Substrates were exposed to various pretreatment methods, as summarized in Table 1, to generate different modified surfaces prior to sol-gel application. The sandblasting was done with 60 grit aluminium oxide particles. Sandblasting was performed at an air pressure of 80 pounds, the distance between the nozzle and the stainless-steel mesh was 30cm.

Table 1. Surface treatment methods applied on stainless steel (SS) and titanium (T) substrates.

Substrate	Treatment	Description
SS[a], Ti	Chemical cleaning (CC)	Meshes were sonicated consecutively in ethanol (5 min) and ketone (5 min).
SS[a]	Hydroxylation (H)	After chemical cleaning, meshes were immersed in a standard piranha solution, H_2SO_4/H_2O_2, 70:30 (v/v), for 30 min.
SS[a]	Passivation (P)	After chemical cleaning, meshes were immersed in 30 % HNO_3, for 1 h.
SS[a]	Hydroxylation-Passivation (H/P)	After hydroxylation, meshes were immersed in 30 % HNO_3, for 1 h.
Ti	Pickling (PC)	After chemical cleaning, meshes were immersed in 5% HF, for either 10 s [PC(10s)] or 20 s [PC(20s)].

[a] Previous to treatment the substrates were sandblasted.

Deionized water and analytical reagents were used in all experiments. To prepare the N-S co-doped TiO_2 sol, a mixture of 5mL of titanium(IV) isopropoxide and 5mL of isopropyl alcohol was added dropwise to 31mL of 0.3M HNO_3 containing 1mL of acetyl acetone. The mixture was stirred for 12h, afterwards 0.033g thiourea was added and the sol was again stirred for an additional 12h. In order to improve the adherence of sol-gel layer on substrates, the sol was aged in the dark for 21 days. Each metallic substrate was dipped into the aged sol, holded for 2min and withdrawed out of the sol at 3cm min^{-1}. After withdrawal, the coated substrates were dried at room temperature and later at 110°C for 1 h. After drying, samples were calcined at 400°C for 1h at a heating rate of 3deg· min^{-1}.

2.2. Characterization of surface modified TiO_2 films and photoelectrocatalytic degradation

XPS data were obtained on a SPECS PHOIBOS 150 spectrometer with a hemispheric analyser using the Kα radiation of an Al anode (1486.6eV). Deconvolution was performed by fitting after Shirley background subtraction using the Casa XPS software package. Surface morphologies of the films were observed by using a FEI QUANTA FEG 650 field emission scanning electron microscope, the analyses were performed at an accelerating voltage of 20kV. Diffuse reflectance spectra (DRS) were obtained with a Shimadzu double beam UV-2401 PC UV/Vis spectrophotometer. Band gaps of the films were determined from the DRS using the Kubelka-Munk function [1]. (Photo)electrochemical measurements were made in a conventional three-electrode cell, using each one of modified TiO_2 films as working electrode. The geometric area of each electrode plate in contact with the electrolyte was 44mmx33mm (22.5 diamonds) for SS and 41mmx34 mm (27 diamonds) for Ti. The counter electrode was a high purity graphite rod, and an Ag/AgCl, KCl (3.0M) electrode (+0.207V vs NHE) supported by a home-made Luggin capillary filled with 3.0M KCl was used as reference electrode. Dissolved oxygen was removed from the solutions prior to the measurements by bubbling nitrogen for 20min. All (photo)electrochemical measurements were performed on an Autolab PGSTAT302N potentiostat. Oxides and hydroxides of iron were determined by voltammetry by reductive dissolution of TiO_2/SS films. A HPL-N 250W UVA-visible lamp (Phillips) was used for the illumination. PEC degradation was performed by holding the potential at 0.65V vs Ag/AgCl. Discoloration of 4.5ppm solution of methyl orange at pH 3.2 (adjusted with H_2SO_4) in 0.1M Na_2SO_4 was followed by measuring the decrease in absorption at 501nm.

3. Results and discussion

3.1. Characterization of TiO_2 films

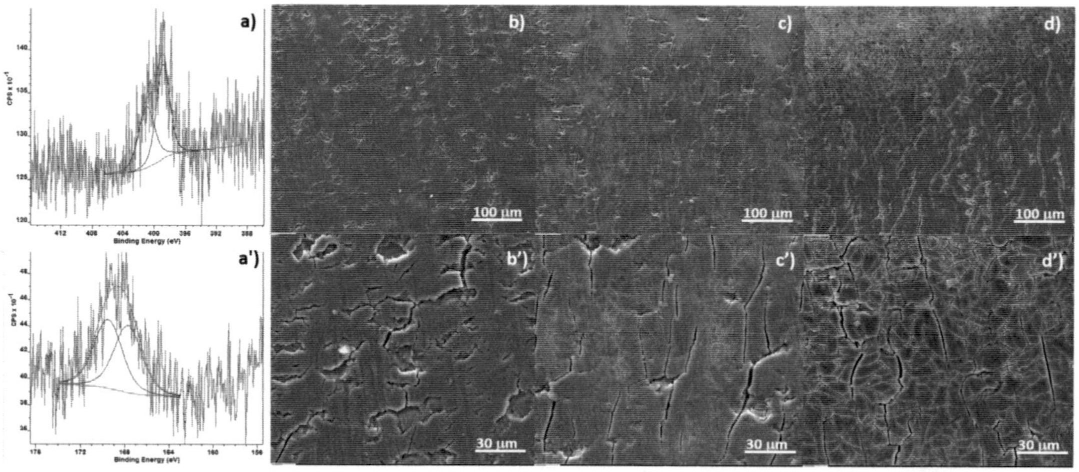

Figure 1. Typical XPS spectra of (a) N 1s and (a') S 2p for a TiO_2 film photoanode. FESEM images of N-S co-doped TiO_2 films supported on titanium pretreated by chemical cleaning (b), (b') and pickling for 10s (c), (c') and 20s (d), (d'). Magnifications: 800X (b)-(d) and 3000X (b')-(d').

XPS measurements were performed in order to investigate the chemical modification by thiourea on the crystal structure of the as-prepared and aged sol-gel TiO_2 films. Figure 1(a), (a') shows representative XPS spectra in the N 1s and S 2p regions. The results of XPS analysis for TiO_2 films supported on stainless steel are given in Table 2. Titanium dioxide films shows peaks in the range of 398.9–402.0eV (N1s) indicating the presence of N–Ti–O, N–O–Ti, and O–N–Ti–O bonds in the bulk of films, ascribed to interstitial and substitutional nitrogen doping [10]. Additional N 1s peaks on TiO_2 film supported on hydroxylated SS surface have been attributed to nitrogen adsorbed compounds [10]. XPS peaks in the range of 167.1–170.2eV have been attributed to the presence of S(IV) and S(VI) species. According to literature, peaks around 167.4eV can be attributed to sulphur doping, while peaks close to 170.1 has been associated to adsorbed surface sulphate species [11]. These results indicate that N and S atoms are co-doped into the bulk phase of TiO_2. The decrease in the band gap energy of modified TiO_2 confirms the doping (see Table 2). The surface morphology of TiO_2 films supported on SS and Ti was investigated by FESEM. Figure 1 shows the surface of the films deposited on the pickled titanium support for 20s was more cracked than that deposited for 10s. It is well-known that formation of cracks in sol-gel films depends on layer thickness [12,13]. Thus, thickness of N-S co-doped TiO_2 films increases with the cracking of the layer in the following order: CC<PC(10s)<PC(20s). As a result of the sandblasting required to improve the adherence of N-S co-doped TiO_2 films on SS, Figure 2 shows that these coatings exhibit a strong non-uniformity, thereby

making it difficult to determine the thickness of the films. However, it is clearly observed that surface roughness of films on passivated stainless steel substrates is greater than in the other pretreated surfaces. Passivated surface was rougher than that hydroxylated/passivated. Thus, roughness increases in the following order: CC < H < H/P < P.

Table 2. Sample labels of substrate surface treatment, XPS binding energies of the doping elements, their stoichiometric ratio, and band gap energies of TiO_2 films supported on stainless steel.

Label	Element	Binding energy (eV)	Atomic ratio N/S	Band gap (eV)
CC	N 1s	399.82, 401.44	1.45	3.13
	S 2p	167.19, 170.15		
H	N 1s	399.22, 400.11, 400.96, 402.03	1.90	2.89
	S 2p	167.42, 170.09		
H/P	N 1s	399.35, 401.67	1.10	2.90
	S 2p	167.70, 169.51		
P	N 1s	398.89, 400.99	1.47	2.97
	S 2p	167.08, 170.12		

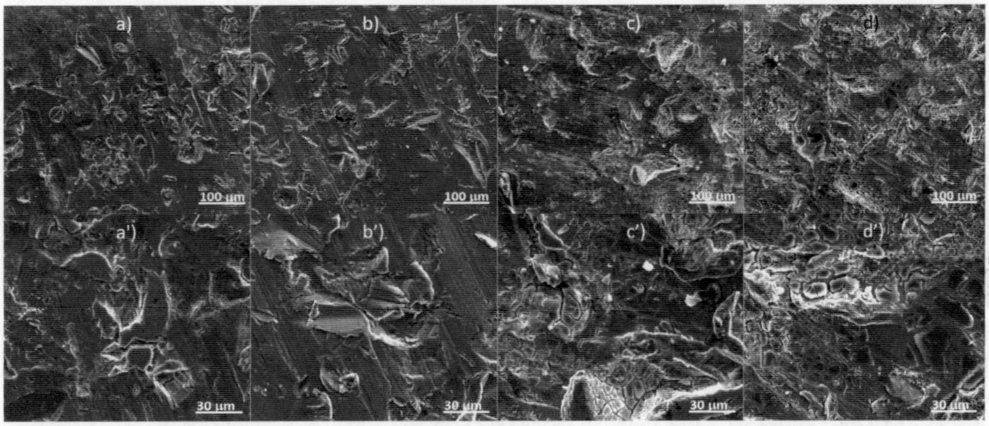

Figure 2. FESEM images of N-S co-doped TiO_2 films supported on stainless steel pretreated by chemical cleaning (a), (a'), hydroxylated (b), (b'), hydroxylated/passivated (c), (c') and passivated (d), (d'). Magnifications: 800X (a)-(d) and 3000X (a')-(d').

3.2. Characterization of the Fe(III) contamination on N,S-TiO_2/stainless steel

Formation of iron oxide in sol-gel TiO_2 films deposited on stainless steel has been reported in the literature [4-7]. Voltammetry is a suitable technique for identification of iron oxides and oxyhydroxides in acid media [14,15]. Figure 3 shows the electrochemical behaviour of SS before and after being coated with N-S co-doped TiO_2 film. Voltammogram of bare stainless steel does not present any current before hydrogen evolution reaction (Figure 3(a)). Two voltammetric peaks are observed around -0.2 and -0.4V (vs Ag/AgCl) for films deposited on hydroxylated and hydroxylated followed by passivated (Figure 3(b), (c)) substrates. According to literature report [14], peaks C_1 and C_2 are assigned to the reductive dissolution of α-FeOOH and α-Fe_2O_3, respectively. In such wise, chemically cleaned and passivated substrates coated with N-S co-doped TiO_2 films were only poisoned by α-Fe_2O_3. Analysis of these results indicates that hydroxylation favours the formation of α-FeOOH in the film, and passivation decreases the contamination of film surface. It is noteworthy that the response in current for passivated substrate is lower than that of stainless steel substrates treated (see Figure 3). These results are in good agreement with XPS analysis (results not shown).

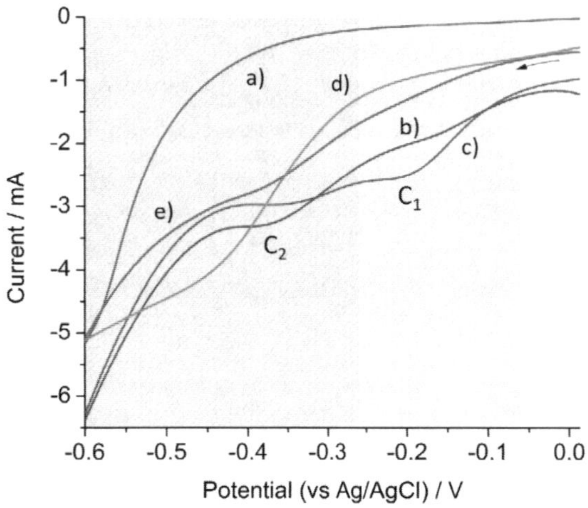

Figure 3. Voltammetric curves of (a) stainless steel and (b)-(e) N-S co-doped TiO_2 films supported on stainless steel. Substrate surface modification by (b) hydroxylation, (c) hydroxylation-passivation, (d) chemical cleaning and (e) passivation treatment. The scans (v=10mVs^{-1}) were initiated in the negative-going direction from the open circuit potential. Electrolyte: 0.1M $HClO_4$.

3.3. Photocurrent measurements and dye discoloration

The photocurrent response depends on type of substrate and on received surface treatment. The photocurrent transient behaviour in Figure 4(i) show that the charge involved in discoloration process (area under the curve), for each of the four surface treatments on SS substrates, increases in the following order (see labels in Table 1): H<H/P<CC<P, according to the following values of charge in mC: 0.033<0.084<0.132<0.182, respectively. The higher photoresponse is obtained for passivated process, which is ascribed to lower iron contamination of the modified TiO_2 films supported on SS. Likewise, the lower photoresponse showed for electrodes that were hydroxylated is associated to high content of iron by the presence of both α-Fe_2O_3 and α-FeOOH in the film. It is noteworthy that photocurrent responses coming from SS are noisy, unstable (photocurrent increases vary over time), and have a lower magnitude than those from Ti substrates; thereby, films supported on Ti show high and stable photocurrents. This can be explained because α-FeOOH or α-Fe_2O_3 can modify the electronic states of TiO_2 acting as electron-hole recombination sites [4,16]. These results are consistent with those of PEC degradation of methyl orange, since the rate of dye discoloration (Figure 4(ii)) increases as photocurrent increases. The increase in photocurrent over time was most likely due to iron dissolution; however, more evidence is needed to confirm this hypothesis. On other hand, dye discoloration on modified TiO_2 films supported on Ti substrates enhances in the following order: PC(10s)<CC<PC(20s). Thus, titanium after 20s of pickling in HF improves both the PCA and PEC discoloration rate with respect to the other electrode materials; it is associated to an increase in surface area of the film (Figure 1).

4. Conclusions

Metallic meshes were coated by N-S co-doped TiO_2 through sol-gel dip-coating technique. It was found that certain surface treatments, depending on the substrate, are required to improve the photocatalytic activity of N- and S-modified TiO_2 films. Hence, surface preparation of substrate is a critical factor in the photocatalytic performance of coatings. N-S co-doped TiO_2 films coated on titanium showed better PCA than on stainless steel, because on this last substrate α-Fe_2O_3 and α-FeOOH poison the films. The solution proposed herein to improve the photocatalytic activity of N-S co-doped TiO_2 films supported on stainless steel is not useful for practical applications in photo(electro)catalysis, due to low PCA of modified TiO_2 photocatalyst, but indeed constitutes a way to solve the detrimental effect of metal diffusion. Accordingly, is suggested the formation of a thermodynamic stable layer (diffusion barrier) between stainless steel support and the film to avoid the iron diffusion. In the case of titanium substrate, depending on the extent of hydrofluoric acid pickling, the photocatalytic activity of modified anodic TiO_2 films can be modulated.

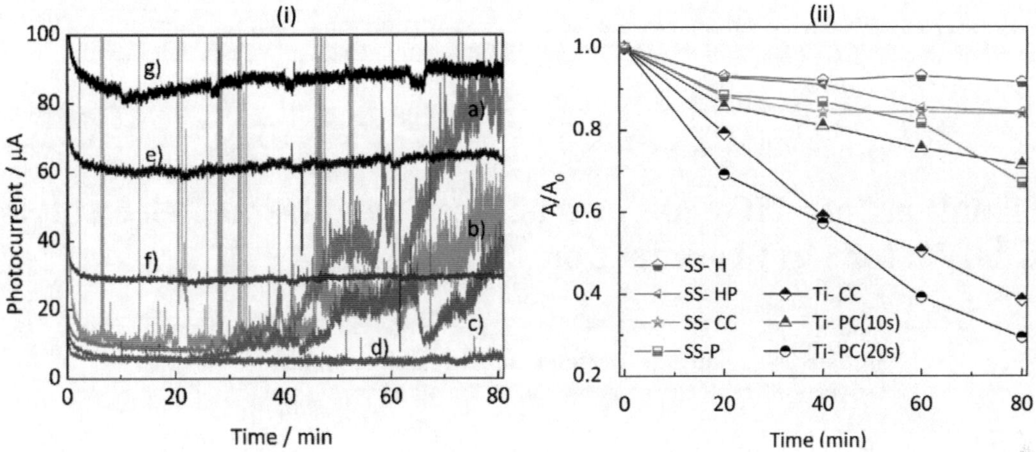

Figure 4. (i) Photocurrent transient behaviour and the corresponding (ii) discoloration of 4.5ppm solution of methyl orange by sol–gel N-S co-doped TiO_2 electrodes in 0.1 M Na_2SO_4. Electrodes: (a) P-SS, (b) CC-SS, (c) H/P-SS, (d) H-SS, (e) CC-Ti, (f) PC(10s)-Ti, and (g) PC(20s)-Ti. Volume of solution 50mL, stirring rate 800rpm, UVA-vis illumination and an applied potential of 0.65V vs Ag/AgCl.

Acknowledgements
This research was supported by the Universidad Industrial de Santander and was funded by the Vicerrectoría de Investigación y Extensión as part of the projects VIE-UIS 1320 and 1336.

References

[1] Gualdrón-Reyes A F, Meléndez A M, Niño-Gómez M E, Rodríguez-González V and Carreño-Lizcano M I 2015 *Rev. Acad. Colomb. Cienc. Ex. Fis. Nat.* **39** 77
[2] Sutter E M M and Goetz-Grandmont G J 1990 *Corrosion Science* **30** 461
[3] Zareidoost A, Yousefpour M, Ghaseme B and Amanzadeh A 2012 *J. Mater. Sci. Mater. Med.* **23** 1479
[4] Pan J H, Lei Z, Lee W I, Xiong Z, Wang Q and Zhao X S 2012 *Catal. Sci. Technol.* **2** 147
[5] Zhu Y, Zhang L, Wang L, Fu Y and Cao L 2001 *J. Mater. Chem.* **11** 1864
[6] Bestetti M, Sacco D, Brunella M F, Franz S, Amadelli R and Samiolo L 2010 *Mater. Chem. Phys.* **124** 1225
[7] Chen Y and Dionysiou D D 2006 *Appl. Catal. B-Environ.* **62** 255
[8] Yoon K H and Chung K S 1992 *J. Appl. Phys.* **72** 5743
[9] Chen L C, Hsieh C T, Lee Y L and Teng H 2013 *ACS Appl. Mater. Interfaces* **5** 11958
[10] Viswanathan B and Krishanmurthy K R 2012 *Int. J. Photoenergy* **2012** 269654
[11] Ohno T, Akiyoshi M, Umebayashi T, Asai K, Mitsui T and Matsumura M 2004 *Appl.Catal. A-General* **265** 115
[12] Garino T J 1990 *Mat. Res. Soc. Symp. Proc.* **180** 497
[13] Kappert E J, Pavlenko D, Malzbender J, Nijmeijer A, Benes N E and Tsai P A 2014 *Soft Matter* **11**, 882
[14] Grygar T, Subrt J and Bohacek J 1995 *Collect. Czech. Chem. C* **60** 950
[15] Grygar T and van Oorschot I H M 2002 *Electroanalysis* **14** 339
[16] Dukes F M, Iuppa E, Meyer B and Shultz M J 2012 *Langmuir* **28** 16933

Formation of TiO_2 nanostructure by plasma electrolytic oxidation for Cr(VI) reduction

D A Torres[1], **F Gordillo-Delgado**[1] **and J Plazas-Saldaña**[1]
[1] Universidad del Quindío, Armenia, Colombia

E-mail: fgordillo@uniquindio.edu.co

Abstract. Plasma electrolytic oxidation (PEO) is an environmentally friendly technique that allows the growth of ceramic coatings without organic solvents and non-toxic residues. This method was applied to ASME SB-265 titanium (Ti) plates (2x2x0.1cm) using voltage pulses from a switching power supply (340V) for 10 minutes at frequency of 1000Hz changing duty cycle at 10, 60 and 90% and the electrolytes were Na_3PO_4 and NaOH. The treated sheets surfaces were analysed by X-ray diffraction and scanning electron microscopy. According to the diffractograms, the duty cycle increase produces amorphous TiO_2 coating on Ti sheets and the thickness increases. After sintering at 900°C during 1 hour, the 10% duty cycle generated a combination of anatase and rutile phases at the sample surface with weight percentages of 13.3 and 86.6% and particle sizes of 32.461±0.009nm and 141.14±0.03 nm, respectively. With this sample, the total reduction of hexavalent chromium was reached at 50 minutes for 1ppm solution. This photocatalytic activity was measured following the colorimetric method ASTM-3500-Cr B.

1. Introduction

The surface modification of metal substrates and alloys using plasma electrolytic oxidation (PEO) for the formation of ceramic coatings with high hardness and large surface area is an environmentally friendly technique, simple for the porous material production and low-cost. In PEO, microdischarges are produced by high potential difference between the metal sheet and another electrode and modifying the frequency or duty cycle of the voltage pulses, which are important parameters for determining the growth characteristics and morphology of the oxide layer [1]. Titanium dioxide (TiO_2) is a semiconductor widely studied due to its large variety of applications such as environmental purification, sewage treatment, metal reduction, biomedical implants and others. TiO_2 has three crystallographic phases: rutile, anatase and brookite, which play an important role according to applications [2,3]. The TiO_2 photocatalytic activity is a solution to attack environmental problems related to contamination by high metals as hexavalent chromium (Cr(VI)); especially, it is one of the most notorious in tanning industry, mining activity and pigment manufacture causing health problems by ingestion, inhalation or absorption through the skin [4,5].

2. Materials and methods

2.1. Preparation of samples and TiO_2 coatings growth

Titanium sheets, 99% purity grade (ASME SB-265), with dimensions 20×20×0.1mm were used in this work. Before to PEO, the plates were polished with 1000 and 1200 grit SiC sandpapers successively and subsequently cleaned by ultrasonic bath (BRANSON 1510, Colombia) in distilled water for 30

IOP Conf. Series: Journal of Physics: Conf. Series **786** (2017) 012046 doi:10.1088/1742-6596/786/1/012046

minutes. A DC pulse power supply designed to deliver a maximum peak voltage of 340V between the electrodes, with variable frequency of 1000Hz and duty cycle of 10, 60 or 90%, was employed in the PEO system. A water cooled stainless steel chamber, containing an aqueous electrolyte 8 g/L Na_3PO_4 and 0.4g/LNaOH, was used as electrode and the Ti sheet was the other electrode. Subsequently, the sintering of the samples was performed at 500, 700 and 900°C for one hour [1].

2.2. Characterization of TiO_2

The coatings' thickness was measured using a micrometer (Mitutoyo 543-460B Digital, Colombia) with an accuracy of ±0.1µm. The thickness was measured at 20 different locations on the sample surface. The phase composition of the coatings was studied by glancing angle XRD measurements in parallel beam geometry using X-ray diffractometer (XBruker D8- Advance, Colombia). The Pictures of the samples' surface were taken using a scanning electron microscopy (SEM, JEOL JSM-6390LV, CICATA, México) [6].

2.3. Photocatalytic activity

The analysis of the photocatalytic activity reduction of Cr(VI) to Cr(III) was done following the colorimetric method (ASTM-3500 B-Cr) with 20 mL aqueous Cr(VI) solution with a concentration of 1ppm. The light from 125W UV-lamp, which has main emission at 253.7nm, was used to radiate the sample in contact with the solution into a beaker; TiO_2/Ti catalyst was placed in the middle of this beaker and the process was completed at two hours.

The TiO_2 coating absorbs UV light so the electrons are excited from valence band to conduction band; this electron-hole pair generates oxidization and radicals OH^- [7,8] as seen in the Equations (1) and (2):

$$O_2 + 4H_2O + 4e^- \xrightarrow{TiO_2/hv} 4OH^- \quad (1)$$

$$CrO_4^{2-} + 4H_2O + 3e^- \rightarrow Cr(OH)_3 + 5OH^- \quad (2)$$

A UV–Vis spectrophotometer (Thermo Scientific Evolution 201/220 UV-Vis system, Colombia) was used to measure the absorbance of the treated solutions under the maximum absorbance wavelength (540nm). In the range of low concentration, the reduction rate in the solution of Cr (VI) to Cr (III) was calculated through the Equation (3):

$$\text{Removal rate } (\%) = \frac{C_0 - C}{C_0} \times 100\,\% \quad (3)$$

Where C is the concentration of Cr(VI) at a certain interval of time (mg/L) and C_0 is the initial concentration of Cr(VI) [5].

3. Results and discussion

3.1. Characterization of TiO_2 coatings

During PEO, white sparks appear when critical voltage is achieved; later the tension stabilizes and orange sparks appear as in Figure 1 [9]. The coating thickness presents a progressive increase with the duty cycle 10, 60 and 90% of 0.32%, 5.4% and 9.49%, respectively at 1000Hz [10]. We assumed that this thickness increase occurs on t_{on} and solidifies during t_{off}; this suggests that the heat is absorbed by the coating and the micro-discharge colour changes. The coatings by PEO were examined using SEM, revealing uniform microstructures by voltage pulse frequency 1000Hz and duty cycle of 10%, as shown in Figure 2. When this frequency is applied the pores density is large enough; because of the high density of the sparks [11]. The pores diameter was approximately between 356 and 1200nm. The weight and atomic chemical composition were determined by EDS [12] giving 48.66% of oxygen, 43.06% of titanium and 8.28% of phosphorus. The phosphorus presence in the films shows that electrolyte enters in discharge channels during the growth process; although SEM analysis shows a significant amount of phosphorus, it was not found by XRD, because it is in amorphous shape or the crystalline component may be below diffractometer detection limit [11].

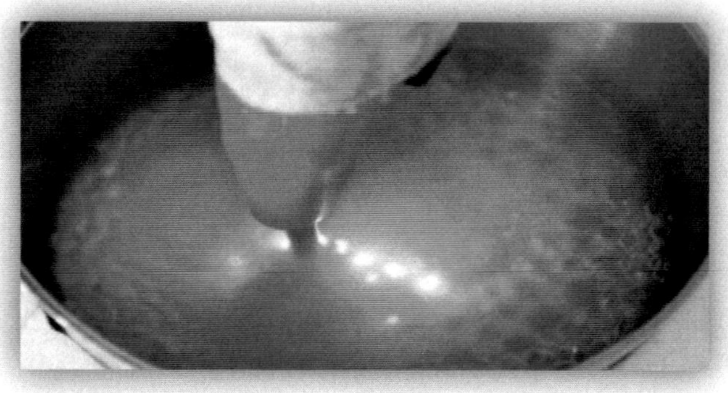

Figure 1. Formation of orange sparks during the TiO_2 superficial coating by PEO on the Ti plate.

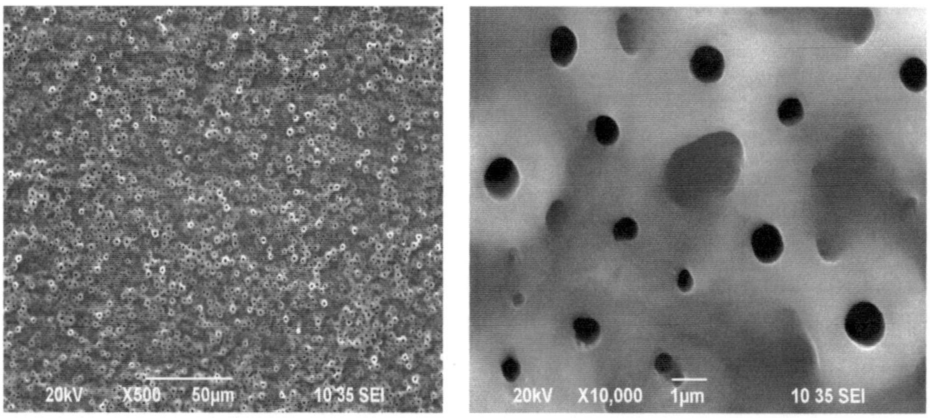

(a) (b)

Figure 2. Surface morphology by SEM of TiO_2 on Ti sheet at 1000Hz, duty cycle of 10% and 500°C: (a) low magnification x500 and (b) high magnification x10000.

The XRD patterns of titanium substrate and coatings (10, 60 and 90% duty cycles) are shown in Figure 3, where peaks of titanium and amorphous TiO_2 can be seen (2θ between 20° and 30°).

Sample diffractograms after thermal treatment at 500, 700 and 900°C can be seen in Figure 4. The average particle size was determined by the Scherrer Equation (4), involving the Bragg angle (2θ) and the Full Width at Half Maximum (FWHM); the values were listed in Table 1 [13].

$$T = \frac{0.9\,\lambda}{\mathrm{FWHM}\cos\theta} \quad (4)$$

From Figure 5, anatase and rutile weight fractions during sintering of TiO_2 coatings were calculated using the Spurr and Myers's Equations, (5) and (6) [14]:

$$X_R = \frac{1}{1+1.26\left(\frac{I_A}{I_R}\right)} \quad (5)$$

$$X_A = 1 - X_R \quad (6)$$

Where X_A and X_R are the phase's percentages and I_A and I_R are the maximum-intensity of the XRD peaks of anatase (101) and rutile (110) phases. The results are shown in Table 1.

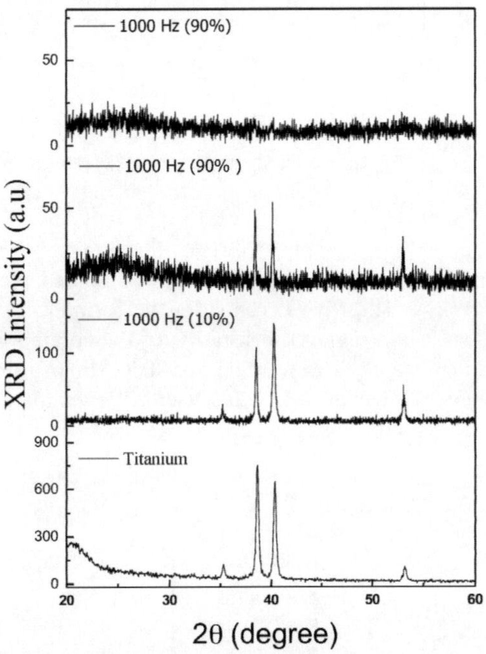

Figure 3. Diffractograms of TiO_2 coatings; PEO with duty cycle: 10, 60 and 90%.

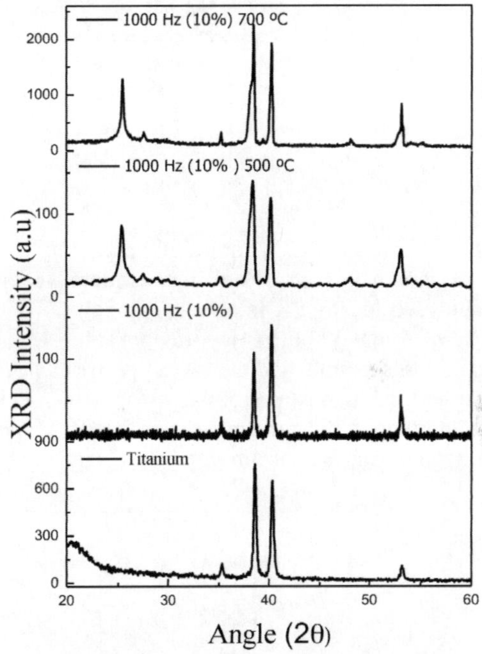

Figure 4. Diffractograms of TiO_2 coatings; PEO with 1000 Hz and 10% duty cycle and different sintering temperature.

Table 1. Particle size; phase percentage and reduction of Cr (VI) using Ti sheets modified by PEO with 1000 Hz and 10 % of duty cycle.

Sintering Temperature (°C)	Anatase average size (nm)	Rutile average size (nm)	Anatase (%)	Rutile (%)	Percentage reduction of Cr (%)
500	16.567±0.004	---	100	---	99.488±0.004
700	24,40±0.01	73.819±0.069	75.2	24.6	98.848±0.004
900	32.461±0.009	141.143±0.029	13.3	86.6	26.442±0.004

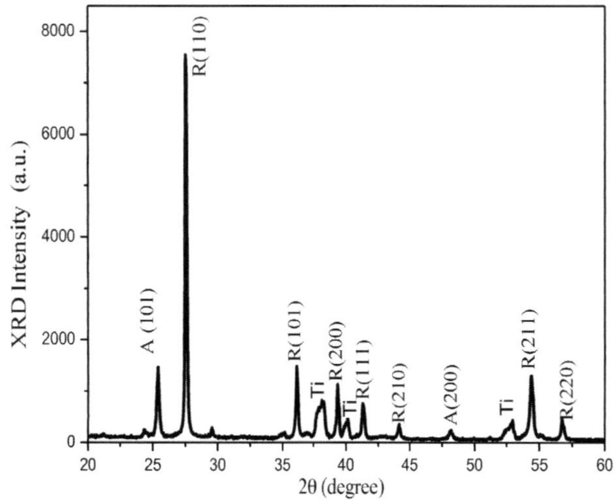

Figure 5. Diffractogram of TiO_2 films sintered at 900 °C and formation of crystalline anatase and rutile phase.

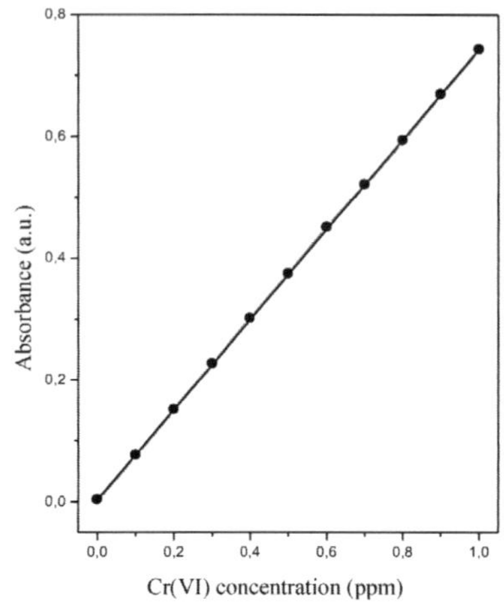

Figure 6. Calibration curve for quantification of Cr (VI) reduction.

3.2. Photocatalytic activity

For the evaluation of the photocatalytic activity of TiO_2 coatings, a 1ppm Cr(VI) calibration curve was done as shown in Figure 6, using Standard Methods 3500-Cr. Absorption capacity of coated plates was analysed locating it into the same solution in darkness. The Cr(VI) reduction performance with UV irradiation at 2 hours is reported in Table 1; for 900°C sintering, a deterioration of the coating was found and this sample had the lowest performance (approximately 26%); while at 500°C sintering, the sample had the highest performance (approximately 99%). However, this value was achieved after 50 and 90 minutes for 1ppm and 10ppm Cr(VI) respectively, according to Figure 7 that shows reduction percentage in function of time.

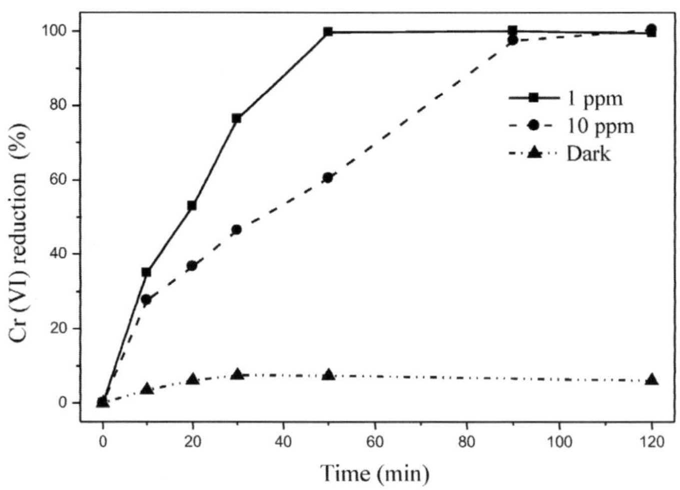

Figure 7. Percent reduction of 1 ppm (squared dots) and 10 ppm (circle dots) Cr(VI) and the effect by solution adsorption in the darkness (triangle dots).

4. Conclusion

TiO_2 ceramic coatings on titanium sheets were accomplished by PEO technique using voltage pulses frequency 1000Hz and changing the duty cycle. These porous films were sintering for growing TiO_2 nanostructures of a mixed phase with particle sizes between 16 and 32nm for anatase and between 74 and 141nm for rutile phase; the SEM pictures showed a topography of microcavities; the EDS analysis shows small phosphorus amounts, which were not detected by XRD, but it could be present in the sample as an impurity.

The equilibrium time of maximum adsorption (7%) was 30 minutes and a reduction almost complete was achieved after 50 minutes of radiation with the TiO_2 coating made with 10% of duty cycle.

Acknowledgments

The authors thank to the University of Quindío for the financial support through the project 805 and the research group's incentive program. Special thanks to Luis Salazar from the Chemistry department for the technical assistance with Uv-Vis absorbance measurements, to Dr. Liliana Tirado and Angélica Marcela Castillo from the Interdisciplinary Institute of Sciences for the XRD measurements and to E. Marín and A. Bedoya from CICATA-IPN, México for the SEM pictures.

References

[1] Arunnellaiappan T & Kishore N 2015 Influence of frequency and duty cycle on microstructure of plasma electrolytic oxidized AA7075 and the correlation to its corrosion behavior *Surface and Coatings Technology* **280** 136-147

[2] Franz S & Perego D 2016 Photoactive TiO_2 coatings obtained by Plasma Electrolytic Oxidation in refrigerated electrolytes *Applied Surface Science* **385** 498-505

[3] Lekphet W & Tsai-Chyuan Ke 2016 Morphology control studies of TiO_2 microstructures via surfactant-assisted hydrothermal process for dye-sensitized solar cell applications *Applied Surface Science* **382** 15-26

[4] Zhou Q & Jing Li Z 2015 Applications of TiO_2 nanotube arrays in environmental and energy fields: A review *Microporous and Mesoporous Materials* **202** 22-35

[5] Zhongping Yao, Fangzhou Jia & Yanli J 2010 Photocatalytic reduction of potassiumchromate by Zn-doped TiO_2/Ti film catalyst *Applied Surface Science* **256** 1793-1797

[6] Mohammad K & Hossein S 2014 Effect of heat treatment temperature on the performance of nano-TiO_2 coating in protecting 316L stainless steel against corrosion under UV illumination and dark conditions *Surface and Coatings Technology* **258** 861-870

[7] Qing S & Xiaolong H 2015 Influence of calcination temperature on the structural adsorption and photocatalytic properties of TiO_2 nanoparticles supported on natural zeolite *Powder Technology* **274** 88-97

[8] Yang J & Lee S 2012 Photocatalytic removal of CrVI with illuminated TiO_2 *Desalination and Water Treatment* **46** 375-380

[9] Venkateswarlu K & Rameshbabu N 2012 Role of electrolyte additives on in-vitro electrochemical behavior of micro arc oxidized titania films on Cp Ti *Applied Surface Science* **258** 6853-6863

[10] Vahid D & Ben L 2013 Effect of duty cycle and applied current frequency on plasma electrolytic oxidation *Surface & Coatings Technology* **226** 100-107

[11] Gowtham S, Arunnellaiappan T & Rameshbabu N 2016 An investigation on pulsed DC plasma electrolytic oxidation of cp-Ti and its corrosion behaviour in simulated body fluid *Surface and Coatings Technology* **301** 63-73

[12] Nenad T & Stevan S 2016 Characterization and photocatalytic properties of tungsten doped TiO_2 coatings on aluminum obtained by plasma electrolytic oxidation *Surface and Coatings Technology* **305** 192-199

[13] Allen, B, Kenneth, O & Thomas, R 2009 On the estimation of average crystallite size of zeolites from the Scherrer equation: A critical evaluation of its application to zeolites with one-dimensional pore systems. *Microporous and Mesoporous Materials* **117** 75-90

[14] Roach M & Williamson R 2016 Tuning anatase and rutile phase ratios and nanoscale surface features by anodization processing onto titanium substrate surfaces *Materials Science and Engineering: C* **58** 213

Institute of Physics
Dirac House, Temple Back
Bristol BS1 6BE UK

ISSN: 1742-6588
ISBN 978-1-5108-3604-4

proceedings.com